Lecture Notes in Mathematics

Edited by A. Dold and B. Eckmann

870

Shape Theory and Geometric Topology

Proceedings of a Conference Held at the Inter-University Centre of Postgraduate Studies, Dubrovnik, Yugoslavia, January 19 – 30, 1981

Edited by S. Mardešić and J. Segal

Springer-Verlag
Berlin Heidelberg New York 1981

Editors

Sibe Mardešić
Department of Mathematics, University of Zagreb
P.O.Box 187, 41001 Zagreb, Yugoslavia

Jack Segal
Department of Mathematics, University of Washington
Seattle, WA 98195, USA

AMS Subject Classifications (1980): 54C56, 55-06, 55M10, 55N07, 55P55, 57Q10

ISBN 3-540-10846-7 Springer-Verlag Berlin Heidelberg New York
ISBN 0-387-10846-7 Springer-Verlag New York Heidelberg Berlin

Library of Congress Cataloging in Publication Data Main entry under title:
Shape theory and geometric topology. (Lecture notes in mathematics; 870)
Bibliography: p. Includes index. 1. Algebraic topology--Congresses.
2. Shape theory (Topology)-- Congresses. I. Mardešić, S. (Sibe),
1927- II. Segal, Jack. III. Winter School and Conference on
Shape Theory and Geometric Topology (2nd: 1981 : Dubrovnik, Croatia) IV. Series:
Lecture notes in mathematics (Springer-Verlag); 870. QA3.L28 vol.870 [QA612]
510s [514'.2] 81-9145 ISBN 0-387-10846-7 (U.S.) AACR2

Printing and binding: Beltz Offsetdruck, Hemsbach/Bergstr.
2141/3140-543210

FOREWORD

From January 19 to January 30, 1981, the University of Zagreb sponsored a Winter school and Conference on Shape Theory and Geometric Topology at the Inter-University Centre of postgraduate studies, Dubrovnik, Yugoslavia. This was the second such school and conference held there. The first was held from January 12 to January 30, 1976 under the title Shape Theory and Pro-homotopy.

The Winter school consisted of a series of lectures devoted to the interaction of shape theory with various areas of geometric topology and also involved the participation of the graduate students. In particular, the aim was to cover the following topics: homotopy and shape domination, pointed and unpointed shape, improving shape representations, dimension and shape dimension, cell-like mappings and hereditary shape equivalences, approximate fibrations and shape fibrations, complement theorems and embeddings up to shape, strong shape.

The contributed papers have been divided into four areas to form the chapters of this volume. These are:

Shape and homotopy domination

Geometric topology and dimension theory

Complement theorems and embeddings up to shape

Shape and strong shape. Steenrod homology

In each area the articles appear in alphabetical order by author.

The addresses of all participants are given at the end of the volume.

S. Mardešić

J. Segal

CONTENTS

FINITELY DOMINATED COMPACTA NEED NOT HAVE FINITE TYPE

Steve Ferry*

In this paper we construct examples of compacta which are homotopy dominated by finite complexes but which fail to be homotopy equivalent to finite complexes. The author discovered two somewhat different constructions in late 1977 and early 1978. The more general construction was published in [F]. There is some interest (see [G]) in the explicit geometry of these compacta. For this reason we are including a revised version of the earlier, more explicit version in these Proceedings.

We recall that a map $d : X \to Y$ is called a *homotopy domination* if there is a map $u : Y \to X$ such that $d \circ u$ is homotopic to the identity. In this case Y is said to be *homotopy dominated* by X.

§1. Recall that the projective class group $\tilde{K}_0(\Lambda)$ of a ring Λ is the Grothendieck group of isomorphism classes of finitely generated projective Λ-modules. If π is a group, $Z\pi$ will denote the integral group ring of π.

The following theorem is basic to our study.

THEOREM (Wall, [Wa]). If a topological space Y is homotopy dominated by a finite complex, then Y has the homotopy type of a finite complex if and only if a certain obstruction $\sigma(Y) \in \tilde{K}_0(Z\pi_1 Y)$ vanishes. If K is a finite complex with $\dim K = n \geq 2$ and $d : K \to Y$ is an n-connected map, then the kernel, $\ker_n d_*$, of $d_* : \pi_n K \to \pi_n Y$ is a finitely generated projective module over $Z\pi_1 K$ and $\sigma(Y) = (-1)^n d_\#[\ker_n d_*]$, where $[\ker_n d_*] \in \tilde{K}_0(Z\pi_1 K)$ and $d_\# : \tilde{K}_0(Z\pi_1 K) \to \tilde{K}_0(Z\pi_1 Y)$ is the induced isomorphism. Moreover, if π is a finitely presented group and $[P] \in \tilde{K}_0(Z\pi)$, then there exist a three-dimensional CW complex L and an isomorphism $i : \pi \to \pi_1 L$ such that $\sigma(L) = i_\#[P]$. ■

The main result of §§1-3 says that the CW complex L above may be replaced by a three-dimensional compact metric space.

THEOREM 1. If π is a finitely presented group and $[P] \in \tilde{K}_0(Z\pi)$, then there exist a compact finitely dominated three-dimensional metric space Y and an isomorphism $i : \pi \to \pi_1 Y$ such that $\sigma(Y) = i_\#[P]$.

We begin the construction of Y. If $\sigma \in \tilde{K}_0(Z\pi)$, choose representatives P and Q for σ and $-\sigma$ so that $P \oplus Q$ is a free $Z\pi$-module on, say, ℓ generators. We write $P \oplus Q = F(x_1, \ldots, x\ )$. Let A be the matrix of the projection $P \oplus Q \to P \oplus 0 \subset F$. A is an $\ell \times \ell$-matrix with entries from $Z\pi$ such that $A^2 = A$.

Let $(K,*)$ be a pointed two-dimensional finite CW complex such that $\pi_1 K$ is isomorphic to π. Let $Y = K \vee \bigvee_{i=1}^{\ell} S_i^2$, the one-point union of K with a bouquet of S^2's.

---

* Partially supported by a National Science Foundation grant.

To conserve notation, we will identify $\pi_1 K$ and $\pi_1 Y$ with π. The retraction $r : (Y, \vee_{i=1}^{\ell} S_i^2) \to (K,*)$ splits the homotopy exact sequence of (Y,K) and shows that $\pi_2(Y)$ is isomorphic as a $Z\pi$-module to $\pi_2(K) \oplus \pi_2(Y,K)$. By passing to the universal cover and using the Hurewicz theorem, we see that $\pi_2(Y,K)$ is a free $Z\pi$-module "generated by the two-spheres $\{S_i^2\}_{i=1}^{\ell}$."

Imbed K^2 in R^n, n large, and consider the subset X' of $[0,\infty] \times R^n$ obtained by attaching a copy of $\vee_{i=1}^{\ell} S_i^2$ to each $\{m\} \times K \subset \{m\} \times R^n$, where m is an integer, $0 \leq m \leq \infty$. This should be done in such a way that the translation $T : [0,\infty] \times R^n \to [0,\infty] \times R^n$ defined by $T(s,x) = (s+1,x)$ carries X' into itself and so that the projection $P_{R^n} : [0,\infty] \times R^n \to R^n$ restricts to an imbedding on $X' \cap (\{0\} \times R^n)$.

Choose characteristic homeomorphisms $f_i : S^2 \to S_i^2 \subset X' \cap (\{0\} \times R^n)$ for $i = 1,\ldots,\ell$. Let $f_i' = T \circ f_i$. f_i' is a characteristic homeomorphism for $S_i^2 \subset X' \cap (\{1\} \times R^n)$. If we regard homotopy groups in X' as based along the path $[0,\infty] \times \{*\}$, then $\pi_2(X' \cap [0,1] \times R^n) \cong \pi_2 K \oplus F_0 \oplus F_1$, where F_0 and F_1 are free $Z\pi$-modules generated by $\{[f_i]\}_{i=1}^{\ell}$ and $\{[f_i']\}_{i=1}^{\ell}$, respectively.

Let P, Q, and A be as before. For each $k = 1,\ldots,\ell$, attach a disc D_k^3 to $X' \cap ([0,1] \times R^n)$ using an attaching map $\partial D_k^3 \to X'$ which represents $A[f_k] + (I-A)[f_k'] \in \pi_2(X' \cap ([0,1] \times R^n))$. These discs may be imbedded in $[0,1] \times R^n$ in such a way that the projection $[0,1] \times R^n \to R^n$ is one-to-one on the union of their interiors. Call the resulting subset of $[0,1] \times R^n$ X''. The effect of these attachings is to kill the part of F_1 corresponding to P and the part of F_2 corresponding to Q.

Let $X = \overline{\bigcup_{m=0}^{\infty} T^m(X'')} \subset [0,\infty] \times R^n$. X consists of copies of X'' between each pair of consecutive integers in $[0,\infty] \times R^n$ together with the projection of X'' into $\{\infty\} \times R^n$. In the following schematic picture, the wiggly lines represent attached discs.

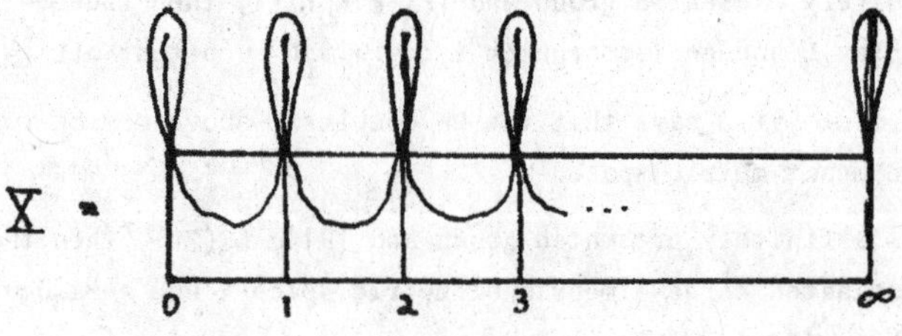

Figure 1

This completes the construction of X. ■

§2. Our goal in the next two sections is to show that X is homotopy dominated by the finite "subcomplex" $X_N^* = [0,\infty] \times K \cup \bigcup_{i=0}^{N-1} T^i(X'') \subset X$ for some large N and that $\sigma(X) = [P]$. Showing that the inclusion $i : X_N^* \hookrightarrow X$ is a homotopy domination is equivalent to constructing a deformation $r_t : X \to X$, $0 \leq t \leq 1$ such that $r_0 = \text{id}$ and $r_1(X) \subset X_N^*$. The construction of such a map requires above all that we be able to construct continuous self-maps of the unpleasant space X. This motivates the following definition and proposition.

DEFINITION. Let $X^- = X \cap ([0,\infty) \times R^n)$. A homotopy $h_t : X^- \to X^-$ is said to be *periodic near infinity* if there is an integer $M \geq 0$ such that

(i) $h_t \mid [0,\infty) \times K = \text{id}$ for all t and

(ii) $h_t \circ T(s,x) = T \circ h_t(s,x)$ for all $(s,x) \in X^-$ with $s \geq M$ and for all $0 \leq t \leq 1$.

PROPOSITION. If $h_t : X^- \to X^-$ is periodic near infinity, then h_t extends to a continuous homotopy $\bar{h}_t : X \to X$.

PROOF. We first extend $\bar{h}_t$ to $(\{\infty\} \times R^n) \cap X$. Let M be as in the definition above. We define $\bar{h}_t(\infty,x) = (\infty, p_{R^n} h_t(p_{R^n}^{-1}(\infty,x) \cap ([M,\infty) \times R^n) \cap X))$ where $p_{R^n} : [0,\infty] \times R^n \to \{\infty\} \times R^n$ is the projection. $\bar{h}_t(\infty,x)$ is well-defined by the periodicity near ∞ of h_t and the construction of X. For each $q \in R^n$, $([0,\infty) \times \{q\}) \cap X$ is either null, contained in $[0,\infty) \times K$, or is a positive semi-orbit of T. Since h_t commutes with T and $p_{R^n} \circ T = p_{R^n}$, $\bar{h}_t$ is well-defined. Moreover $p_{R^n}|(X^- \cap ([M,M+1] \times R^n))$ is a quotient map, so $\bar{h}_t|(X \cap (\{\infty\} \times R^n))$, at least, is continuous.

It remains to prove that $\bar{h}_t$ is continuous. It suffices to prove that $p_{[0,\infty]} \circ \bar{h}_t$ and $p_{R^n} \circ \bar{h}$ are continuous. The first is obvious - the continuity of $\bar{h}_t$ is only in doubt at points of $X \cap (\{\infty\} \times R^n)$ and the periodicity guarantees that points near ∞ stay near ∞ throughout the homotopy. To show that $p_{R^n} \circ \bar{h}_t$ is continuous, let $\{(s_k,x_k,t_k)\}$ be a sequence in $X \times I \subset [0,\infty] \times R^n \times I$ which converges to (∞,x,t). For large k we have $p_{R^n} \circ \bar{h}_{t_k}(s_k,x_k) = p_{R^n} \circ \bar{h}_{t_k}(\infty,x_k)$. That this limit is $\bar{h}_t(\infty,x)$ follows immediately from the continuity of $\bar{h}_t|(X \cap (\{\infty\} \times R^n))$. ∎

We now construct the required deformation. Let Z be the quotient space of X^- obtained by identifying x with T(x). The resulting space is a copy of $K \times S^1$ with a bouquet $\bigvee_{i=1}^{\ell} S_i^2$ of two-spheres and a family $\{D_i^3\}_{i=1}^{\ell}$ of three-discs attached. See Figure 2 for a schematic picture. Again, the wiggly lines represent discs.

Z = 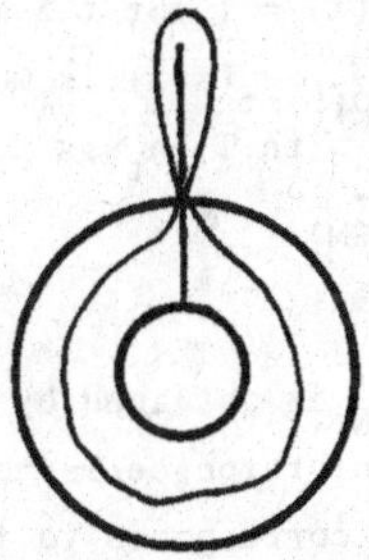

Figure 2

We will first show that $K \times S^1$ is a strong deformation retract of Z. For this, we need only show that $\pi_*(Z, K \times S^1) = 0$. The inclusion $K \times S^1 \hookrightarrow Z$ induces an isomorphism on π_1, so the matter can be settled by showing that $\pi_*(\tilde{Z}, \widetilde{K \times S^1}) = 0$, where $\tilde{Z}$ and $\widetilde{K \times S^1}$ are the universal covers. By the relative Hurewicz theorem, it suffices to show that $H_*(\tilde{Z}, \widetilde{K \times S^1}) = 0$. This is computed from a chain complex $0 \to C_3 \to C_2 \to 0$, where C_2 and C_3 are free $Z\pi_1(K \times S^1)$-modules generated by the cells $\{S_i^2\}$ and $\{D_i^3\}$. The matrix of the boundary operator is $A + t(I - A)$, where A is the matrix used in the construction of X and t is the element of $\pi_1(K \times S^1)$ corresponding to the clockwise loop $* \times S^1$. This matrix is invertible, its inverse being $A + t^{-1}(I - A)$. (Multiplications involving A are simplified by the fact that $A^2 = A$.) This proves that $K \times S^1$ is a strong deformation retract of Z.

Now let $r_t : Z \to Z$ be a strong deformation retraction to $K \times S^1$. Let $\bar{r}_t : \bar{Z} \to \bar{Z}$ be a lifting of r_t to the infinite cyclic covering of Z which corresponds to $\pi_1 K \subset \pi_1(Z)$. We choose $\bar{r}_t$ so that $\bar{r}_0 = \text{id}$. See Figure 3 for a schematic picture of $\bar{Z}$.

$\bar{Z}$ =
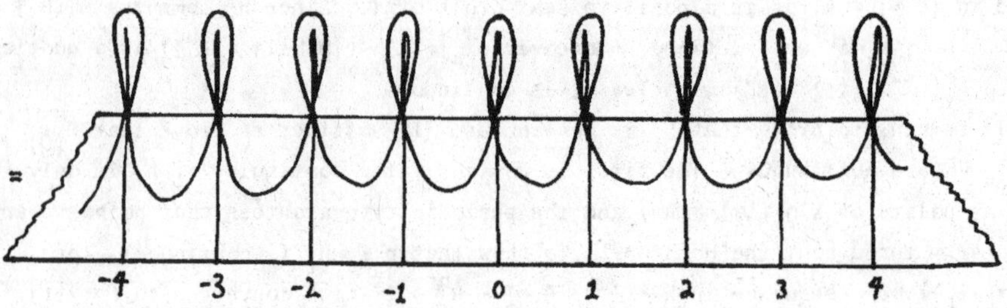

Figure 3

We will consider $\bar{Z}$ to be a subset of $(-\infty,\infty) \times R^n$ in such a way that $X^- = ([0,\infty) \times R^n) \cap \bar{Z}$ and $\bar{Z}$ is invariant under the obvious extension of T to $(-\infty,\infty) \times R^n$. Note that $T \circ \bar{r}_t = \bar{r}_t \circ T$. Since Z is compact, there is an integer M' such that $|p_{(-\infty,\infty)} \circ \bar{r}_t(s,x) - s| < M'$ for all $(s,x) \in \bar{Z}$ and $0 \le t \le 1$. Here, $p_{(-\infty,\infty)} : (-\infty,\infty) \times R^n \to (-\infty,\infty)$ is the projection. Let $\rho : (-\infty,\infty) \to [0,1]$ be a function such that $\rho(t) = 0$ for $t \le M'$ and $\rho(t) = 1$ for $t \ge M' + 1$. Then the function $h_t : X^- \to X^-$ defined by $h_t(s,x) = \bar{r}_{\rho(s)\cdot t}(s,x)$ is well-defined. Since $\bar{r}_t$ is a strong deformation retraction and commutes with T, h_t is periodic near infinity. The extension $\bar{h}_t$ of h_t to X deforms X into $X^*_{2M'+1}$. ∎

§3. It remains to compute $\sigma(x)$. X_N^* is obtained by attaching $N + 1$ bouquets $\vee_{i=1}^{\ell} S_i^2$ of two-spheres and N families $\{D_i^3\}$ of three-cells to $K \times [0,\infty]$. It is evident as in §1 that $\pi_2 X_N^* \cong \pi_2 K \oplus Q \oplus P$ where Q corresponds to the part of the first bouquet missed by the first family of discs and P corresponds to the part of the last bouquet missed by the last family of discs.

The deformation of X into X_N^* is rel the first bouquet. Therefore, if an element of Q bounds a singular disc D in X, then it bounds a singular disc $h_1(D)$ in X_N^*. Thus, Q injects into $\pi_2 X$. P, on the other hand, clearly bounds in X, so $\ker_2(i_* : \pi_2 X_N^* \to \pi_2 X) = P$. Unfortunately, X_N^* is three-dimensional, so our situation does not satisfy the hypotheses of Wall's theorem. This can be remedied, however, by attaching 3-cells to X_N^* to kill P. The resulting space X_N^{**} is three-dimensional and dominates X. The kernel $\ker_3 d_*$ is stably isomorphic to Q. Thus, $\sigma(X) = (-1)^3[Q] = [P] \neq 0$.

This completes the proof. ∎

REFERENCES

[E-G] D.A. Edwards and R. Geoghegan, Shapes of complexes, ends of manifolds, homotopy limits, and the Wall obstruction, Annals of Math. 101 (1975), 521-535 with correction 104 (1976) 389.

[F] Steve Ferry, Homotopy, simple homotopy, and compacta, Topology 19(1980),101-110.

[G] R. Geoghegan, Fixed points in finitely dominated compacta: the geometric meaning of a conjecture of H. Bass, these proceedings.

[Wa] C.T.C. Wall, Finiteness conditions for CW complexes, Annals of Math. 81 (1965), 55-69.

University of Kentucky
Department of Mathematics
Lexington, Kentucky 40506

Fixed points in finitely dominated compacta: the geometric meaning of a conjecture of H. Bass

by

Ross Geoghegan

§1. Introduction

The conjecture of Bass mentioned in the title appears in [B]. We will begin by stating it.

Let G be a group, G_1 the set of conjugacy classes in G, $\mathbb{Z}[G_1]$ the free abelian group generated by G_1, and $p\colon \mathbb{Z}[G]\to\mathbb{Z}[G_1]$ the obvious quotient of the additive group of the ring $\mathbb{Z}[G]$. If A is an $n \times n$ $\mathbb{Z}[G]$-matrix the Hattori-Stallings trace of A is $T(A) = \sum_{i=1}^{n} p(a_{ii}) \in \mathbb{Z}[G_1]$. Denote the conjugacy class $\{1\}$ by $C(1)$.

Strong Conjecture 1.1: If A is idempotent (i.e. $A^2 = A$) then $T(A) = (\text{integer}).C(1)$.

Accompanying this is the

Weak Conjecture 1.2: If A is idempotent and $T(A) = \Sigma\{n_C . C \mid C \in G_1,\ n_C \in \mathbb{Z}\}$ then $\Sigma\{n_C \mid C \neq C(1)\} = 0$.

In either case, if the conjecture is true for all finitely generated groups G, it is true for all groups. If one thinks of the conjectures as dependent on G (i.e. statements for all idempotent A, given G) then the Strong Conjecture is known to be true for various classes of groups including finite groups, abelian groups and linear groups, while the Weak Conjecture is known to be true for residually finite groups. See [B].

This lecture is about the geometric meaning of the Strong Conjecture 1.1 when G is finitely presented. (A complete set of analogous statements exists concerning the Weak Conjecture 1.2, but we will not give them.) We wish to show that the Strong Conjecture 1.1 is equivalent to a geometrically plausible conjecture about fixed point theory on finitely dominated compact metric spaces (≡ compacta). We maintain that the geometrical conjecture has been around <u>implicitly</u> for about fifty years, and that the recent theorem of Ferry [F_1] stating that every finitely dominated space has the homotopy type of a compactum, makes it reasonable to hope for a proof. Our main theorem is

<u>Theorem 6.4</u>: Let G be a finitely presented group and A an idempotent $\mathbb{Z}[G]$-matrix. Then the following are equivalent:

(i) the Strong Conjecture 1.1 is true for A ;

(ii) there exists a connected finitely dominated compactum Z such that $\pi_1(Z,z) = G$, $\sigma(Z) = [A]$, and $(Z, \text{identity}_Z)$ supports fixed point theory;

(iii) whenever Z is a connected finitely dominated compactum such that $\pi_1(Z,z) = G$, $\sigma(Z) = [A]$ and $\chi(Z) = 0$, then $(Z, \text{identity}_Z)$ supports fixed point theory.

In the above theorem, "finitely dominated" means homotopically dominated by finite complex, $\sigma(Z) \in \tilde{K}_0(\mathbb{Z}[G])$ is the Wall finiteness obstruction of Z , and $[A]$ is the element of $\tilde{K}_0(\mathbb{Z}[G])$ defined by A . Much of the lecture will consist of a review of concepts needed in order to define the statement that (Z,f) "supports fixed point theory", where Z is compact and $f : Z \to Z$ is a map. These concepts are: the Reidemeister trace of a map on a finite complex (§2), the extension of this to finitely dominated spaces (§3), homotopically essential fixed points (§5) and the way in which the Reidemeister trace detects homotopically essential fixed points (§6). The connection between the Strong Conjecture 1.1 and the Reidemeister trace is made in Theorem 4.1. In Appendix I we note a useful algebraic fact. Appendix II explains how these matters are related to previous work.

It is our hope that the program organized here will lead to a geometric proof of the Strong Conjecture 1.1 for finitely presented G . If, however, the Strong Conjecture turns out to be false (in which case a counterexample would likely be found algebraically) then we feel the discussion here still has interest. For it reveals the obstruction to the existence of a compactum Z, having given Wall obstruction, such that $(Z, \text{identity}_Z)$ supports fixed point theory. What this means is explained in §6.

We end this introduction by establishing some notation. $\simeq$ denotes homotopy. If $J : f \simeq g : Z \to Z$ is a homotopy, $z \in Z$ is a base point, and $\omega = J(z,.)$ is the loop at z traced out by z, we write $J : f \underset{\omega}{\simeq} g$. If ω is constant we write $J : f \underset{z}{\simeq} g$. A <u>CW polyhedron</u> is a CW complex whose underlying space is a polyhedron, each closed cell being a subpolyhedron; this ensures that all base points are non-degenerate. If g is an element of a group G, $T_g : G \to G$ denotes the conjugation $x \longmapsto g^{-1}.x.g$.

§2. The Reidemeister trace of a map on a finite complex

If $f : X \to X$ is a map on a finite complex the Lefschetz number, $L(f)$, is a widely understood invariant. In the 1930's, Reidemeister [R_1] studied the analogue of $L(f)$ in the universal cover, i.e. the corresponding invariant defined by $\tilde{f} : \tilde{X} \to \tilde{X}$ when the chains in $\tilde{X}$ are regarded as finitely generated $\mathbb{Z}[\pi_1]$-modules. The invariant so obtained, more delicate then $L(f)$, is called the "Reidemeister trace" of f. Just as $L(f) \neq 0$ implies all maps homotopic to f have fixed points, so the Reidemeister trace was shown by Wecken [We] to give information about the minimum number of fixed points of maps homotopic to f (see §6).

In this section we describe the Reidemeister trace, since we know of no suitable exposition in the literature. However, a special case (when f induces the identity on π_1) is well explained in [S]. We omit proofs of propositions in this section because they can be obtained from the corresponding proofs in [S] by routine changes.

Let X be a connected finite CW polyhedron, $x \in X$, and

$(\tilde{X},\tilde{x})$ the pointed universal cover. Write $\pi_1(X,x) = G$ and identify G with the group of covering transformations of $\tilde{X}$ in the usual way, namely: if $[\omega] \in \pi_1(X,x)$ and $\tilde{\omega} : I \to \tilde{X}$ is the lift of ω such that $\tilde{\omega}(0) = \tilde{x}$, then $h_{[\omega]}$ is the covering transformation such that $h_{[\omega]}(\tilde{x}) = \tilde{\omega}(1)$. Thus $h_{[\omega'][\omega]} = h_{[\omega']} \circ h_{[\omega]}$.

Denote the integral group ring $\mathbb{Z}[G]$ by Λ.

For each n, regard the abelian group of cellular n-chains (see [C p.7]) as a right Λ-module: if c is a cellular n-chain and $[\omega] \in G$ define $c.[\omega] = (h_{[\omega]}-1)_{\#}(c)$, and extend this action of G $\mathbb{Z}$-linearly to an action of Λ. The boundary homomorphism respects this action. Denote the right Λ-chain complex of cellular chains in $\tilde{X}$ by $(C_*(X), \partial)$; this is a good notation because the n-chains in $\tilde{X}$ are finitely generated (over Λ) by the oriented n-cells of X; see [C p.11].

Identify G with the natural set of generators in $\Lambda \equiv \mathbb{Z}[G]$. A group homorphism $\phi : G \to G$ has a natural extension to a ring homomorphism $\phi : \Lambda \to \Lambda$. If M is a right Λ-module, a ϕ-<u>endomorphism</u> of M is an additive morphism α with $\alpha(m,\lambda) = \alpha(m).\phi(\lambda)$ for all $m \in M$, $\lambda \in \Lambda$. (An ordinary endomorphism corresponds to $\phi = 1 \equiv \text{identity}_G$). We come at once to the important example of a ϕ-endomorphism.

Let $f : X \to X$ be a cellular map such that $f(x) = x$, and let $\phi = f_{\#} : G \to G$. The unique pointed lift of f, $\tilde{f} : (\tilde{X},\tilde{x}) \to (\tilde{X},\tilde{x})$, induces a ϕ-endomorphism $\tilde{f}_{\#} : C_n(X) \to C_n(X)$ for each n.

We wish to define the trace of a ϕ-endomorphism $\alpha : F \to F$ on a finitely generated free right Λ-module. The naive definition would be: pick a basis for F and add the diagonal elements of the corresponding matrix of α, but in general this "naive trace" depends on the choice of basis. The correct definition, which we now give, is due to Reidemeister $[R_1]$. The Hattori-Stallings trace in §1 is a special case.

If $\phi : G \to G$ is a homomorphism, elements x and y of G are ϕ-conjugate if there exists z such that $x = z.y.\phi(z)^{-1}$. Let G_ϕ be the set of ϕ-conjugacy classes in G, $\mathbb{Z}[G_\phi]$ the free abelian group generated by G_ϕ, and $p : \mathbb{Z}[G] \to \mathbb{Z}[G_\phi]$ the obvious quotient of the additive group of $\mathbb{Z}[G]$. The p-image of the "naive trace" of α is well-defined (independent of basis). Extending the notation of §1 (when $\phi = 1$), we denote this trace by $T(\alpha) \in \mathbb{Z}[G_\phi]$.

Returning to the ϕ-endomorphisms $\tilde{f}_\# : C_n(X) \to C_n(X)$ we define the Reidemeister trace of the pointed map $f : (X,x) \to (X,x)$ to be

$$R(f,x) = \sum_{n\geq 0} (-1)^n T(\tilde{f}_\# : C_n(X) \to C_n(X)) .$$

$R(f,x)$ lies in $\mathbb{Z}[G_\phi]$.

We now discuss the homotopy invariance of $R(f,x)$. Let $J : f \underset{\omega}{\simeq} g$ be cellular where $g : (X,x) \to (X,x)$. If $\psi = g_\# : G \to G$ then $\psi = T_{[\omega]} \circ \phi$. The relationship between ϕ-conjugacy cl sses and ψ-conjugacy classes in G is simple: right multiplication by $[\omega]$ turns a ϕ-conjugacy class into a ψ-conjugacy class.

Proposition 2.1: If $R(f,x) = \Sigma\{n_C.C \mid C \in G_\phi\} \in \mathbb{Z}[G_\phi]$ (n_C being the $\mathbb{Z}$-coefficient of C), then $R(g,x) = \Sigma\{n_C.(C.[\omega]) \mid C \in G_\phi\} \in \mathbb{Z}[G_\psi]$.

Next, suppose (Y,y) is another pointed connected finite CW polyhedron and $(X,x) \underset{h}{\overset{g}{\rightleftarrows}} (Y,y)$ are pointed cellular maps. We wish to compare $R(h \circ g,x)$ with $R(g \circ h,y)$. Let $G' = \pi_1(Y,y)$ and $\Lambda' = \mathbb{Z}[G']$. Let $\mu = (h \circ g)_\# : G \to G$ and $\nu = (g \circ h)_\# : G' \to G'$. Then $[\alpha]$ and $[\beta]$ in G are μ-conjugate if and only if $g_\#([\alpha])$ and $g_\#([\beta])$ are ν-conjugate. Thus g induces a bijection $g_\# : G_\mu \to G'_\nu$.

Proposition 2.2: If $R(h \circ g,x) = \Sigma\{n_C.C \mid C \in G_\mu\}$ then $R(g \circ h,y) = \Sigma\{n_C.(g_\# C) \mid C \in G_\mu\} \in \mathbb{Z}[G'_\nu]$.

We have defined $R(f,x)$ when g is cellular and $f(x) = x$. We extend the definition to non-cellular maps using the Cellular

Approximation Theorem as follows. Let y be a vertex of the CW complex X. Pick a homotopy J from f to a cellular map g such that $g(y) = y$. Let σ be the path $J(y,.)$ from $f(y)$ to y, and let γ be any path from x to y. If $H = \pi_1(X,y)$ and $\psi = g_{\#}$ then $[\alpha]$ and $[\beta]$ are ψ-conjugate if and only if $[\gamma.\alpha.\sigma^{-1}.f\gamma^{-1}]$ and $[\gamma.\beta.\sigma^{-1}.f\gamma^{-1}]$ are ϕ-conjugate. Let the induced bijection be $\Gamma : H_\psi \to G_\phi$. Γ is independent of γ. If the (already defined) $R(g,y) = \Sigma\{n_C.C|C\epsilon H_\psi\}$, we define

$$R(f,x) = \Sigma\{n_C.\Gamma(C)|C\epsilon H_\psi\} \in \mathbb{Z}[G_\phi] .$$

The reader can check that this is well-defined, and that Propositions 2.1 and 2.2 continue to hold.

As a special case of the above, we observe that if $f(x) = x$ and $f(y) = y$ then $R(f,x)$ and $R(f,y)$ determine each other in a simple way. Read the above with $g = f$ and J the constant homotopy to get:

<u>Proposition 2.3</u>: If $R(f,y) = \Sigma\{n_C.C|C\epsilon H_\psi\}$ then $R(f,x) = \Sigma\{n_C.\Gamma(C)|C\epsilon H_\psi\}$.

There is little of fixed points in all this, apart from the fact that x is a fixed point of f. The connection with fixed points will come in §6. For now we simply observe two immediate consequences of the definitions:

<u>Proposition 2.4</u>: (a) If $R(f,x) = \Sigma\{n_C.C|C\epsilon G_\phi\}$ then $\Sigma\{n_C\} = L(f)$, the Lefschetz number of f. (b) $R(\text{identity}_X,x) = \chi(X).C(1)$, where $C(1)$ is the ϕ-conjugacy class of $1 \in G$ and $\chi(X)$ is the Euler characteristic of X.

§3. <u>Extension to finitely dominated spaces</u>

A connected space Z is <u>finitely dominated</u> if there exist $Z \underset{d}{\overset{u}{\rightleftarrows}} X$ such that X is a connected finite CW polyhedron, and $d \circ u \simeq \text{identity}_Z$. Let $f : Z \to Z$ be a map on a finitely dominated space, and let $f(z) = z$. We wish to define a Reidemeister trace $R(f,z)$ extending the definition given in §2 when Z is a finite complex.

We can certainly choose X, u and d so that for some $x \in X$, $u(z) = x$ and $d(x) = z$. Using z as base point, let $d \circ u \underset{\alpha}{\simeq} \text{identity}_Z$. We write $H = \pi_1(Z,z), \psi = f_{\#}$, $G = \pi_1(X,x)$, $\phi = (u \circ f \circ d)_{\#}$. Then $[\beta]$ and $[\gamma]$ are ϕ-conjugate if and only if $d_{\#}([\beta]).[\alpha]$ and $d_{\#}([\gamma]).[\alpha]$ are ψ-conjugate. Since $d_{\#}$ is onto, there is an induced bijection $\Delta : G_\phi \to H_\psi$. If $R(u \circ f \circ d, x) = \Sigma\{n_C . C | C \in G_\phi\}$, we define $R(f,z) = \Sigma\{n_C . \Delta(C) | C \in G_\phi\} \in \mathbb{Z}[H_\psi]$. By a long but straightforward calculation using Propositions 2.1 and 2.2, one checks that this definition is independent of choices (compare $R(u \circ d' \circ u' \circ f \circ d, x)$, $R(u' \circ f \circ d \circ u \circ d', x')$, $R(u \circ f \circ d, x)$ and $R(u' \circ f \circ d', x)$, where $Z \underset{d'}{\overset{u'}{\rightleftarrows}} X'$ is another choice).

All the Propositions in §2 extend immediately except for Proposition 2.4(b). It is far from clear that $R(\text{identity}_Z, z) = \chi(Z).C(1)$. In fact Theorem 4.1 will say that this is equivalent to the Strong Conjecture 1.1.

§4. Finitely dominated spaces and the Strong Conjecture

Let Z be a connected finitely dominated space, $z \in Z$, and $G = \pi_1(Z,z)$. The Wall finiteness obstruction of Z is an element $\sigma(Z) \in \tilde{K}_0(\mathbb{Z}[G])$ which is zero if and only if Z is homotopy equivalent to a finite complex [W]. [See Appendix I for the definition of $\tilde{K}_0(\mathbb{Z}[G])$, and see the proof of Lemma 4.3 for the definition of $\sigma(Z)$].

If A is an idempotent $\mathbb{Z}[G]$-matrix representing an endomorphism α of a free $\mathbb{Z}[G]$-module F, then $\text{image}(\alpha : F \to F)$ is projective and its class in $\tilde{K}_0(\mathbb{Z}[G])$ will be denoted by $[A]$.

The connection between the Strong Conjecture, Reidemeister traces and the Wall obstruction can now be stated:

<u>Theorem 4.1</u>: Let G be a finitely presented group and A an idempotent $\mathbb{Z}[G]$-matrix. Then the following are equivalent:

(i) the Strong Conjecture 1.1 is true for A;

(ii) for some connected finitely dominated space Z such that $\pi_1(Z,z) = G$ and $\sigma(Z) = [A]$, $R(\text{identity}_Z, z) = \chi(Z).C(1)$;

(iii) for all connected finitely dominated spaces Z such that $\pi_1(Z,z) = G$ and $\sigma(Z) = [A]$, $R(\text{identity}_Z, z) = \chi(Z).C(1)$.

In preparation for the proof we need two lemmas.

<u>Lemma 4.2</u>: Let A and A' be idempotent $\mathbb{Z}[G]$-matrices. If $[A] = [A']$ in $\tilde{K}_0(\mathbb{Z}[G])$ then $T(A) - T(A') = (\text{integer}).C(1)$.

The proof of 4.2 is easy, and is omitted.

<u>Lemma 4.3</u>: Let Z be connected and finitely dominated, $z \in Z$, and $\pi_1(Z,z) = G$. Then there is an idempotent $\mathbb{Z}[G]$-matrix A such that $\sigma(Z) = [A]$ and $R(\text{identity}_Z, z) = T(A) + (\text{integer}).C(1)$.

<u>Proof</u>: By [W] we may assume Z is a CW complex of finite type (i.e. finite k-skeleton for all k). Extending the notation and concepts of §2, consider the cellular chain complex $C_*(Z)$. Let $Z \underset{d}{\overset{u}{\rightleftarrows}} X$ be such that X is a finite complex, $d \circ u \underset{\omega}{\simeq} \text{identity}_Z$, and $\pi_1(Z,z) \underset{d_\#}{\overset{u_\#}{\rightleftarrows}} \pi_1(X,x)$ are isomorphisms, where $x = u(z)$. It is well-known that this can be arranged. Let $n = \dim X$. Then $\sigma(Z)$ is represented by the projective $\mathbb{Z}[G]$-module $C_n(Z)/B_n(Z) \equiv D_n(Z)$, as is explained in [G] where it is also shown (implicitly - see Theorem 1.4 of [G] and adapt the proof) that $C_*(Z)$ is chain homotopy equivalent to the finite projective chain complex

$$\mathcal{D}_* : \ldots \to 0 \to D_n(Z) \xrightarrow{\bar{\partial}} C_{n-1}(Z) \xrightarrow{\partial} C_{n-2}(Z) \xrightarrow{\partial} \ldots$$

Now $D_n(Z)$ is a direct summand of a free module F. $\alpha \equiv$ (projection on $D_n(Z)$) $: F \to F$ is an idempotent. Form the finite free chain complex

$$\mathcal{E}_* : \ldots \to 0 \to F \xrightarrow{\partial \circ \alpha} C_{n-1}(Z) \xrightarrow{\partial} \ldots$$

Clearly there are chain maps $\mathcal{D}_* \underset{\overline{d}}{\overset{\overline{u}}{\rightleftarrows}} \mathcal{E}_*$ such that $\overline{d}\circ\overline{u}$ = identity, and $\overline{u}\circ\overline{d}$ is the chain map

$$\begin{array}{ccccccc} 0 & \to & F & \to & C_{n-1}(Z) & \to & \cdots \\ & & \downarrow{\scriptstyle\alpha} & & \downarrow{\scriptstyle id} & & \\ 0 & \to & F & \to & C_{n-1}(Z) & \to & \cdots \end{array}$$

The Reidemeister trace of this chain map is

$$(-1)^n T(\alpha) + \left(\sum_{i=0}^{n-1} (-1)^i \text{ rank } C_i(Z)\right).C(1)$$

By the discussion in §3 (extended to chain maps) this Reidemeister trace is also $R(\text{identity}_Z, z)$. Choosing n even, we get the required result. ||

<u>Proof of Theorem 4.1</u>: (ii) → (i): given Z as in (ii), Lemma 4.3 gives A' with $\sigma(Z) = [A']$ and $R(\text{identity}_Z, z) = T(A') + (\text{integer}).C(1)$; apply Lemma 4.2. (iii) → (ii): by [W] there exists Z such that $\pi_1(Z,z) = G$ and $\sigma(Z) = [A]$; apply (iii). (i) → (iii): let Z be such that $\pi_1(Z,z) = G$ and $\sigma(Z) = [A]$; Lemma 4.3 gives A' with $\sigma(Z) = [A']$ and $R(\text{identity}_Z, z) = T(A') + (\text{integer}).C(1)$; by Lemma 4.2 and (i), $T(A') = (\text{integer}).C(1)$; so $R(\text{identity}_Z, z) = (\text{integer}).C(1)$; so $R(\text{identity}_Z, z) = \chi(Z).C(1)$, by §2 and §3. ||

§5. <u>The pure topology of removing fixed points</u>

Let $f : Z \to Z$ be a map on a topological space and let $\Phi(f)$ be the set of fixed points of f . We partition $\Phi(f)$ by the equivalence relation $x \sim y$ iff for some path α from x to y $\alpha \simeq f\alpha$ rel{0,1}. Following [Br] we dènote the set of equivalence classes (called <u>fixed point classes</u>) by $\Phi'(f)$.

Now let $f, f' : Z \to Z$ and let $J : f \simeq f'$. Fixed points x of f and x' of f' are J-<u>related</u> if there is a path σ from

x to x' such that the paths σ and $J(\sigma(.),.)$ are homotopic rel{0,1} . Fixed point classes F of f and F' of f' are J-related if some (equivalently any) $x \in F$ is J-related to some (equivalently any) $x' \in F'$.

Here are some easily checked properties of J-relation; see [Br, pp. 87-92]:

If F is J-related to F' then F' is J^{-1}-related to F . If $J' : f' \simeq f''$, then F J-related to F' and F' J'-related to F'' imply $F(J*J')$-related to F'' . If $f = f'$ and J is the constant homotopy then F is J-related to F' iff $F = F'$. Each $F \in \Phi'(f)$ is J-related to at most one $F' \in \Phi'(f')$.

$F \in \Phi'(f)$ is homotopically essential if for every homotopy $J : Z \times I \to Z$ such that $J(.,0) = f$, F is J-related to some fixed point class of $J(.,1)$. Otherwise F is homotopically inessential. Note that identity_Z has only one fixed point class, so if $f \simeq \text{identity}_Z$ then f has at most one homotopically essential fixed point class. Moreover, if $f \underset{z}{\simeq} \text{identity}_Z$ then that class contains z .

If $Z \underset{h}{\overset{g}{\rightleftarrows}} Z'$ are maps, then $g(\Phi(h \circ g)) = \Phi(g \circ h)$ and one easily checks that there is an induced bijection $g_{\#} : \Phi'(h \circ g) \to \Phi'(g \circ h)$. However on this level of generality (non-compact spaces are allowed) the property of being homotopically essential is not preserved by this bijection.

§6. The Reidemeister trace as detector of fixed points

First consider a map $f : X \to X$ where X is a connected finite CW polyhedron, and let $f(x) = x$. $R(f,x)$ detects fixed points of f in the following way.

Write $G = \pi_1(X,x)$ and $\phi = f_{\#} : G \to G$. Each fixed point class $F \in \Phi'(f)$ determines a ϕ-conjugacy class $C(F) \in G_{\phi}$ by the rule: pick a path α from z to some $y \in F$ and let $C(F)$ be the class of $[\alpha.f\alpha^{-1}] \in G$. The function $F \longmapsto C(F)$ is well-defined and injective ([Br. page 104]) .

Theorem 6.1 ([We]): Let $R(f,x) = \Sigma\{n_C.C|C\in G_\phi\}$. Whenever $n_C \neq 0$, $C = C(F)$ for some fixed point class F of f .

Corollary 6.2: The fixed point classes F detected by $R(f,x)$ in Theorem 6.1 are homotopically essential.

Proof: Consider a homotopy $J : f \underset{\omega}{\simeq} f'$ where $f'(x) = x$. If C has non-zero coefficient in $R(f,x)$ then $C.[\omega]$ has non-zero coefficient in $R(f',x)$, by Proposition 2.1. Theorem 6.1 gives classes $F \in \Phi'(f)$ and $F' \in \Phi'(f')$ corresponding to these coefficients. It is an exercise in the definitions to show that F and F' are J-related. In the same spirit it is easy to show the existence of F' such that F is J-related to F' , even when f' does not fix x . ||

The number, $N(f)$, of non-zero coefficients in $R(f,x)$ is called the Nielsen number of f . By the discussion in §2, $N(f)$ is a homotopy invariant. Corollary 6.2 implies that every map homotopic to f has at least $N(f)$ homotopically essential fixed point classes. Note that the class of x might not be among them.

Now let $f : Z \to Z$ be a map on a connected finitely dominated space, and let $f(z) = z$. As in §3 we may choose $Z \underset{d}{\overset{u}{\rightleftarrows}} X$ where X is a connected finite CW polyhedron, $u(z)=x$, $d(x) = z$ and $d \circ u \simeq \text{identity}_Z$ in order to define $R(f,z)$. Call the data (X,u,d) a finite domination of Z . Theorem 6.1 implies that if $R(u \circ f \circ d,z) = \Sigma\{n_C.C|C\in G_\phi\}$ then, whenever $n_C \neq 0$, $C = C(F)$ for some fixed point class F of $u \circ f \circ d$. Write $H = \pi_1(Z,z)$ and $\psi = f_\# : H \to H$. Then, by §3 and §5, d induces bijections $\Delta : G_\phi \to H_\psi$ and $d_\# : \Phi'(u \circ f \circ d) \to \Phi'(d \circ u \circ f)$. We defined $R(f,z) = \Sigma\{n_C.\Delta(C)|C\in G_\phi\} \in \mathbb{Z}[H_\psi]$ and saw that this definition is independent of the finite domination used. Theorem 6.1 clearly implies:

Theorem 6.3: Let $R(f,z) = \Sigma\{n_C.C|C\in H_\psi\}$. Then every finite domination (X,u,d) of Z has the property that whenever $n_C \neq 0$, $C = C(F)$ for some fixed point class F of $d \circ u \circ f$.

We cannot expect an analogue of Corollarly 6.2 in general. For example, if $Z = [0,\infty)$, X = point, and $d(X) = 0$ then $d \circ u$ is homotopic to a fixed point free map, so the fixed point class detected by $R(\text{identity}_Z, z)$ is not homotopically essential. But what if Z is compact?

Only recently has it been shown by Ferry $[F_1]$ that <u>every finitely dominated space has the homotopy type of a compact metric space</u> ($\equiv$ compactum). A neat construction of compacta realizing every Wall obstruction is given elsewhere in this volume $[F_2]$.

It is a very old intuition in topology that the Lefschetz Fixed Point Theorem, and related notions, ought to carry over from compact polyhedra to some large class of finitely dominated compacta. This intuition motivated the work of Lefschetz himself over a long period; see [Wi] for an interesting discussion. With this in mind we make the following definition. Given a map $f : Z \to Z$ on a finitely dominated compactum, (Z,f) <u>supports fixed point theory</u> if for some (hence any) finite domination (X,u,d) of Z, the fixed point classes of $d \circ u \circ f$ detected by $R(f,z)$ (in the sense of Theorem 6.3) are homotopically essential. We are picking out the class of (Z,f) for which Corollary 6.2 generalizes.

There is a famous example of a contractible compactum Z for which $(Z, \text{identity}_Z)$ does not support fixed point theory. It is the "can with roll of toilet paper" of Kinoshita [K] (named by Bing!): From the compact spiral $\Sigma \equiv \{\frac{\theta}{1+\theta} e^{i\theta} \mid \theta \geq 0\} \cup S^1$ form $Z = (\Sigma \times I) \cup (B^2 \times \{0\})$. This contractible compactum admits fixed point free maps. (In comparing with 6.4 below, note that $\chi(Z) \neq 0$).

Very few such examples are known, and how they occur is not understood. See [Wi; page 388] for a conjecture. It would be very interesting to know if, for each possible Wall obstruction σ, one of Ferry's compacta Z having $\sigma(Z) = \sigma$ has the property that $(Z, \text{identity}_Z)$ supports fixed point theory, or whether such compacta could be found. Because to decide this question would be to settle the Strong Conjecture 1.1, as the following theorem shows:

Theorem 6.4: Let G be a finitely presented group and A an idempotent $\mathbb{Z}[G]$-matrix. Then the following are equivalent:

(i) the Strong Conjecture 1.1 is true for A ;

(ii) there exists a connected finitely dominated compactum Z such that $\pi_1(Z,z) = G$, $\sigma(Z) = [A]$, and $(Z, \text{identity}_Z)$ supports fixed point theory;

(iii) whenever Z is a connected finitely dominated compactum such that $\pi_1(Z,z) = G$, $\sigma(Z) = [A]$, and $\chi(Z) = 0$, then $(Z, \text{identity}_Z)$ supports fixed point theory.

Proof: (i) $\rightarrow$ (ii): By $[F_1]$ or $[F_2]$ there exists a finitely dominated compactum Z for which $\pi_1(Z,z) = G$ and $\sigma(Z) = [A]$. By wedging on spheres of dimension > 1 we may assume $\chi(Z) = 0$. By Theorem 4.1, $R(\text{identity}_Z, z) = 0$. So no fixed points are detected, so $(Z, \text{identity}_Z)$ supports fixed point theory. (ii) $\rightarrow$ (iii): Let Z be as in (ii) and let (X,u,d) be a finite domination of Z . As is well known, we can arrange that $d \circ u \underset{z}{\simeq} \text{identity}_Z$. By remarks in §5 the only possible homotopically essential fixed point class of $d \circ u$ is the class of z . Since $(Z, \text{identity}_Z)$ supports fixed point theory, $R(\text{identity}_Z, z) = \chi(Z).C(1)$. Hence, by Theorem 4.1, $R(\text{identity}_{Z'}, z') = 0$ whenever Z' is as in (iii), so $(Z', \text{identity}_{Z'})$ supports fixed point theory. (iii) $\rightarrow$ (i): By $[F_1]$ or $[F_2]$ there exists a connected finitely dominated compactum Z such that $\pi_1(Z,z) = G$, $\sigma(Z) = [A]$ and $\chi(Z) = 0$. If the Strong Conjecture 1.1 is false for A , then, by Theorem 4.1, $R(\text{identity}_Z, z)$ detects a fixed point class of $d \circ u$ other than that of z , where (X,u,d) is a finite domination such that $d \circ u \underset{z}{\simeq} \text{identity}_Z$. As explained above, such a class cannot be homotopically essential, so $(Z, \text{identity}_Z)$ does not support fixed point theory. ||

Appendix I: $\tilde{K}_0$ is infinite whenever the Strong Conjecture is false

Let P be the set of isomorphism classes of finitely generated projective right $\mathbb{Z}[G]$-modules, $\mathbb{Z}[P]$ the free abelian group generated by P, $W[P]$ the subgroup of $\mathbb{Z}[P]$ generated by $\{P \oplus Q - P - Q\}$ and $K_0(\mathbb{Z}[G])$ the quotient abelian group $\mathbb{Z}[P]/W[P]$. Define rank $P = \dim_{\mathbb{Z}}(P \otimes_{\mathbb{Z}[G]} \mathbb{Z})$. This induces an epimorphism, rank: $K_0(\mathbb{Z}[G]) \to \mathbb{Z}$, whose cokernel is defined to be $\tilde{K}_0(\mathbb{Z}[G])$.

If A is an idempotent $n \times n$ $\mathbb{Z}[G]$-matrix, let $\alpha : F \to F$ be the corresponding idempotent endomorphism on a free $\mathbb{Z}[G]$-module of rank n. Image(α) is projective, hence has a representative $\bar{A} \in K_0(\mathbb{Z}[G])$. We have a commutative diagram with the top line exact:

$$
\begin{array}{ccccccccc}
0 & \to & \mathbb{Z} & \to & K_0(\mathbb{Z}[G]) & \to & \tilde{K}_0(\mathbb{Z}[G]) & \to & 0 \\
 & & & \searrow^{s} & \downarrow t & & & & \\
 & & & & \mathbb{Z}[G_1] & & & &
\end{array}
$$

where $t(\bar{A}) = T(A)$ and $s(1) = C(1)$. If for some A, $T(A) \neq$ (integer)$.C(1)$, then the induced homomorphism $\tilde{K}_0(\mathbb{Z}[G]) \to \mathbb{Z}[G_1 \setminus \{C(1)\}]$ is non-zero, so $\tilde{K}_0(\mathbb{Z}[G])$ maps onto a non-trivial free abelian group.

Hence we have proved

<u>Proposition</u>: If A is an idempotent $\mathbb{Z}[G]$ matrix for which the Strong Conjecture 1.1 is false, then $\tilde{K}_0(\mathbb{Z}[G])$ has a $\mathbb{Z}$-summand.

We draw attention to this because $\tilde{K}_0(\mathbb{Z}[G])$ is often known to be finite, in particular (Swan's Theorem) whenever G is finite.

Appendix II. Connections with a previous paper

This lecture was given to an audience of geometric topologists with a special interest in shape theory. We will explain the connection.

We have discussed the fixed point theory of finitely dominated compacta. There are natural bijections between: homotopy types of finitely dominated compacta, shape types of compacta shape dominated by finite complexes (≡ FANR's), and shift equivalence classes of homotopy idempotents on finite complexes (f and f' are "shift-equivalent" if $f = h \circ g$ and $f' = g \circ h$); see the paper of Hastings in this volume, and also $[F_1]$. Hence, by Theorem 4.1, the element $T(A) - n_{C(1)} \cdot C(1)$ which measures the failure of the Strong Conjecture is a shape invariant of FANR's. However we believe it is more useful to look on it as a homotopy invariant of finitely dominated compacta.

There is a point worth noting here. Before Hastings and Heller proved that every homotopy idempotent on a finite complex splits (see Hastings' paper in this volume) we had attempted to prove that result for homotopy idempotents $f : X \to X$ for which the Nielsen number $N(f) > 0$. The main part of our work was the following theorem [Ge]:

<u>Theorem</u>: Let $f : X \to X$ be a homotopy idempotent on a finite complex, $f(x) = x$, and let $n_{C(1)} \neq 0$ in $R(f,x)$. Let $H : f \underset{\omega}{\simeq} f^2$. Then there are integers $0 < m < n$ and a loop σ based at x such that, in $\pi_1(X,x)$, $[\omega]^m = [\sigma].f^n_{\#}([\sigma]^{-1})$.

This theorem and Bass' theorem [B; §9] that the Strong Conjecture is true for linear groups, appear to be essentially the same, in the sense that the proof of either could (it seems) be altered easily to give the other. The above Theorem is a generalization of a theorem of Gottlieb, which in turn is known to be relevant to [S], which in turn is known to be relevant to the Strong Conjecture, so this is not altogether surprising.

Finally we point out that Theorem 4.1 can be restated entirely in the language of classical fixed point theory. A homotopy idempotent on a finite complex, $f : X \to X$, always splits; i.e. there exist $Z \underset{d}{\overset{u}{\rightleftarrows}} X$ with $d \circ u \simeq \text{identity}_Z$ and $u \circ d \simeq f$. See Hastings' paper for details. Then $\sigma(f) \equiv u_{\#}(\sigma(Z)) \in \tilde{K}_0(\mathbb{Z}[\pi_1(X,x)]$ can be interpreted as the obstruction to splitting f through a finite complex. Clearly we have:

Theorem 4.1$'$: Let G be a finitely presented group and A an idempotent $\mathbb{Z}[G]$-matrix. Then the following are equivalent:

(i)$'$ the Strong Conjecture 1.1 is true for A;

(ii)$'$ for some homotopy idempotent on a connected finite complex $f : X \to X$ such that $\pi_1(X,x) = G$ and $\sigma(f) = [A]$, $N(f) = 0$ or 1;

(iii)$'$ for all homotopy idempotents on connected finite complexes $f : X \to X$ such that $\pi_1(X,x) = G$ and $\sigma(f) = [A]$, $N(f) = 0$ or 1.

State Univeristy of New York at Binghamton
Binghamton, NY 13901
USA

References

[B] H. Bass, Euler characteristics and characters of discrete groups, Inventiones Math. 35 (1976) 155-196.

[Br] R. F. Brown, The Lefschetz fixed point theorem, Scott Foresman, Glenview and London, 1971.

[C] M. M. Cohen, A course in simple homotopy theory, Graduate Texts in Mathematics, Springer-Verlag, New York, Heidelberg, Berlin, 1973.

[F_1] S. Ferry, Homotopy, simple homotopy and compacta, Topology 19 (1980) 101-110.

[F_2] S. Ferry, Finitely dominated compacta need not have finite type, this volume.

[Ge] R. Geoghegan, The homomorphism on fundamental group induced by a homotopy idempotent having essential fixed points, Pacific J. Math. (to appear).

[G] S. Gersten, A product formula for Wall's obstruction, Amer. J. Math. 88 (1966) 337-346.

[K] S. Kinoshita, On some contractible continua without the fixed point property, Fund. Math. 40 (1953) 96-98.

[R_1] K. Reidemeister, Automorphismen von Homotopiekettenringen, Math. Ann. 112 (1936), 586-593.

[R_2] K. Reidemeister, Complexes and homotopy chains, Bull. Amer. Math. Soc. 56 (1950), 297-307.

[S] J. R. Stallings, Centerless groups - an algebraic formulation of Gottlieb's theorem, Topology 4 (1965) 129-134.

[W] C. T. C. Wall, Finiteness conditions for CW complexes, Annals of Math., 81 (1965), 55-69.

[We] F. Wecken, Fixpunktklassen, I, II, III, Math. Ann. 117 (1941) 659-671; 118 (1942) 216-234 and 544-577.

[Wi] R. L. Wilder, Some mapping theorems with applications to non-locally connected spaces, pp. 378-388 of Algebraic Geometry and Topology, a symposium in honor of S. Lefschetz, edited by R. H. Fox, D. C. Spencer and A. W. Tucker, Princeton University Press, Princeton, 1957.

SPLITTING HOMOTOPY IDEMPOTENTS

Harold M. Hastings
Department of Mathematics
Hofstra University
Hempstead, NY 11550 USA
and
Alex Heller
Department of Mathematics
CUNY Graduate Center
33 West 42nd Street
New York, NY 10036 USA

We shall discuss the question of splitting homotopy idempotents and its applications in pro-homotopy and shape theory. A map $f:X \dashrightarrow X$ is a homotopy idempotent if f^2 is homotopic to f ($f^2 \simeq f$); f is said to split if there are maps $d:X \dashrightarrow Y$ and $u:Y \dashrightarrow X$ with $du \simeq 1_Y$ and $ud \simeq f$. We consider the case where X is a pointed, connected CW complex, and f a pointed map. It is well known (D. A. Edwards and R. Geoghegan, 1975*; the result is also implicit in P. Freyd, 1966) that pointed homotopy idempotents split. We sketch a proof in Section 2 to establish the role of the fundamental group.

The next three sections contain the main results on unpointed homotopy idempotents. We begin by sketching a proof of the following result in Section 3. The first part was proven independently by J. Dydak (1977a) and P. Minc and P. Freyd and A. Heller (1978); the second part is due to Freyd and Heller.

Theorem A (Dydak (1977a) - Minc; Freyd-Heller, 1978). There is a group G whose $K(G,1)$ admits an unsplit, unpointed homotopy idempotent f. Also, G is universal in the following sense. Suppose a complex X admits an unsplit, unpointed homotopy idempotent f'. Then there is an injection $G \dashrightarrow \pi_1 X$, equivariant with respect to f_* and f'_*.

Dydak (1977b) found several applications to shape theory; however the space $K(G,1)$ is infinite-dimensional so the tower

$$K(G,1) \overset{f}{\longleftarrow} K(G,1) \overset{f}{\longleftarrow} \dots$$

---

* We use this format for references to emphasize the historical development of the study of homotopy idempotents.

does not yield an interesting compactum in the limit (cf. Edwards and Geoghegan, 1975).

We next consider finite-dimensional complexes. In 1978, Dydak and the first-named author proved the following.

Theorem B (Dydak-Hastings, 1978). (Unpointed) homotopy idempotents on two-dimensional complexes split.

Two years later we proved the following theorem.

Theorem C (Hastings-Heller, 1980). (Unpointed) homotopy idempotents on finite-dimensional complexes split.

We review the proofs in Sections 4 and 5, respectively. The proof of Theorem B is included for historical reasons and because it is simpler than the proof of Theorem C.

The concluding section, Section 6, contains geometric applications and a brief discussion of open problems.

2. Pointed homotopy idempotents.

We prove the well-known result that pointed homotopy idempotents split, largely following Freyd (1966), and discuss its consequences. From now on, all (CW) complexes are assumed to be pointed and connected, and all maps are assumed to preserve basepoints. It is easy to reduce all splitting questions to this case. Homotopies need not preserve basepoints unless otherwise stated. We shall write $\simeq$ for unpointed homotopies and $\simeq_*$ for pointed homotopies.

(2.1) Theorem. Let X be a complex and f a pointed homotopy idempotent. Then f splits.

Proof. We try to construct the "image" of f by forming the telescopes (infinite mapping cylinders)

(2.2) $T = \mathrm{Tel}\,(X \xrightarrow{f} X \xrightarrow{f} X \xrightarrow{f} \ldots)$,

$\tilde{T} = \mathrm{Tel}\,(\tilde{X} \xrightarrow{\tilde{f}} \tilde{X} \xrightarrow{\tilde{f}} \tilde{X} \xrightarrow{\tilde{f}} \ldots)$.

Define maps d: X--->T by inclusion at the base of the telescope and u: T--->X using the pointed-homotopy commutative diagram

(2.3) $$\begin{array}{l} X \xrightarrow{f} X \xrightarrow{f} X \xrightarrow{f} \cdots \\ \quad f \searrow \;\; f \downarrow \;\; f \swarrow \\ \qquad X \end{array}$$

By construction, $f \simeq_* ud$. It is easy to check that

$$\pi_1 T = \operatorname{colim}\,(\pi_1 X \xrightarrow{f_*} \pi_1 X \xrightarrow{f_*} \ldots), \text{ and}$$

$$H_* \tilde{T} = \operatorname{colim}\,(H_* \tilde{X} \xrightarrow{f_*} H_* \tilde{X} \xrightarrow{f_*} \ldots),$$

and finally that du induces the identity on π_1 and H_*. Therefore, du is a pointed homotopy equivalence.

To show that ud is the identity, let g be a (pointed) homotopy inverse to ud. Then

$$f^2 \simeq_* f,$$

$$ud \cdot ud \simeq_* ud,$$

$$gd \cdot ud \cdot ud \cdot ug \simeq_* gd \cdot ud \cdot ug,$$

$$gdu \cdot du \cdot dug \simeq_* gdu \cdot dug,$$

$$du \simeq_* 1,$$

as required. Therefore f splits. //

We observe that if f were an unpointed homotopy idempotent, the above proof would break down at only one place: the map du (which could always be pointed) need not induce an isomorphism on the fundamental group.

About five years ago D. A. Edwards and R. Geoghegan (1975) constructed strange compacta in shape theory (compacta shape dominated by finite complexes, but not of the shape of finite complexes). They also constructed strange ends in proper homotopy theory by applying the (T.A.) Chapman (1972) complement theorem. (S. Ferry (1980) recently constructed strange compacta in homotopy theory, using similar techniques and a clever compactification.) The Edwards-Geoghegan result required a preliminary theorem.

(2.4) Theorem (Edwards-Geoghegan, 1975). Idempotents split in pro-categories.

The proof is similar to the first part of the proof of Theorem (2.1), and is omitted excepted for some remarks. Edwards and Geoghegan replaced the telescope (2.2) by the tower (countable inverse system)

(2.5) $X \xleftarrow{f} X \xleftarrow{f} X \xleftarrow{f} \cdots$.

This tower splits f in pro-Ho (CW_o). They then took the homotopy limit to split f in Ho (CW_o). They then took the homotopy limit to split f in Ho (CW_o) (note that the telescope (2.2) is a homotopy colimit), and the (ordinary) limit to split f in pointed shape theory. Thus, every pointed FANR (compactum shape dominated by a finite complex) has the shape of a possibly infinite complex. As usual, there is a possible Wall (1965) obstruction.

For analogues in unpointed shape theory, we need a splitting theorem for unpointed homotopy idempotents. However, an example of Dydak (1977a) and Minc, and Freyd and Heller (1978) shows that additional hypotheses are needed to split unpointed homotopy idempotents. We now consider this example.

3. Proof of Theorem A (Dydak-Minc, Freyd-Heller).

We outline a proof of Theorem A. It is convenient to define a homotopy category of groups, Ho (Groups) as in Freyd-Heller (1978). Two maps $g, g' : H \rightrightarrows H'$ are called homotopic if they are conjugate. Pointed maps of pointed, connected spaces which are unpointed homotopic induce homotopic maps on fundamental groups. Conversely, homotopic maps of groups induce unpointed homotopic maps on $K(\pi,1)$'s, see, e.g., Spanier (1966). Therefore, a homomorphism g is a homotopy idempotent on a group H if and only if the induced map K(g,1) on K(H,1) is an unpointed homotopy idempotent. Also, g splits if and only if the induced map K(g,1) splits.

A group G with an unsplit homotopy idempotent f must have at least two generators x_o and x_1 (otherwise G is cyclic, hence abelian, and f is an honest idempotent which splits). Therefore let

$G = \langle x_o, x_1 \mid \text{relations}\rangle$
$fx_o = x_1$

where the relations will be defined in order to make f a homomorphism, homotopy idempotent, and, as we shall see, not splittable. For f to be a homotopy idempotent we need

$$fx_1 = f^2x_o, \text{ and } fx_o = x_1,$$

so we define

$$fx_1 = x_1^{x_o}(=x_o^{-1}x_1x_o).$$

In order for f to also be a homomorphism, relations of the following form are needed.

$$f^2x_1 = f(fx_1) = (fx_1)^{x_o} \text{ (homotopy idempotent property);}$$
$$f^2x_1 = f(fx_1) = f(x_o^{-1}x_1x_o) = x_1^{-1}fx_1 = (fx_1)^{x_1} \text{ (homomorphism property).}$$

Therefore,

$$(fx_1)^{x_o} = (fx_1)^{x_1}.$$

If we inductively define $x_{n+1} = x_n^{x_o}$, $n>1$, then all relations of the form

$$x_n^{x_m} = x_{n+1}, \; m<n$$

must hold. (Dydak, 1977a, and Minc; Heller, (1978). Also, these relations are consequences of the two relations

$$x_2^{x_o} = x_2^{x_1}$$
$$x_3^{x_o} = x_3^{x_1}$$

(Freyd-Heller, 1978; see Dydak-Hastings, 1978, for a proof). Therefore define

$$G = \langle x_o, x_1 | \; x_2^{x_o} = x_2^{x_1}, x_3^{x_o} = x_3^{x_1} \rangle,$$
$$fx_o = x_1, fx_1 = x_2(=x_1^{x_o}).$$

The f is a homotopy idempotent on the group G. We now sketch a proof that f does not split, following Dydak (1977a,b), Freyd-Heller (1978), Geoghegan (1978), and Dydak-Hastings (1978).

(3.1) <u>Proposition</u>. Let g be a homotopy idempotent on a group H. Then the following are equivalent.

(i) g splits in Ho(Groups).
(ii) g is equivalent in Ho(Groups) to an idempotent in Groups.
(iii) $\mathrm{Im}\, g^{k+1} = \mathrm{Im} g^k$ for some $k>1$.

Outline of proof. (i) ---> (iii)

Since g splits in Ho(Groups), there are maps d: G--->G' and u: G' --->G with $g \simeq ud$ and $du \simeq 1_G$. Then for some map d': G--->G' with $d' \simeq d$, $d' \cdot u = 1_G$. The required idempotent is given by g' = ud'.

(ii) ---> (iii)

The three observations Im g' = Img, $g'^2 = g'$, and $g' \simeq g$ imply that Im g^3 = Im g^2. In fact, (iii) holds for all $k \geq 1$.

(iii) ---> (i)

For some element y_o in H, the restriction g| Img is conjugation by y_o. Thus, (iii) implies that Img^2 = Img, or, equivalently, that g| Img is surjective. Since $g^2 \simeq g$, g| Img is already injective. Thus, g| Img is an isomorphism. It is easy to check that g is homotopic to the identity on Img (as in the proof of Theorem (2.1) above). Thus, g splits in Ho(Groups) as required. //

(3.2) Corollary. Any of the above conditions is equivalent to the condition Img = Img^2.

(3.3) Proposition. (Freyd-Heller, 1978; see also Geoghegan, 1978). Let g: X--->X be an unpointed homotopy idempotent (and a pointed map on a pointed, connected complex). Then g splits if and only if the induced map g_* on $\pi_1 X$ splits.

Proof. (--->) This is immediate.

(<---) If g_* splits, then the restriction g_*| Im g_* is an isomorphism. Now follow the proof of Theorem (2.1). The conclusion follows. //

We now continue with the proof of Theorem A.

(3.4) Proposition. For the map f: G ---> G defined above, Imf = Imf^2.

Outline of proof (Freyd-Heller, 1978). First, verify that the group G admits a normal form in which each element except for the identity e has a unique representation of the form

(3.5) $$x_{i_1} x_{i_2} \cdots x_{i_r} x_{j_s}^{-1} \cdots x_{j_2}^{-1} x_{j_1}^{-1},$$

in which

$i_{m+1} = i_m$ or $i_m \pm 1$,

$j_{m+1} = j_m$ or $j_m \pm 1$, and

$j_s \neq i_r$.

Next, verify that the image of f^k contains precisely those normal forms with $i_1 \geq k$ and $j_1 \geq k$, together with e. The conclusion follows. //

Corollary (3.2) and Proposition (3.4) imply that f does not split in Ho(Groups). This yields the first part of Theorem A.

For the second part (universality), consider a (pointed, connected) complex X with an (unpointed) homotopy idempotent g. Let $H = \pi_1 X$, then $g_*{}^2$ is conjugate to g_* on H, say via an element y_o. Then (Proposition (3.3)), g splits if and only if g_* splits in Ho(Groups). By Proposition (3.1), this is equivalent to the condition $\mathrm{Im} g^3 = \mathrm{Im} g^2$, where, for simplicity, we write g in place of g_*. We now follow the proof in Dydak-Hastings (1978) - Freyd and Heller (1978) gave an earlier proof.

(3.6) Proposition (Dydak-Hastings, 1978). Let g: H ---> H be a homomorphism with $g^2(y) = y_o^{-1} g(y) y_o$ and $\mathrm{Im} g^3 = \mathrm{Im} g^2$. Then there is a monomorphism α : G ---> H defined by $\alpha(x_n) = g^n(y_o)$.

P r o o f. We first show that for any k>0,

(3.7) $g(y_o{}^k) \notin \mathrm{Im}\, g^2$,

Assume, otherwise, that for some k>0, that $g(y_o{}^k) = g^2(z)$. Define a homomorphism $h: \mathrm{Im}\, g^2$ ---> $\mathrm{Im}\, g^2$ by $h(t) = g(y_o{}^{-1}) t\, g(y_o)$. Then $h.g^2(w) = g(y_o{}^{-1} g(w) y_o)$ $= g^3(w) \in \mathrm{Im}\, g^3$, that is, $\mathrm{Im}\, h \subset \mathrm{Im}\, g^3$. However, $h^k(t) = g(y_o{}^k)^{-1} t g(y_o{}^k)$.

Since $g(y_o{}^k) \in \mathrm{Im}\, g^2$ by assumption, the map h^k: $\mathrm{Im}\, g^2$ ---> $\mathrm{Im}\, g^2$ is an isomorphism. In particular, $\mathrm{Im}\, h = \mathrm{Im}\, g^2$. Thus $\mathrm{Im}\, g^3 = \mathrm{Im}\, g^2$, a contradiction. Therefore, condition (3.7) holds.

Now suppose that $\alpha(x) = e$ for some x in G. Recall the normal form (3.5), which allows us to write

(3.8) $x = x_s{}^a w x_s{}^{-b}$, where a+b>0, a>0, b>0 and w belongs to the subgroup of G generated by x_{s+1}, x_{s+2}, ...

If a>b, let

(3.9) $x' = x_s^{-b} x x_s^{b} = x_s^{a-b} w$, so that $\alpha(x') = 1$.

If k = a-b, we obtain

$g^s(y_o{}^k) = (x_s{}^k) = (w^{-1}) \in \mathrm{Im}\, g^{s+1}$.

If $s < 1$, this yields a contradiction, thus $\alpha(x) \neq 1$. If $s>1$ since $g|\operatorname{Im} g$ is clearly a monomorphism,

(3.10) $g(y_o^k) \in \operatorname{Im} g^2$,

a contradiction.

If $a<b$, replace x' by $(x')^{-1}$, following (3.9) to obtain (3.10) If $a=b$, then $f(w) = 1$. By (3.8), w is shorter than x. Continue, either obtaining the contradiction (3.10) or $x = e$. The conclusion follows. //

This yields universality, the second part of Theorem A. We complete this section with a technical lemma, again first proven by Freyd and Heller (1978).

(3.11) <u>Proposition</u>. The subgroup H of G generated by $x_0x_1^{-1}$, $x_2x_3^{-1}$, ... is free abelian and infinitely generated.

<u>P r o o f</u>. (Dydak and Hastings, 1978) Let us verify that H is abelian. Let $1<k+1<n$. Then

$$x_k(x_{k+1}^{-1}x_nx_{n+1}^{-1}x_{k+1})x_k^{-1}x_{n+1}x_n^{-1}$$
$$=(x_kx_{n+1}x_{n+2}^{-1}x_k^{-1})x_{n+1}x_n^{-1} = x_nx_{n+1}^{-1}\ x_{n+1}x_n^{-1} = 1, \text{ i.e.}$$
$$x_kx_{k+1}^{-1}.x_nx_{n+1}^{-1} = x_nx_{n+1}^{-1}.x_kx_{k+1}^{-1}.$$

Thus H is abelian.

Now recall (see Dydak and Segal, 1978a, pp. 82-83) that there is an epimorphism $\varphi : G \dashrightarrow G_o$ such that $\varphi(x_n) = g_{n+1}$, where G_o is the subgroup of all bijections of Z x N (Z is the set of integers and N is the set of positive integers) which is generated by $g_n : Z \times N \dashrightarrow Z \times N$, $n = 1,2,\ldots$, defined as follows

$$g_n\ (j,k) = \begin{cases} (j,k) & \text{for } j < n \\ (n,2k) & \text{for } j = n \\ (n,\ 2k+1) & \text{for } j = n+1 \\ (j-1,\ k) & \text{for } j > n+1. \end{cases}$$

Now $(g_ng_{n+1}^{-1})^k(n+1,0) = (n,2^{k-1})$ for $k>1$ and $(g_mg_{m+1}^{-1})(n+1,0) = (n+1,0)$ for $m>n+1$. Thus $(g_ng_{n+1}^{-1})k$ cannot be expressed as a product of $g_mg_{m+1}^{-1}$ for $m>n+1$. Hence H is free and infinitely generated. //

(3.12) <u>Corollary</u>. The group g is infinite - dimensional.

4. Proof of Theorem B (Dydak and Hastings, 1978).

In 1978, Dydak and Hastings proved Theorem B: homotopy idempotents on two-dimensional complexes split. We briefly review the proof. Many details are omitted because this result was subsumed last year by a more general result (Hastings and Heller, 1980): homotopy idempotents on finite dimensional complexes split.

Let X be a two-dimensional CW complex, and let f: X ---> X be a pointed map and an unpointed homotopy idempotent. The f_*^2 is conjugate to f_* on $\pi_1(X)$. Let y_o be the conjugating element. Assume that f does not split. Then there is a canonical equivariant injection G --->$\pi_1(X)$ (x_o --->y_o, x_1 ---> f_*y_o, see Proposition (3.5) Form the universal cover X and lift f to a map f: X ---> X. We regard $H_2(X)$ as a (free) ZG module via the injection G --->$\pi_1(X)$. For any element a in $H_2(X)$, compute f_*^3 a in the following two ways:

$$f_*^3a = (f_*^2a)(x_o) = (f_*a)^{(x_o^2)};$$
$$f_*^3a = f_*(f_*^2a) = f_*(f_*a)^{x_o}) = (f_*a)^{x_o x_1}$$

Since $H_2(X)$ is contained in the free $Z\pi_1(X)$ - module $H_2(X,X^1)$ (Whitehead, 1949) and $x_o x_1^{-1}$ has infinite order, we infer that $f_*a=o$. Thus, $f_*=o$ on $H_2(X)$.

Now form the telescopes (infinite mapping cylinders)

$T = tel(X\overset{f}{-\!-\!>}X\overset{f}{-\!-\!>}X\overset{f}{-\!-\!>}\ldots)$, and $\tilde{T} = tel(\tilde{X}\overset{\tilde{f}}{-\!-\!>}\tilde{X}\overset{\tilde{f}}{-\!-\!>}\tilde{X}\overset{\tilde{f}}{-\!-\!>}\ldots)$

Then $\tilde{T}$ is a covering space of T, and $\pi_1(\tilde{T}) = 0$ and $H_2(\tilde{T}) = 0$ by construction. Hence T is contractible, so T is a $K(\pi,1)$ for some group π containing G, again by construction. Thus π and G are finite-dimensional; the latter fact contradicts corollary (3.12). Consequently, f splits, as required. //

5. Proof of Theorem C (Hastings and Heller, 1980).

We now prove Theorem C: homotopy idempotents on finite-dimensional CW complexes split. Let X be a finite-dimensional CW complex, and let f:X ---> X be a pointed map and an unpointed homotopy idempotent. As in Section 4, let f_*^2 be conjugate to f_* on $\pi_1(X)$ by an element y_o in $\pi_1(X)$. Assume f does not split, and construct the canonical equivariant injection G --->π_1 (Theorem A, Prop. (3.5). Also, form the telescopes T and $\tilde{T}$ as in Section 5.

Compute

$\pi_1 T = G^1 = \langle \ldots, x_{-1}, x_o, x_1, \ldots | x_n x_m = x_{n+1}$

for all m<n>.

For the nth copy of X in T, n=0,1,2,..., we have the map

$$x_i (\text{in } \pi_1 X) \dashrightarrow x_{i-n} \ (\text{in } G^1).$$

For each n>0, let A_n be the subgroup of G^1 generated by the n elements $x_{-3}^{-1}x_{-2}, x_{-6}^{-1}x_{-5}, \ldots x_{-3n}^{-1}x_{-3n+1}$.

As in the proof of Proposition (3.11), A_n is a free abelian group on n generators. As in Section 4, A_n acts trivially on the image of the 0th injection.

$$H_*\tilde{X} \dashrightarrow H_*\tilde{T} = \operatorname{colim} (H_*\tilde{X}, f_*).$$

If $H_r(T) = 0$ for all r, then T is contractible, a contradiction (see Section 4). Otherwise, consider the largest r with $H_r(\tilde{T}) \neq 0$. Since $f_*{}^2 a = (f_* a) x_o$ for all a in $H_*(\tilde{T})$, $\tilde{f}_*| \operatorname{Im}\tilde{f}_*$ is a monomorphism, and the image of $H_r\tilde{X}$ in $H_r\tilde{T}$ is non-zero. Let T_n be the covering space T with fundamental group A_n, above. Consider the spectral sequence of the covering $\tilde{T} \dashrightarrow T_n$ (see Hilton and Wylie, 1962, or equivalently, use the Serre spectral sequence of the fibration sequence $\tilde{T} \dashrightarrow T_n \dashrightarrow BZ_n$)

$$E^2{}_{p,q} = H_p(A_n, H_q(\tilde{T})) \Rightarrow H_{p+q}(T_n).$$

In this sequence $E^2{}_{p,q} = 0$ if p>n or q>r. Thus,

$E^\infty{}_{n,r} = E^2{}_{n,r} = H_n(A_n, H_r(\tilde{T})) = H_r(\tilde{T})^{A_n}$, the subgroup of $H_r\tilde{T}$ fixed under the action of A_n. But the image $H_*(\tilde{X}) \dashrightarrow H_*(\tilde{T})$ is a non-trivial subgroup of $(H_q\tilde{T})^{A_n}$. Therefore, $E^\infty{}_{p,q} \neq 0$ and $H_{n+r}(T_n) \neq 0$. Since n was arbitrary, and T_n covers T, the dimensions of T is infinite. This contradicts the hypothesis that X is finite-dimensional. Therefore, f splits, as required. //

6. Geometric applications and questions.

We briefly discuss several geometric applications of the main splitting result, Theorem C, and its relevance to the following question.

(6.1) <u>Question</u>. Is every shape equivalence a strong shape equivalence?

Strong shape theory was introduced by Edwards and Hastings (1976), see also T. Porter (1973). Many authors have since given equivalent definitions. In particular, we note the definition of Dydak and Segal (1978b): an inclusion of compacta f: X--->Y is a strong shape equivalence if for all maps X--->Z the induced map $Z \dashrightarrow Z \cup_X Y$ is a shape equivalence.

The Dydak-Minc, Freyd-Heller example, Theorem A, above, shows that the analogous question in pro-homotopy has a negative answer. Here is a sketch.

(6.2) Proposition. There is a map of towers of complexes which is invertible in towers-Ho(Top), but not Ho(towers-Top).

Sketch of proof. Consider the unpointed homotopy idempotent K(f,1) on K(G,1) of Theorem A. Split this idempotent in towers-Ho(Top) (Edwards-Geoghegan, 1975, see Theorem (2.4) above):

$$K(G,1) \xrightarrow{d} Y \xrightarrow{u} K(G,1),$$

where Y is the tower

$$K(G,1) \xleftarrow{f} K(G,1) \xleftarrow{f} \cdots$$

Replace Y by a tower of fibrations and d and u by maps in towers-Top (Edwards-Hastings 1976; see also Edwards-Geoghegan, 1975):

$$K(G,1) \xrightarrow{d'} Y' \xrightarrow{u'} K(G,1).$$

Then the composite u'd': Y'--->Y' is an isomorphism in towers-Ho(Top), but not in Ho(towers-Top). Otherwise, application of homotopy limits (Bousfield and Kan, 1973) would yield an isomorphism in Ho(Top), and split f there. See Edwards- Geoghegan (1975), and Edwards-Hastings (1976). //

On the other hand, (strong) shape theory corresponds to the study of towers of finite complex, so Theorem C suggests a possible affirmative answer. See Dydak and Geoghegan (1980) for further discussion. Finally, an affirmative answer, together with results of Calder and Hastings (1981) would yield a pleasant description of the strong shape category. It would be W. Holsztynski's (1971) universal shape category, the localization of metric compacta at shape equivalences.

Here are more applications.

(6.3) Theorem. Let X be a pointed, (strong shape) connected, compact metric space. Then the following are equivalent.

(i) X is a FANR.

(ii) X is a pointed FANR.

(iii) X is shape dominated by a complex.

(iv) X is pointed shape dominated by a (pointed) complex.

(v) X has the shape of a complex.

(vi) X has the pointed shape of a (pointed) complex.

Sketch of proof. Of course, (i) and (iii), also (ii) and (iv), are equivalent. The remaining equivalences for pointed spaces and maps are due to Edwards and Geoghegan (1975). To obtain unpointed equivalences, combine their results, Theorem C, and the observation that a pointed map of complexes which is an unpointed homotopy equivalence is also a pointed homotopy equivalence. Details are omitted. //

There are also analogues in proper homotopy theory; see Chapman (1972), Edwards and Geoghegan (1975), Chapman and Siebenmann (1976).

References

A. K. Bousfield and D. M. Kan, 1973, Homotopy limits, completions, and localizations, Lecture Notes in Math. 304, Springer, Berlin-Heidelberg-New York.

A. Calder and H. M. Hastings, 1981. Realizing strong shape equivalences, J. Pure Appl. Alg., (to appear).

T. A. Chapman, 1972. On some applications of infinite dimensional topology to the theory of shape, Fund. Math. 76, 181-193.

T. A. Chapman and L. Siebernmann, 1976. Finding a boundary for a Hilbert cube manifold, Acta. Math. 139, 171-208.

J. Dydak, 1977a. A simple proof that pointed, connected FANR spaces are regular fundamental retracts of ANR's Bull. Acad. Polon. Sci. ser. sci. math. phys. 25, 55-62.

J. Dydak, 1977b. 1-movable continua need not be pointed 1-movable, ibid., 485-488.

J. Dydak and R. Geoghegan, 1980. The behavior on fundamental group of a free pro-homotopy equivalence, (preprint).

J. Dydak and H. M. Hastings, 1978. Homotopy idempotents on two-dimensional complexes split, (Proc. Intern. Conf. Geom. Topology, Warszaw). (preprint, 1978).

J. Dydak and J. Segal, 1978a. Shape theory - an introduction, Lecture Notes in Math. 688, Springer, Berlin-Heidelberg-New York.

J. Dydak and J. Segal, 1978b. Strong shape theory, Dissertationes Mathematicae.

D. A. Edwards and R. Geoghegan, 1975. Shapes of complexes, ends of manifolds, homotopy limits, and the Wall obstruction, Ann. of Math. 101 (1975), 521-535, and 104 (1976), 379.

D. A. Edwards and H. M. Hastings, 1976. Čech and Steenrod homotopy theory, with applications to geometric topology, Lecture Notes in Math. 542, Springer, Berlin-Heidelberg-New York, 1976.

S. Ferry, 1980. Homotopy, simple homotopy, and compacta, Topology 19, 101-110.

P. Freyd, 1966. Stable homotopy theory, in Proc. Conf. on categorical Algebra, La Jolla (1965), Springer, Berlin-Heidelberg-New York.

P. Freyd and A. Heller, 1978. Splitting homotopy idempotents, I, II, (submitted for publication).

R. Geoghegan, 1978. (In Proc. Oklahoma Top. Conf.)

H. M. Hastings and A. Heller, 1980. Homotopy idempotents on finite-dimensional complexes split, Proc. Amer. Math. Soc., (to appear).

P. J. Hilton and S. Wiley, 1962. Homology theory: an introduction to algebraic topology, Cambridge, England: University Press.

W. Holsztynski, 1971. An extension and axiomatic characterization of Borsuk's theory of shape, Fund. Math. 70, 57-68.

T. Porter, 1973. Cech homotopy, I, J. London Math. Soc. (2) 6, 429-436.

E. Spanier, 1966. Algebraic Topology, McGraw-Hill, New York.

C. T. C. Wall, 1965. Finitenes conditions for CW complexes, Annals of Math. (2) 81, 55-69.

J. H. C. Whitehead, 1949. Combinatorial homotopy I, II, Bull. Amer. Math. Soc. 55 (1949), 213-245, and 453-496.

Approximate Fibrations-A Geometric Perspective

by Donald S. Coram

In the study of mappings on manifolds, cell like mappings have been important. (A mapping is cell like if its point inverses have trivial shape. See [L_4] for an exposition of this topic). In order to study mappings with non-trivial point inverses, Coram and Duvall invented the concept of approximate fibration [C-D_1]. This was intended to generalize Hurewicz fibrations in roughly the same way that cell-like mappings generalize homeomorphisms. Thus one of the main themes of the work is to merge earlier results about fibrations and cell like mappings. Particular topics where this theme has been explored are 1) relationships between domain, range and fibers, 2) approximation by nicer maps, 3) detection by observing point inverses, and 4) applications to manifold topology.

One of the significant developments after the introduction of approximate fibrations has been the invention of a theory of shape fibrations. Now approximate fibrations seem to be a hybrid between Hurewicz fibrations for ANR's and shape fibrations for FANR's. However, for questions about arbitrary continuous mappings between manifolds (or more generally ANR's) approximate fibrations seem to be just the right tool. To expand on this perspective let us state the two applications with which this exposition will end.

Theorem 5.1 Let $p : S^3 \to S^2$ be a surjection such that each point inverse has the shape of S^1. Then p can be approximated arbitrarily closely by Seifert fiber maps.

Theorem 5.2 Let G be an upper semicontinuous decomposition of S^3 in which every element has the shape of S^1. If (i) for each $g \in G$ there are saturated neighborhoods $\tilde{U}$ and $\tilde{V}$ of g in S^3 such that every decomposition element g' in $\tilde{V}$ is essential in $\tilde{U}$, and (ii) S^3/G is an ANR, then S^3/G is homeomorphic to S^2.

In making the choices of theorems and proofs to include in this exposition, we will be guided by the idea of making the proofs of these two theorems understandable. There are many interesting and important ideas for higher dimensional manifolds, ANR's and even metric spaces but it is necessary to omit all but the briefest mention of these for the sake of a sense of continuity. It is true moreover that the path to the above applications does pass through each of the topics mentioned in the first paragraph. Also details of proofs are generally omitted in order to emphasize the basic ideas. References are given for those wanting to see the details.

The following notation and terminology will be used. Euclidean space, the real line, the unit interval and the n-sphere are denoted R^n, R, I, and S^n respectively. An n-manifold is a separable metric space locally homeomorphic to R^n. An ANR is a locally compact, separable, metric, absolute neighborhood retract. A mapping is a continuous function. A <u>surjection</u> is a mapping onto its range. A mapping is <u>proper</u> if it is closed and each point inverse is compact. The notations H_i and H^i refer to singular homology and cohomology with integral coefficients, while $\check{\pi}_i$, $\check{H}_i$, and $\check{H}^i$ refer to the shape (≡Čech ≡ inverse limit) groups.

1. <u>Approximate fibrations</u>. In the study of fibrations the homotopy lifting property is encountered [Du], [Sp]. A surjection $p : E \to B$ has the <u>homotopy lifting property</u> for a space X provided that given maps g and G in the commutative diagram

$$\begin{array}{ccc} X \times \{0\} & \xrightarrow{\ g\ } & E \\ \cap & & \downarrow p \\ X \times I & \xrightarrow{\ G\ } & B \end{array}$$

there is a map $\tilde{G} : X \times I \to E$ which extends g such that $p\tilde{G} = G$. A mapping is a <u>Serre fibration</u> if it has the homotopy lifting property for I^n for all n. A mapping is a <u>Hurewicz fibration</u> if it has the homotopy lifting property for all spaces.

In [A-P], [K_1], and [L_1] cell like mappings were shown to satisfy an approximate version of the homotopy lifting property for polyhedra. Thus we are led to the following definition. A surjection $p : E \to B$ has the <u>approximate homotopy lifting property</u> for a space X provided that given any open cover ε of B and maps g and G in the commutative diagram

$$\begin{array}{ccc} X \times \{0\} & \xrightarrow{\ g\ } & E \\ \cap & & \downarrow p \\ X \times I & \xrightarrow{\ G\ } & B \end{array}$$

there is a map $\tilde{G} : X \times I \to E$ which extends g such that $p\tilde{G}$ is ε-close to G. (The last statement means that for each $(x, t) \in X \times I$, there is a U in ε such such that both $\tilde{G}(x, t)$ and $p\tilde{G}(x, t)$ lie in U.) A mapping is an <u>approximate fibration</u> if it has the approximate homotopy lifting property for all spaces. For any type of fibration a <u>fiber</u> is a point inverse, denoted $F_b = p^{-1}(b)$.

In the original paper [$C\text{-}D_1$] the term approximate fibration included the assumption that the domain and range be ANR's and that the map be proper. In light of subsequent developments it seems better to retain flexibility in the definition and to make such assumptions explicitly where they are needed.

<u>Examples</u>. Obviously any Hurewicz fibration is an approximate fibration. In fact any Serre fibration between ANR's is an approximate fibration (Theorem 4.2). Also any cell like map between ANR's is an approximate fibration ([L_1] and Theorem 4.2). Furthermore if B is a compact ANR and $p : E \to B$ is the limit of a uniformly convergent sequence of Hurewicz fibrations, then p is an approximate fibration [$C\text{-}D_1$, Prop 1.1 and Prop 1.4]. More specifically there is an approximate fibration $p : S^1 \times S^1 \to S^1$ such that for some $x_0 \in S^1$, $p^{-1}(x_0)$ is a "$\sin(\frac{1}{x})$ circle" but over $S^1 - \{x_0\}$ p is topologically a product projection [$C\text{-}D_1$]. There is another approximate fibration $q : S^1 \times S^1 \to S^1$ such that $q^{-1}(x)$ is homeomorphic to S^1

for every $x \in S^1$ but q is not a Hurewicz fibration [C-D_4]. This is accomplished by putting "zig-zags" into a sequence of fibers converging to a "straight" fiber. Any approximate fibration $p : S^1 \times S^1 \to S^1$ with S^1 fibers can be used to construct a map $\tilde{P}_{k,\ell} : S^3 \to S^2$ using the join structure $S^3 = S^1 * S^1$ such that the fibers are (k, ℓ) curves on the intermediate tori [C-D_4]. Such examples may fail to be Hurewicz fibrations over large sets, but they will be approximate fibrations except at two points perhaps.

In the definition of approximate fibration it is equivalent in ANR spaces to allow inexactness in the commutativity of the original diagram of maps or the upper triangle of the conclusion: [C-D_1, Lem 1.2] and [C-D_2, Lem 2.5]. Another variation we will often use is "regular lifting". The map $\tilde{G} : X \times I \to E$ in either exact or approximate homotopy lifting is a <u>regular lifting</u> of G provided that whenever $G(x, t) = G(x, 0)$ for all t, then $\tilde{G}(x, t) = \tilde{G}(x, 0) = g(x, 0)$ for all t.

<u>Lemma</u> 1.1 (see [C-D_1, Prop 1.5] and [Du, Ch.XX, Cor 2.4]). If $p : E \to B$ is an approximate fibration between metric spaces, then all approximate homotopy liftings may be chosen to be regular.

The proof is very similar to the proof for Hurewicz fibrations and involves path lifting functions.

2. <u>Relationships between domain, range, and fiber</u>. The first relationship is the equivalence of the fibers. Fibers of cell like maps are shape equivalent by definition. Fibers of Hurewicz fibrations over a path connected space are homotopy equivalent.

<u>Theorem</u> 2.1 [C-D_3] If $p : E \to B$ is an approximate fibration between metric spaces and B is path connected, then any two fibers are shape equivalent.

Proof. Let $b \in B$ and A be any arc containing b. We wish to show that the inclusion $i : F_b \to p^{-1}(A)$ is a shape equivalence. By the Mardesic-Kozlowski formulation of shape theory [M_2], [K_2], i is a shape equivalence provided $i^* : [p^{-1}(A), L] \longrightarrow [F_b, L]$ is a bijection for every ANR L. (The notation [X, L] means the set of homotopy classes of maps form X to L.) To see that i^* is surjective, let $f : F_b \to L$ be given. Since L is an ANR, there is an extension $\bar{f} : U \to L$ where U is a neighborhood of F_b in E. Applying regular lifting to a contraction of A to b rel b gives a deformation H_t of $p^{-1}(A)$ into U holding F_b fixed. Then $f H_1 : p^{-1}(A) \to L$ and $i_*(f H_1) = f H_1 | F_b = f$. To see that i^* is injective, let $f, g : p^{-1}(A) \to L$ be maps such that $f | F_b \simeq g | F_b$. Extend f and g to $\bar{f}$ and $\bar{g}$: $V \to L$ where V is a neighborhood of $p^{-1}(A)$. Then extend to a homotopy $H_t : W \to L$ where W is a neighborhood of F_b in V, $H_0 = \bar{f}|W$, and $H_1 = \bar{g}|W$. As before there is a homotopy $G_t : p^{-1}(A) \to V$ such that $G_0 = 1$, $G_t | F_b = 1$, and $G_1(p^{-1}(A)) \subset W$. Now define a homotopy $F_t : p^{-1}(A) \to L$ by

$$F_t(x) = \begin{cases} \bar{f}\, G_{3t}(x) & 0 \le t \le \frac{1}{3} \\ H_{3t-1}G_1(x) & \frac{1}{3} \le t \le \frac{2}{3} \\ \bar{f}\, G_{3-3t}(x) & \frac{2}{3} \le t \le 1 \end{cases}$$

Then $f \simeq g$ and i_* is a bijection as desired.

The second relationship concerns properties. If $p : F \to B$ is a Hurewicz fibration between metric spaces and any two of E, B, and F are ANR's, then the third is also [C-F], [Fa], [Fe_2], [A-F]. Similarly if $p : F \to B$ is a proper cell like map, and E is an ANR then B is an ANR provided that B is finite dimensional [L_1]. Of course cell like sets are fundamental absolute neighborhood retracts FANR's [B_2], [M_1] not ANR's. Hence for an approximate fibration $p : F \to B$ we consider the following three statements: 1) E is an ANR, 2) B is an ANR, 3) F is an FANR. We ask which two imply the other.

Theorem 2.2 [C-D_1, Cor 2.] If $p : E \to B$ is a proper approximate fibration and both E and B are ANR's, then F_b is an FANR.

The idea of the proof is to use local contractibility of B and regular lifting to construct a "strong shape retraction" to F_b.

Theorem 2.3 [C-D_7] If $p : E \to B$ is a proper approximate fibration, E is an ANR, F_b is an FANR, and B is finite dimensional, then B is an ANR.

Note that the finite dimensionality assumption was needed for the corresponding result for Hurewicz fibrations until only recently and still is needed for cell like mappings. In fact there are some important counterexamples and open questions here for cell like mappings [Ge].

To prove Theorem 2.3 we first prove that B satisfies an approximate version of local n-connectivity. Given $b \in B$ and $\varepsilon > 0$ choose small neighborhoods $W \subset V$ so that $p^{-1}(V)$ deforms into $p^{-1}N(b, \frac{\varepsilon}{3})$ rel $p^{-1}(W)$. Now consider $f : S^n \to V$ with $b \in f(S^n)$. Write $S^n = D_1 \cup D_2$ where D_1, D_2 are n-disks, $D_1 \cap D_2 = \Sigma$ is an (n-1) sphere, and $f(D_1) \subset W$. Consider $f|D_2$ to be a homotopy of Σ and lift starting with a constant to a map $g_2 : D_2 \to f^{-1}(v)$ so that $g_2(\Sigma) \subset f^{-1}(w)$. By the choice of V and W, g_2 is homotopic rel to a map $g_1 : D_1 \to N(b, \frac{\varepsilon}{3})$. The result is a null-homotopic map $g : S^n \to f^{-1}(V)$ for which $d(pg, f) < \varepsilon$. Thus f can be arbitrarily closely approximated by maps which are null homotopic. We now construct an extension of f to the $(n + 1)$-disk. By local compactness we may assume that $V \subset Q$, the Hilbert cube. We may also assume inductively that B is LC^{n-1} (locally n-connectivity). First extend f to $\tilde{f} : S^n \cup \frac{1}{2}D^{n+1} \to V$ using g on $\frac{1}{2}D^{n+1}$, the disk of radius $\frac{1}{2}$. Next extend over the product $[\frac{1}{2}, 1]$ cross the $(n - 1)$ skeleton of a trangulation of S^n, the map still into V. Finally extend to $f_1 : D^{n+1} \to Q$ using the local contractibility of Q. Now $f_1(D^{n+1})$ has some "bubbles" which stick out into Q, but we can repeat the process to get smaller bubbles closer to V modifying the map only on the preimage of the bubbles. The limit of a sequence of such maps gives the desired extension $f : D^{n+1} \to V$, so B is locally n-connected. Since B

is finite dimensional, B is an ANR [Be], [Hu].

The third relationship concerns dimension. If $p : E \to B$ is a proper Hurewicz fibration where E is a finite dimensional separable metric space and B is a finite dimensional manifold, then $\dim E = \dim B + \max\{\dim F_b\}_{b \in B}$ [Li]. Also if $p : E \to B$ is a proper cell like mapping between finite dimensional manifolds of dimensional $\neq 3,4$ then $\dim E = \dim B$ since E is homeomorphic to B [S]. This agrees with the first result if we use fundamental dimension, since $Fd(F_b) = 0$. We present the following theorem as a sample, but refer the reader to [D-H_2] and section 3 for further results.

Theorem [D-H_1]. Let $p : E \to B$ be a proper approximate fibration where E is a manifold and B is a polyhedron. Then $\dim E = \dim B + FdF_b$ for each $b \in B$.

To see the idea of the proof consider the case where F has the shape of a finite complex and $\dim E - \dim B \geq 6$. Let $m = \dim E$ and $n = \dim B$. By [H_2] $F_b \times T^{n+1}$ has the shape of a closed $(m+1)$-mainfold where T^{n+1} is the $(n+1)$-fold product of circles. Hence $Fd(Fb \times T^{n+1}) = m + 1$. On the other hand $Fd(Fb \times T^{n+1}) = Fd(Fb) + n + 1$ by [N]. Hence $m = n + Fd(Fb)$.

Finally we mention the exact sequence of an approximate fibration. This generalizes the exact sequence of a fibration [Sp] and the Vietoris mapping theorem for cell like mappings [L_4].

Theorem 2.4 If $p : E \to B$ is a proper approximate fibration between ANR's, $b \in B$ and $e \in F_b$, then there is an exact sequence.

$$\cdots \to \check{\Pi}_q(F_b, e) \to \check{\Pi}_q(B,b) \to \check{\Pi}_{q-1}(F_b, e) \to \cdots$$

The proof is a fairly routine generalization of earlier proofs and we omit we.

3. Approximating approximate fibrations by fibrations. A cell like mapping between manifolds can be approximated by homeomorphisms in many cases [A] [S] [C_1] [Mo]. Thus one would hope to show that an approximate fibration can be approximated by some kind of fibrations. Husch [H_1], Coad [G_1], [G_2], Chapman [C_1] [C_2], and Quinn [Q] have positive results along this line. However Husch [H_1], and Ferry [Fe_1] have examples to show that such approximations do not always exist. We present here the following special case of [H_1, Th. A].

Theorem 3.1 Let $p : E \to B$ be an approximate fibration where E is an open subset of S^3, B is a 2-manifold and F_b has the shape of S^1 for each $b \in B$. Then p can be approximated arbitrarily closely by locally trivial fiber map.

The idea of the proof is as follows. Let U be an open 2-cell in B. Using the exact sequence it can be proven that π_1 of the end of $p^{-1}(U)$ is stable and in fact is $Z \oplus Z$. Hence $p^{-1}(U)$ is the interior of a compact 3-manifold [H-P] which must be $S^1 \times D^2$ since $E \subset S^3$. Then the projection $S^1 \times D^2$ onto D^2 gives a trivial fibration $p^{-1}(U) \to U$ which differs from p by no more than the diameter of U. This argument is applied to open 2-cells which are the interiors of stars of simplices in a fine triangulation of B. These must be fit together carefully using the same philosophy: preimages of open subsets under p behave as if p were locally trivial.

4. Detection of approximate fibrations. Our aim is to show that very little needs to be known about a map to show that it is an approximate fibration. First we reduce the variety of spaces for which the AHLP needs to be checked. Secondly we find a condition on the point inverses which implies lifting. And finally we show that for certain maps this condition occurs automatically almost everywhere. The first theorem generalizes one about exact homotopy lifting; the proof is similar too.

Theorem 4.1 [C-D_1] Let $p : E \to B$ be a surjection between metric spaces. If p has the AHLP for metric spaces, then p is an approximate fibration.

The next theorem generalizes [U, Cor. 1].

Theorem 4.2 [C-D_2] Let $p : E \to B$ be a surjection between ANR's. If p has the AHLP for all k-cells, then p is an approximate fibration.

In the proof one first proves that p has the AHLP for finite polyhedra by stepwise extension. Next prove p has the AHLP for any countable, locally finite polyhedron by partitioning it into compact pieces. Then one can show that p has the AHLP for all separable metric spaces by using nerves of covers. This isn't quite enough for Theorem 4.1 as is but its proof works because E and B are separable.

We now turn to the important concept of movability of mappings for detecting approximate fibrations. For cell like mappings the precedents were UV^∞ and UV^k [L_2]. For fibrations there was complete regularity, homotopy n-regularity, strong regularity, etc. Survey articles on these topics can be found in [Mc]; particularly the ones by Hamstrom, Addis and Ungar.

Let F be a compactum in the space E and $V \subset U$ be neighborhoods of F. If the inclusion induced map $\check{\Pi}_k(F, x) \to \Pi_k(U, x)$ is a monomorphism whose image equals the image of the inclusion induced map $\Pi_k(V, x) \to \Pi_k(U, x)$ for every $x \in F$, then we say $\check{\Pi}_k F$ is realized as the image of $\Pi_k V$ in $\Pi_k U$. A proper surjection $p : E \to B$ between metric spaces is a k-movable mapping provided that for each $b \in B$ and each neighborhood U_0 of F_b, there exist neighborhoods $V \subset U$ of F_b in U_0 such that if F_c is any point inverse in V, then $\Pi_i F_c$ is realized as the image of $\Pi_i V$ in $\Pi_i U$ for all i, $0 \le i \le k$. A proper surjection $p : E \to B$ is completely movable provided that for each $b \in B$ and each neighborhood U of F_b there exists a neighborhood V of F_b in U such that if F_c is any point inverse in V and W is any neighborhood of F_c in V, then there is a homotopy $H_t : V \to U$ such that $H_0 = 1$, $H_1(V) \subset W$, and $H_t | F_c = 1$. It is instructive to check 1-movability for the maps $\tilde{p}_{k,\ell} : S^3 \to S^2$ constructed in the first section. If b is the "north pole", U and V are open solid tori neighborhoods of the join circle F_b and $F_c \subset V$, $c \neq b$, then $\check{\Pi}_1 F_c \to \Pi_1 U$ has image $k \cdot \Pi_1 U$ while $\Pi_1 V \to \Pi_1 U$ is an isomorphism.

Theorem 4.3 [C-D_2] Let $p : E \to B$ be a proper surjection between ANR's. The following are equivalent.

(1) p is completely movable,

(2) p is k-movable for all $k \geq 0$,

(3) p is an approximate fibration.

To prove that (1) implies (2) one uses the given homotopies to move representative singular spheres around. To prove (2) implies (3) one uses the given algebraic information to build liftings of homotopies of k-cells by stepwise extension. Then Theorem 4.2 gives the result. Finally (3) implies (1) is proven by lifting the local contractibility of B.

Corollary. (Restriction) Suppose $p : E \to B$ is a proper approximate fibration between ANR's. If U is an open set in U, then $p|p^{-1}(U) : p^{-1}(U) \to U$ is an approximate fibration.

Corollary. (Uniformization) Suppose $p : E \to B$ is a proper surjection between ANR's and C is an open covering of B. If $p|p^{-1}(U) : p^{-1}(U) \to U$ is an approximate fibration for every $U \in C$, then p is an approximate fibration.

Often it is not necessary to check k-movability for all k but only up to some finite stage. Compare [L_2] [U] for precedents. Results of this type can be found in [C-D_2] and [C-D_5]. Here is a sample result.

Theorem 4.4 [C-D_2] Let $p : E \to B$ be a proper surjection between ANR's with each fiber having the shape of S^1. If p is a 1-movable mapping then p is an approximate fibration.

The proof of this is easy since each F_b is k-UV (approximately k-connected) for $k \geq 2$ which in the presence of 1-movability implies k-movability for all k.

Finally we consider some finiteness results following [L_3]. The philosophy is that any map on a manifold should be an approximate fibration almost everywhere. Such results occur in [C-D_3] for S^3, in [C-D_4] for S^{2k+1}, and in [C-D_6] for manifolds and ANR's in general.

Theorem 4.5 [C-D_6] Let $p : E \to B$ be a surjection between ANR's where F_b has the shape of S^1 for each $b \in B$. Then there is a dense open set $B_0 \subset B$ such that p is an approximate fibration over B_0.

The main idea of the proof is to measure how much winding about a fiber there is by nearby fibers. Using the fact that each F_b is an FANR, there is a homomorphism $r : \Pi_1(p^{-1}(V)) \to \check{\Pi}_1(F_b)$ such that

$$\begin{array}{ccc} \Pi_1(p^{-1}(V)) & \longrightarrow & \Pi_1(p^{-1}(U)) \\ & {\scriptstyle r} \searrow \quad \nearrow & \\ & \check{\Pi}_1(F_b) & \end{array}$$

commutes. (The unlabeled homomorphisms are inclusion induced.) Let K be the set of points $b \in B$ such that each neighborhood V of b contains a point c such that $\check{\Pi}_1(F_c) \to \Pi_1(p^{-1}(V))$ is not monic. It can be shown that K is closed and nowhere dense. Now given $b \in B$ and neighborhoods U and V as above, let $x \in V$ and let φ_x be the composition $\check{\Pi}_1 F_x \to \Pi_1 p^{-1}(V) \to \check{\Pi}_1 F_b$. Define a winding function $\alpha_b : V \to R$ by $\alpha_b(x) = |\varphi_x(1)|$ thinking of φ_x as a homomorphism on the integers. It is easy to show that $\alpha_b(x) = \alpha_b(y) \cdot \alpha_y(x)$ for any y sufficiently close to x.

Hence α_b is lower semi continuous on $V-K$. Standard point set arguments then say that the points of continuity of α_b form a dense open set C_b in V. Then p is 1-movable over C_b. The theorem now follows by setting B_0 equal to the union of the C_b's for all $b \in B$ since 1-movability is a local property.

5. Applications. We now indicate the proofs of the applications stated in the introduction. For other results see [C-D$_4$], [C-D$_6$].

Theorem 5.1 [C-D$_3$] Let $p : S^3 \to S^2$ be a surjection where each point inverse has the shape of S^1. The p can be approximated arbitrarily closely by Seifert fiber maps.

In the proof we again use the winding functions. If the set of discontinuities were uncountable we could find an arc A such that the winding functions are continuous on the interior of A but discontinuous and equal at the two endpoints, say $A = [d_0, d_1]$ and $\alpha(d_0) = \alpha(d_1) = k$ where $c \in \text{Int } A$. Now $p^{-1}(d_i) \subset p^{-1}(A)$ is a shape equivalence by an argument similar to the one for Theorem 2.1. Similarily $\check{\Pi}_1 F_c \to \check{\Pi}_1 p^{-1}(A)$ is multiplication by k. One then calculates $\check{H}^2(p^{-1}(A)) \cong Z_k$ if $k > 0$ and $\check{H}^2(p^{-1}(A)) \cong Z$ if $k = 0$. The first case is ruled out by Alexander duality; the second since A does not separate to S^2 and p is monotone. Hence the set of discontinuities must be countable and no winding number occurs more than once. In fact it is finite by lower semi continuity. Furthermore the same kind of argument shows that if $K \neq \emptyset$, then there is exactly one discontinuity. On the other hand suppose $K = \emptyset$ but there are at least three discontinuities. Consider a triod Y whose endpoints are discontinuities and whose interior contains only points of continuity. Let U be a 2-cell neighborhood of Y which contains no other discontinuities. As in the proof of Theorem 3.1 $p^{-1}(U)$ is the interior of a cube with knotted hole. One can calculate its fundamental group and discover a contradiction to a result from classical knot theory. In summary there are at most two points d_1, d_2 in S^2 where p is not 1-movable. Therefore p is an approximate fibration over $S^2 - \{d_1, d_2\}$. We We now apply Theorem 3.1 to obtain a locally trivial (in fact trivial) fibration over $S^2 - \{d_1, d_2\}$ approximating p. Modify this over small neighborhoods of d_1 and d_2 so that it extends to a Seifert fiber map.

Theorem 5.2 [Cö] Let G be an upper semi continuous decomposition of S^3 in which every element has the shape of S^1. If (i) for each $g \in G$ there are saturated open neighborhoods $\tilde{U}$ and $\tilde{V}$ of g in S^3 such that every decomposition element g' which lies in $\tilde{V}$ is essential in $\tilde{U}$, and (ii) S^3/G is an ANR, then S^3/G is homemorphic to S^2.

Hypothesis (i) gives us that the set K defined in the proof of Theorem 4.5 is empty. Then the proof of Theorem 4.4 goes through to the point of showing that the set of discontinuities of winding functions is finite. We wish to apply the Kline sphere characterization [Bi]. No pair of points separates S^3/G since no pair of fibers can separate S^3. To prove that any simple closed curve J separates S^3/G one calculates $\check{H}^2(p^{-1}(J)) \cong Z$ in a manner similar to the calculation of $\check{H}^2 p^{-1}(A)$ above. Then $H_0(S^3 - p^{-1}(J)) \cong Z$ so $p^{-1}(J)$ separates S^3. But p is monotone so J separates S^3/G. Hence $S^3/G \cong S^2$.

A-F G. Allaud and E. Fadell, A fiber homotopy extension theorem, Trans. Amer. Math. Soc., 104(1962), 239-251.

A S. Armentrout, Cellular decomposition of 3-manifolds that yeild 3-manifolds, Memoirs Amer. Math. Soc., 107(1971).

A-P S. Armentrout and T. M. Price, Decompositions into compact sets with UV-properties, Trans. Amer. Math. Soc., 141(1969), 433-442.

Bi R. H. Bing, The Kline sphere characterization problem, Bull. Amer. Math. Soc., 52(1946) 644-653.

B_1 K. Borsuk, Theory of Retracts, Polish Scientific Publishers, Warsaw, 1967.

B_2 K. Borsuk, Theory of Shape, Lecture Notes Series NO. 28, Matematisk Inst. Aarhaus. Univ, 1971.

C_1 T. A. Chapman, Lectures on Hilbert Cube Manifolds, C. B. M. S. Regional Conference Series in Math., No. 28, 1976.

C_2 T. A. Chapman, Approximating maps into fiber bundles by homeomorphisms, Rocky Mt. J. of Math. 10(1980), 333-350.

C_3 T. A. Chapman, Carving up manifolds into block bundles. Preliminary Version.

C-F T. Chapman and S. Ferry, Hurewicz fiber maps with ANR fibers, Topology 16 (1977), 131-143.

C-D_1 D. Coram and P. Duvall, Approximate fibrations, Rocky Mt. J. of Math. 7(1977), 275-288.

C-D_2 D. Coram and P. Duvall, Approximate fibrations and a movability condition for maps., Pac. J. of Math, 72(1977), 41-56.

C-D_3 D. S. Coram and P. F. Duvall, Jr., Mappings from S^3 to S^2 whose point inverses have the shape of a circle., Gen. Top. and Appl., 19(1979), 239-246.

C-D_4 D. S. Coram and P. F. Duvall, Jr., Non-degenerate k-sphere mappings, Topology Proceedings, Vol. 4, 1979.

C-D_5 D. Coram and P. Duvall, A Hurewicz-type theorem for approximate fibrations, Proc. Amer. Math. Soc. 78(1980), 443-448.

C-D_6 D. S. Coram and P. F. Duvall, Finiteness theorems for approximate fibrations, preprint.

C-D_7 D. S. Coram and P. F. Duvall, Local n-connectivity and approximate lifting, preprint.

Co D. S. Coram, Decompositions of S^3 into circles, talk given at Austin Topology Summer Conference, preprint.

Du J. Dugundji, Topology, Allyn and Bacon, Inc., Boston, 1966.

D-H_1 P. F. Duvall, Jr. and L. S. Husch, Fundamental dimension of fibers of approximate fibrations, Topology Proceedings, Vol. 3 (1978).

D-H_2 P. F. Duvall, Jr. and L. S. Husch, Fundamental dimension and suspension of approximate fibrations, Proc. Amer. Math. Soc. 79(1980), 122-126.

Fa E. Fadell, On fiber spaces, Trans. Amer. Math. Soc. 90(1959), 1-14.

Fe_1 S. Ferry, Approximate fibrations with non-finite fibers, Proc. Amer. Math. Soc. 64(1977), 335-345.

Fe_2 S. Ferry, Strongly regular mappings with compact ANR fibers are Hurewicz fibrings, Poc. J. Math. 75(1978), 373-382.

Ge R. Geoghogan (ed.) Open problems in infinite dimensional topology, Topology Proceedings, Vol. 4, 1979, 287-338.

G_1 R. E. Goad, Local homotopy properties of maps and approximation by fiber bundle projection, Thesis Univ. of Ga. (1976).

G_2 R. E. Goad, Approximate torus fibrations of high dimensional manifolds can be approximated by torus bundle projections, preprint.

Hu S. T. Hu, Theory of Retracts, Wayne State University Press, Detroit, 1965.

H_1 L. Husch, Approximating approximate fibrations by fibrations, Can. J. Math. 29(1977), 897-913.

H_2 L. Husch, Fibres of Hurewicz and approximate fibrations, Math. Scand. 43(1978), 44-48.

H-P L. S. Husch and T. M. Price, Finding a boundary for a 3-manifold, Ann. Math. 91(1970), 223-235.

K_1 G. Kozlowski, Factorization of certain maps up to homotopy, Proc. Amer. Math. Soc. 21(1969), 88-92.

K_2 G. Kozlowski, Images of ANR's, Trans. Amer. Math. Soc., (to appear).

L_1 R. C. Lacher, Cell-like mappings I, Pac. J. of Math. 30(1969), 717-731.

L_2 R. C. Lacher, Cellularity criteria for maps, Mich. Math. Jour. 17(1970), 385-396.

L_3 R. C. Lacher, Finiteness theorems in the study of mappings between manifolds, Proc. of the Univ. of Okla. Topo. Conf. (1972), 79-96.

L_4 R. C. Lacher, Cell-like mappings and their generalizations, Bull. Amer. Math. Soc. 83(1977), 495-552.

Li S. D. Liao, Some theorems on the dimension of fiber spaces, Amer. J. Math. 71(1949), 231-240.

M_1 S. Mardešić, Strongly movable compacta and shape retracts, Proc. Intern. Sym.

on Top. and its Appl. (Budva, 1972), 163-166.

M$_2$ S. Mardešić, Shapes for topological spaces, Gen. Top. and Appl. 3(1973), 265-282.

Mc L. McAuley, Proceedings of the Conf. on Monotone mappings and open mappings, State Univ. Of N. Y. at Binghamton, 1970.

Mo R. L. Moore, Concerning upper semicontinuous collections of continua, Trans. Amer. Math. Soc. 27(1925), 416-428.

N S. Nowak, On the fundamental dimension of approximately 1-connected compacta, Fund. Math. 89(1975), 61-79.

Q Frank Quinn, Ends of maps and applications, Ann. of Math. 119(1979), 275-331.

S L. C. Siebenmann, Approximating cellular maps with homeomorphisms, Topology, 11(1972), 271-294.

Sp E. H. Spanier, Algebraic Topology, McGraw-Hill Book Co., New York (1966).

U G. Unger, Conditions for a mapping to have the slicing structure property, Pac. J. of Math., 30(1969), 549-553.

LOCAL N-CONNECTIVITY OF QUOTIENT SPACES AND ONE-POINT COMPACTIFICATIONS

Jerzy Dydak (Warsaw)

1. Introduction.

In $[D_3]$ the author studied the following problem:

Let $A \subset X \in LC^n$ $(n \geqslant 1)$ be compacta. Find neccessary and sufficient conditions for the quotient space X/A to be locally k-connected for $k \leqslant n$.

It turns out that the condition $X/A \in LC^n$ is an internal property of A and can be characterized by using some algebraic shape invariants of A.

Then it was suggested by R. Geoghegan [G] that the results of $[D_3]$ can be reformulated in a more general setting. Namely, one can give analogues necessary and sufficient conditions for the one-point compactification ωX of a locally compact, separable and metrizable space X to be locally k-connected for $k \leqslant n$.

In this paper we go much further (and the proofs presented here are simpler) by giving necessary and sufficient conditions for a metrizable space X to be locally k-connected for $k \leqslant n$ at a point $x_0 \in X$ such that $X - \{x_0\} \in LC^n$.

We are going to use notions and results from $[D_1]$, [D-S], [E], [Hu] and [S]. For the sake of completeness, we reprove some results scattered throughout the literature.

2. Algebraic preliminaries.

For the notions of Mittag-Leffler property and stability of pro-groups see [D-S] (pp. 17 and 77).

All results of this section concern the following situation:

Inverse sequences $\underline{G}^k = \left(G_n^k, q_{n,n+1}^k\right)$, $1 \leqslant k \leqslant 4$, of groups and homomorphisms $p_n^{k-1,k}: G_n^k \to G_n^{k-1}$, $n \geqslant 1$ and $2 \leqslant k \leqslant 4$, of groups are given such that

$$p_n^{k-1,k} \cdot q_{n,n+1}^k = q_{n,n+1}^{k-1} \cdot p_{n+1}^{k-1,k}$$

and each finite sequence $\underline{G}_n = (G_n^k, p_n^{k-1,k})$ is exact.

2.1. Lemma. If $\underline{G}^1$ is stable and $\underline{G}^i$ satisfies the Mittag-Leffler condition for $i = 2,4$, then $\underline{G}^3$ satisfies the Mittag-Leffler condition.

Proof. Recall that the fact that a pro-group $(A_n, p_{n,m})$ satisfies the Mittag-Leffler condition (is stable) can be formulated in the following way:

there is an increasing sequence $n_1 < n_2 < \cdots$ of positive integers such that

$$p_{n_k,n_{k+1}}/\text{im } p_{n_{k+1},n_{k+2}} : \text{im } p_{n_{k+1},n_{k+2}} \to \text{im } p_{n_k,n_{k+1}}$$

is an epimorphism (isomorphism).

Therefore, without loss of generality, we may assume that

$$q_{n,n+1}^k / \text{ im } q_{n+1,n+2}^k : \text{im } q_{n+1,n+2}^k \to \text{im } q_{n,n+1}^k$$

is an epimorphism (isomorphism) for all $n \geqslant 1$ and $k = 2,4$ $(k = 1)$.

We are going to prove that

$$q_{n,n+1}^3 / \text{ im } q_{n+1,n+3}^3 : \text{im } q_{n+1,n+3}^3 \to \text{im } q_{n,n+2}^3$$

is an epimorphism.

So, suppose $x_n \in \text{im } q_{n,n+2}^3$, i.e., $x_n = q_{n,n+2}^3(x_{n+2})$ for some $x_{n+2} \in G_{n+2}^3$. Then

$$q^2_{n+1,n+2}\cdot p^{2,3}_{n+2}(x_{n+2}) = q^2_{n+1,n+4}(y_{n+4})$$

for some $y_{n+4} \in G^2_{n+4}$ and

$$q^1_{n+1,n+4}\cdot p^{1,2}_{n+4}(y_{n+4}) = p^{1,2}_{n+4}\cdot q^2_{n+1,n+4}(y_{n+4})$$

$$= p^{1,2}_{n+1}\cdot q^2_{n+1,n+2}\cdot p^{2,3}_{n+2}(x_{n+2}) = q^1_{n+1,n+2}\cdot p^{1,2}_{n+2}\cdot p^{2,3}_{n+2}(x_{n+2}) = 1.$$

Therefore $p^{1,2}_{n+3}\cdot q^2_{n+3,n+4}(y_{n+4}) = q^1_{n+3,n+4}\cdot p^{1,2}_{n+4}(y_{n+4}) = 1$ and there is $x_{n+3} \in G^3_{n+3}$ with $p^{2,3}_{n+3}(x_{n+3}) = q^2_{n+3,n+4}(y_{n+4})$. Now

$$p^{2,3}_{n+1}\cdot q^3_{n+1,n+3}(x_{n+3}) = q^2_{n+1,n+3}\cdot p^{2,3}_{n+3}(x_{n+3}) = q^2_{n+1,n+4}(y_{n+4})$$

$$= q^2_{n+1,n+2}\cdot p^{2,3}_{n+2}(x_{n+2}) = p^{2,3}_{n+1}\cdot q^3_{n+1,n+2}(x_{n+2}).$$

Therefore $q^3_{n+1,n+2}(x_{n+2})\cdot q^3_{n+1,n+3}(x^{-1}_{n+3}) = p^{3,4}_{n+1}(z_{n+1})$ for some $z_{n+1} \in G^4_{n+1}$. Take $z_{n+3} \in G^4_{n+3}$ with $q^4_{n,n+3}(z_{n+3}) = q^4_{n,n+1}(z_{n+1})$. Then

$$q^3_{n,n+3}\cdot(p^{3.4}_{n+3}(z_{n+3})\cdot x_{n+3}) = p^{3,4}_{n}\cdot q^4_{n,n+3}(z_{n+3})\cdot q^3_{n,n+3}(x_{n+3})$$

$$= p^{3,4}_{n}\cdot q^4_{n,n+1}(z_{n+1})\cdot q^3_{n,n+3}(x_{n+3})$$

$$= q^3_{n,n+1}(p^{3,4}_{n+1}(z_{n+1})\cdot q^3_{n+1,n+3}(x_{n+3}))$$

$$= q^3_{n,n+1}\cdot q^3_{n+1,n+2}(x_{n+2}) = q^3_{n,n+2}(x_{n+2}).$$

Thus $q^3_{n,n+1}/\text{im }q^3_{n+1,n+3}$ is an epimorphism and $\underline{G}^3$ satisfies the Mittag-Leffler condition.

Remark. Lemma 2.1 was proved by the author in $[D_5]$. The proof given here is much simpler.

2.2. Lemma. If $\underline{G}^i$ is stable for $i = 1,3$, and satisifies the Mittag-Leffler condition for $i = 2,4$, then $\underline{G}^2$ is stable.

Proof. Without loss of generality we may assume that

$$q^k_{n,n+1}/ \operatorname{im} q^k_{n+1,n+2}: \operatorname{im} q^k_{n+1,n+2} \to \operatorname{im} q^k_{n,n+1}$$

is an isomorphism (epimorphism) for all $n \geq 1$ and $k = 1,3$ $(k = 2,4)$.

We are going to prove that the above homomorphism is a monomorphism for $k = 2$ and $n \geq 2$.

So, suppose $x_{n+1} \in (\operatorname{im} q^2_{n+1,n+2}) \cap (\ker q^2_{n,n+1})$. Take $x_{n+3} \in \operatorname{im} q^2_{n+3,n+4}$ such that $x_{n+1} = q^2_{n+1,n+3}(x_{n+3})$. Then $p^{1,2}_{n+3}(x_{n+3}) \in \operatorname{im} q^1_{n+3,n+4}$ and $q^1_{n,n+3}\cdot p^{1,2}_{n+3}(x_{n+3}) = 1$. Therefore $p^{1,2}_{n+3}(x_{n+3}) = 1$ and there exists $y_{n+3} \in G^3_{n+3}$ with $x_{n+3} = p^{2,3}_{n+3}(y_{n+3})$. Since $p^{2,3}_n\cdot q^3_{n,n+3}(y_{n+3}) = 1$, there exists $z_n \in G^4_n$ with $p^{3,4}_n(z_n) = q^3_{n,n+3}(y_{n+3})$. Take $z_{n+3} \in G^4_{n+3}$ with $q^4_{n-1,n+3}(z_{n+3}) = q^4_{n-1,n}(z_n)$. Then

$$q^3_{n-1,n+2}\cdot p^{3,4}_{n+2}\cdot q^4_{n+2,n+3}(z_{n+3}) = p^{3,4}_{n-1}\cdot q^4_{n-1,n+3}(z_{n+3})$$

$$= p^{3,4}_{n-1}\cdot q^4_{n-1,n}(z_n) = q^3_{n-1,n}\cdot p^{3,4}_n(z_n)$$

$$= q^3_{n-1,n}\cdot q^3_{n,n+3}(y_{n+3}) = q^3_{n-1,n+2}\cdot q^3_{n+2,n+3}(y_{n+3}).$$

Hence $q^3_{n+2,n+3}(y_{n+3}) = p^{3,4}_{n+2}\cdot q^4_{n+2,n+3}(z_{n+3})$ and therefore

$$x_{n+1} = q^2_{n+1,n+3}(x_{n+3}) = q^2_{n+1,n+3}\cdot p^{2,3}_{n+3}(y_{n+3}) = p^{2,3}_{n+1}\cdot q^3_{n+1,n+3}(y_{n+3})$$

$$= p^{2,3}_{n+1}\cdot q^3_{n+1,n+2}\cdot p^{3,4}_{n+2}\cdot q^4_{n+2,n+3}(z_{n+3})$$

$$= q^2_{n+1,n+2}\cdot p^{2,3}_{n+2}\cdot p^{3,4}_{n+2}\, q^4_{n+2,n+3}(z_{n+3}) = 1.$$

Thus $\underline{G}^2$ is stable.

Recall (see [B-K], p. 251) that $\varprojlim^1 G^4 = *$ means that for any sequence $\{a_n\}_{n=1}^{\infty} \in \prod_{n=1}^{\infty} G^4_n$ there is a sequence $\{b_n\}_{n=1}^{\infty} \in \prod_{n=1}^{\infty} G^4_n$ with

$a_n = b_n q^4_{n,n+1}(b^{-1}_{n+1})$. This is a weaker condition than the Mittag-Leffler property (see [D-S], p. 78).

2.3. <u>Lemma</u>. If $\varprojlim^1 \underline{G}^4 = *$, then the sequence

$$\varprojlim \underline{G}^1 \leftarrow \varprojlim \underline{G}^2 \leftarrow \varprojlim \underline{G}^3$$

is exact.

<u>Proof</u>. It is obvious that the image of $\varprojlim \underline{G}^3 \to \varprojlim \underline{G}^2$ is contained in the kernel of $\varprojlim \underline{G}^2 \to \varprojlim \underline{G}^1$. So, suppose the sequence $\{x_n\}_{n=1}^{\infty}$ is contained in the kernel of $\varprojlim \underline{G}^2 \to \varprojlim \underline{G}^1$. Then, for each n there exists $y_n \in G^3_n$ with $p^{2,3}_n(y_n) = x_n$. Since $p^{2,3}_n(y^{-1}_n \cdot q^3_{n,n+1}(y_{n+1})) = p^{2,3}_n(y^{-1}_n) \cdot q^2_{n,n+1} \cdot p^{2,3}_{n+1}(y_{n+1}) = x^{-1}_n \cdot q^2_{n,n+1}(x_{n+1}) = 1$, there exists $z_n \in G^4_n$ with $p^{3,4}_n(z_n) = y^{-1}_n \cdot q^3_{n,n+1}(y_{n+1})$. Hence, in view of $\varprojlim^1 \underline{G}^4 = *$, there exist elements $t_n \in G^4_n$, $n \geqslant 1$, with $z_n = t_n \cdot q^4_{n,n+1}(t^{-1}_{n+1})$. Put $a_n = y_n \cdot p^{3,4}_n(t_n) \in G^3_n$ for each $n \geqslant 1$. Then

$$q^3_{n,n+1}(a_{n+1}) = q^3_{n,n+1}(y_{n+1} \cdot p^{3,4}_{n+1}(t_{n+1}))$$

$$= q^3_{n,n+1}(y_{n+1}) \cdot q^3_{n,n+1} \cdot p^{3,4}_{n+1}(t_{n+1}) = y_n \cdot p^{3,4}_n(z_n) \cdot p^{3,4}_n \cdot q^4_{n,n+1}(t_{n+1})$$

$$= y_n \cdot p^{3,4}_n(t_n) = a_n$$

and $p^{2,3}_n(a_n) = p^{2,3}_n(y_n \cdot p^{3,4}_n(t_n)) = p^{2,3}_n(y_n) = x_n$.

Thus $\{a_n\}_{n=1}^{\infty} \in \varprojlim \underline{G}^3$ and the image of it is $\{x_n\}_{n=1}^{\infty}$ which concludes the proof.

<u>Remark</u>. Lemma 2.3 is usually stated for short exact sequences (see [B-K] for example). Our form of it has the advantage that it can be applied without first reducing a given exact sequence to short exact sequences.

3. Properties of LC^n-spaces.

In the sequel we will use the following categories:

$\mathcal{T}$ (topological spaces and maps),

$H\mathcal{T}$ (the homotopy category of $\mathcal{T}$),

$\mathcal{W}$ (the full subcategory of $H\mathcal{T}$ whose objects are spaces homotopy equivalent to CW complexes),

$\mathcal{G}r$ (groups and homomorphisms).

3.1. Lemma If X is a closed subset of a metrizable space $Y \in LC^n$ $(n \geq 0)$, then for each $k \geq 0$ there exists a natural morphism of pro-$\mathcal{G}r$

$$\underline{h}_k : (H_k(U) : X \subset \text{int } U) \to \text{pro-}H_k(X)$$

which is an isomorphism for $k \leq n$ and an epimorphism for $k = n + 1$.

Remark. By the naturality of the morphism $\underline{h}_k$, we mean the following:

If X' is a closed subset of a metrizable space $Y' \in LC^n$ and $f: Y \to Y'$ is a map with $f(X) \subset X'$, then

$$\text{pro-}H_k(Sf)\ \underline{h}_k = \underline{h}_k'\ (\text{pro-}H_k(\underline{f})),$$

where $\underline{h}_k' : (H_k(V) : X' \subset \text{int } V) \to \text{pro-}H_k(X')$, Sf is the shape morphism induced by f and $\underline{f}$: $(U : X \subset \text{int } U) \to (V : X' \subset \text{int } V)$ is the morphism of pro-$\mathcal{T}$ induced by $f_V = f/f^{-1}(V) : f^{-1}(V) \to V$, $X' \subset \text{int } V$.

Proof of Lemma 3.1. Embed Y as a closed subset of $Z \in ANR$ (see [Hu], Proposition 6.1 on p. 95). By Theorem 3.3.4 in [D-S] (p. 31) and by Corollary 3.1.6 in [D-S] (p. 25) there is a natural isomorphism of pro-$\mathcal{W}$ $\underline{p} : (V : X \subset \text{int}_Z V) \to \check{C}(X)$, where $\check{C}(X)$ is the Čech system of X (see [D-S], p. 21). Let

$$\underline{i} : (U : X \subset \text{int}_Y U) \to (V : X \subset \text{int}_Z V)$$

be the morphism of pro-HT induced by homotopy classes of inclusions $Y \cap V \to V$ for all $V \subset Z$ containing X in its interior. Then we put $\underline{h}_k = \text{pro-}H_k(\underline{p}\cdot\underline{i})$ and it is obvious that $\underline{h}_k$ is natural (in view of Theorem 3.1.5 in [D-S] on p. 24).

To complete the proof of Lemma 3.1, it suffices to show that pro-$H_k(\underline{i})$ is an isomorphism for $k \leq n$ and an epimorphism for $k = n + 1$.

Observe that Lemma 10.1.1 in [D-S] (p. 121) and Theorem 1.2 in [Hu] (p. 112) imply the following:

Claim 1. For any neighborhood V of X in Z there exists a neighborhood $W \subset V$ of X in Z such that if $g:P \to W$ is a map of an $(n+1)$-dimensional polyhedron such that $g(R) \subset W \cap Y$ for some subpolyhedron R of P, then g is homotopic rel.R in V to a map whose values lie in $V \cap Y$.

Then, if $i(A,B)$ denotes the inclusion for any $A \subset B$, under the conditions of Claim 1, we have

$$\text{im } H_k(i(W,V) \subset \text{im } H_k(i(V \cap Y,V)) \quad \text{for } k \leq n+1$$

and

$$\ker H_k(i(W \cap Y,W)) \subset \ker H_k(i(W \cap Y,V \cap Y)) \quad \text{for } k \leq n.$$

Indeed, for any element $d \in H_k(W)$, $k \leq n+1$, there exists a k-dimensional polyhedron P and a map $g:P \to W$ such that $d \in \text{im } H_k(g)$ and by deforming g we get $d \in \text{im } H_k(i(V \cap Y,V))$. Also, if $d \in \ker H_k(i(W \cap Y,W))$, then there exists a $(k+1)$-dimensional polyhedron P and a map $g:P \to W$ such that for some subpolyhedron R of P and $d_1 \in H_k(R)$, there is $g/R:R \to W \cap Y$, $H_k(g/R)(d_1) = d$ and $H_k(i(R,P))(d_1) = 0$. Then we can homotop g rel.R and get $d \in \ker H_k(i(W \cap Y,V \cap Y))$.

By Theorem 2.6 in [D_1] (p. 8) $\underline{h}_{n+1}$ is an epimorphism. To prove that $\underline{h}_k$ is an isomorphism for $k \leq n$, take for any neighborhood U

of X in Z neighborhoods $W \subset V$ of X in U such that

$$\operatorname{im} H_k(i(V,U) \subset \operatorname{im} H_k(i(U \cap Y,U))$$

$$\ker H_k(i(V \cap Y,V)) \subset \ker H_k(i(V \cap Y,U \cap Y))$$

$$\operatorname{im} H_k(i(W,V) \subset \operatorname{im} H_k(i(V \cap Y,V))$$

$$\ker H_k(i(W \cap Y,W)) \subset \ker H_k(i(W \cap Y,V \cap Y)).$$

Then $\alpha:H_k(W) \to H_k(U \cap Y)$ defined by $\alpha(d) = H_k(i(V \cap Y,U \cap Y))(d_1)$ where $d_1 \in H_k(V \cap Y)$ satisfies $H_k(i(W,V))(d) = H_k(i(V \cap Y,V))(d_1)$, is a homomorphism such that $H_k(i(U \cap Y,U))\cdot\alpha = H_k(i(W,U))$ and $\alpha\cdot H_k(i(W \cap Y,W)) = H_k(i(W \cap Y,U \cap Y))$.

By Theorem 2.3.4 in [D-S] (p. 14) $\underline{h}_k$ is an isomorphism.

3.2. Lemma. Let $X \in LC^0$ be a metrizable space such that $X - \{x_0\} \in LC^n$ $(n \geq 1)$ for some point $x_0 \in X$. Then $X \in LC^n$ iff for each neighborhood U of x_0 there exists a neighborhood $V \subset U$ of x_0 such that for any map $f:\partial B^k \to V$ from the boundary of a k-dimensional $(2 \leq k \leq n+1)$ PL-ball B^k and for any neighborhood W of x_0 there exist PL-balls $B_1^k,\cdots,B_m^k \subset B^k$ and an extension $f':B^k - \bigcup_{i=1}^m \operatorname{int} B_i^k \to U$ of f such that $(\operatorname{int} B_i^k) \cap (\operatorname{int} B_j^k) = \emptyset$ for $i \neq j$ and $f'(\partial B_i^k) \subset W$ for $i \leq m$.

Proof. The necessity of the above condition is obvious, because if $X \in LC^n$, then we can choose V in such a way that any map $f:\partial B^k \to V$ is null-homotopic in U for $k \leq n+1$.

So we are going to prove the sufficiency of it.

Suppose U is a neighborhood of x_0 in X. By induction, we can choose a decreasing sequence $\{U_m\}_{m=1}^{\infty}$ of open neighborhoods of x_0 satisfying the following conditions:

a. $U_1 = U$,

b. $\operatorname{diam} U_m < 1/m$ for $m \geq 2$,

c. for any map $f:\partial B^k \to U_{m+1}$ $(m \geq 1,\ 2 \leq k \leq n+1)$ there exist

PL-balls $B_1^k, \cdots, B_p^k \subset B^k$ and an extension

$$f' : B^k - \bigcup_{i=1}^{p} \operatorname{int} B_i^k \to U_m$$

such that $(\operatorname{int} B_i^k) \cap (\operatorname{int} B_j^k) = \emptyset$ for $i \neq j$ and $f'(\partial B_i^k) \subset U_{m+2}$.

Now, suppose $f : \partial B^k \to U_2$ is a map. By induction, we can find finite collections of PL-balls $\{B_{m,i}^k\}_{i=1}^{p_m}$ and maps

$$f_m : B^k - \bigcup_{i=1}^{p_m} \operatorname{int} B_{m,i}^k \to U$$

such that the following conditions are fulfilled:

d. $(\operatorname{int} B_{m,i}^k) \cap (\operatorname{int} B_{m,j}^k) = \emptyset$ for $i \neq j$,

e. for every $i \leq p_{m+1}$ there exists $a(i) \leq p_m$ with $B_{m+1,i}^k \subset B_{m,a(i)}^k$,

f. f_{m+1} is an extension of f_m,

g. for any $i \leq p_m$ $(m \geq 1)$, $f_{m+1}(B_{m,i}^k - \bigcup_{j=1}^{p_{m+1}} \operatorname{int} B_{m+1,j}) \subset U_{m+2}$,

h. $f_m(\partial B_{m,i}^k) \subset U_{m+3}$ for any $i \leq p_m$,

i. f_1 is an extension of f.

Then it is obvious how to construct a map $f' : B^k \to U$ extending f. Thus, $X \in LC^n$.

4. Adding one point to LC^n-spaces.

Recall that the end $e(X)$ of a locally compact Hausdorff space X is the inverse system $(X-A : A \subset X$ is compact$)$ bonded with inclusions.

We need the following generalization of this notion:

4.1. Definition. The end $e(X,x_0)$ of a topological space X at a non-isolated point $x_0 \in X$ is the inverse system

$$(U - \{x_0\} : x_0 \in \operatorname{int} U)$$

bonded with inclusions.

It is clear that if X is locally compact, non-compact and Hausdorff, then $e(X) = e(\omega X, x_0)$, where ωX is the one-point compactification of X and $x_0 \in \omega X - X$.

The reader should have this in mind when reading the results of this section.

4.2. <u>Theorem</u>. Let X be a metrizable space such that $X - \{x_0\} \in LC^0$ for some non-isolated point $x_0 \in X$. Then $X \in LC^0$ iff pro-$H_0(e(X,x_0))$ satisfies the Mittag-Leffler condition.

<u>Proof</u>. Take a decreasing sequence $\{U_n\}_{n=1}^{\infty}$ of open neighborhoods of x_0 in X such that diam $U_n < 1/n$ for $n \geq 1$. The space $U_n - \{x_0\}$ is denoted by U'_n for $n \geq 1$ and the inclusion from U'_m to U'_n is denoted by $i_{n,m}$ for $m > n$.

If $X \in LC^0$ we may assume that each U_n is arcwise-connected. Then $H_0(U_m, U_{m+1}) = 0$ for each m and by the excision property of the singular homology (see [S], p. 189) we get

$$H_0(U'_m, U'_{m+1}) = 0$$

for each $m \geq 1$.

Now the homology exact sequence

$$H_0(U'_{m+1}) \to H_0(U'_m) \to H_0(U'_m, U'_{m+1}) = 0$$

implies that $H_0(i_{m,m+1})$ is an epimorphism. Since $(H_0(U'_m))_{m=1}^{\infty}$ is a cofinal subsystem of pro-$H_0(e(X,x_0))$, we get that pro-$H_0(e(X,x_0))$ satisfies the Mittag-Leffler condition.

If pro-$H_0(e(X,x_0))$ satisfies the Mittag-Leffler condition, we may assume without loss of generality that

$$\text{im } H_0(i_{n,n+1}) = \text{im } H_0(i_{n,m})$$

for $m > n$.

Suppose $x_1 \in U'_m$ for some $m \geq 2$ and let C_m be the arc-component of U'_m containing x_1. By induction, we can choose arc-conponents

C_k of U'_k for $k > m$ such that C_{k+1} and C_k are contained in the same component D_{k-1} of U'_{k-1}. Then D_k is contained in D_{k-1} for each $k > m$. Choose $x_p \in D_{m+p}$ for $p \geq 2$. Then, for each $p \geq 1$, there exists a map

$$f_p: [1/p+1, 1/p] \to D_{m+p}$$

such that $f_p(1/p+1) = x_{p+1}$ and $f_p(1/p) = x_p$ Hence, $f:[0,1] \to U_{m-1}$ defined by $f(x) = f_p(x)$ for $1/p+1 < x \leq 1/p$ and $f(0) = x_0$ is continuous.

Thus, $X \in LC^0$.

To state the next result, we need the following notion.

Suppose $\underline{X} = (X_\alpha, i_{\alpha,\beta}, A)$ is an inverse system of subsets of a space X bonded with inclusions (e.g., we take all neighborhoods of a subset Y of X). $\underline{X}$ is called <u>nearly 1-movable</u> provided for each $\alpha \in A$ there exists $\beta \geq \alpha$ such that for any map $f: S^1 = \partial B^2 \to X$ and for any $\gamma \geq \alpha$ there exist 2-discs $B_1^2, \cdots B_k^2 \subset B^2$ and an extension $f': B^2 - \bigcup_{i=1}^{k} \operatorname{int} B_i^2 \to X_\alpha$ of f such that $(\operatorname{int} B_i^2) \cap (\operatorname{int} B_j^2) = \emptyset$ for $i \neq j$ and $f'(\partial B_i^2) \subset X_\gamma$ for each $i \leq k$.

The above notion was introduced by D.R. McMillan [M] in a case where $\underline{X}$ is the system of all open neighborhoods of a compactum X lying in the Hilbert cube Q.

It is easy to see that the near 1-movability is a hereditary invariant in pro-*HT* (i.e., if $\underline{X}$ is nearly 1-movable and $\underline{X}$ dominates $\underline{Y}$ in pro-*HT*, then $\underline{Y}$ is nearly 1-movable). Also, Corollary 8.4 in [D_1] (p. 39) implies that a compactum X is nearly 1-movable iff the system of all neighborhoods of X in any metrizable space $Y \in LC^1$ containing X is nearly 1-movable.

4.3. <u>Theorem</u>. Let X be a metrizable space such that $X - \{x_0\} \in LC^n$ $(n \geq 1)$ for some non-isolated point $x_0 \in X$. Then, $X \in LC^n$ iff $e(X, x_0)$ is nearly 1-movable and $\text{pro-}H_k(e(X, x_0))$ is stable for $k < n$ and satisfies the Mittag-Leffler condition for $k = n$.

Proof. Take a decreasing sequence $\{U_k\}_{k=1}^{\infty}$ of open neighborhoods of x_0 in X such that $\operatorname{diam} U_k < 1/k$ for $k \geq 1$. Then, $(U'_k)_{k=1}^{\infty}$ is a cofinal subsystem of $e(X,x_0)$, where $U'_k = U_k - \{x_0\}$.

Suppose $X \in LC^n$. For any $k \geq 1$ consider the homology exact sequence

$$H_{p+1}(U_k) \to H_{p+1}(U_k,U'_k) \to H_p(U'_k) \to H_p(U_k) \to H_p(U_k,U'_k)$$

Since $(H_p(U_k,U'_k))_{k=1}^{\infty}$ is stable for each p (by excision it is isomorphic to $H_p(X,X - \{x_0\})$), we infer by Lemmata 2.1 and 2.2 that $(H_p(U'_k))_{k=1}^{\infty}$ is stable for $p < n$ and satisfies the Mittag-Leffler property for $p = n$.

Thus, pro-$H_k(e(X,x_0))$ is stable for $k < n$ and satisfies the Mittag-Leffler condition for $k = n$.

To prove that $e(X,x_0)$ is nearly 1-movable, we may assume that any loop in U_{k+1} is null-homotopic in U_k.

Suppose $f:S^1 = \partial B^2 \to U'_{k+1}$ is a loop and $m > k+1$. Take an extension $f_1:B^2 \to U_k$ of f and let $A = f_1^{-1}(x_0)$. Since $A \cap S = \emptyset$, there exists a neighborhood W of A in $\operatorname{int} B^2$ such that $f_1(W) \subset U_m$. Now, it is clear that there exist disjoint discs $\{D_i\}_{i=1}^{p}$ such that $f_1(\partial D_i) \subset U'_m$ for each i and $A \subset \bigcup_{i=1}^{p} \operatorname{int} D_i$. Consequently, $e(X,x_0)$ is nearly 1-movable.

Now suppose $e(X,x_0)$ is nearly 1-movable, pro-$H_k(e(X,x_0))$ is stable for $k < n$ and satisfies the Mittag-Leffler condition for $k = n$.

We may assume that for each $k+1 < m$ and for any loop $g:S^1 = \partial B^2 \to U'_{k+1}$, there exist discs $D_1,\cdots,D_p$ with mutually disjoint interiors and an extension

$$g':B^2 - \bigcup_{i=1}^{p} \operatorname{int} D_i \to U'_k$$

of g such that $g'(\partial D_i) \subset U'_m$ for each i. Also, we may assume that two components of U'_{m+1} contained in the same component of U'_k are contained in the same component of U'_m (this is possible because pro-$H_0(e(X,x_0))$ is stable).

By Theorem 4.2, $X \in LC^0$.

Suppose $f:S^1 = \partial B^2 \to U_{k+1}$ is a loop and $m > k+1$. We are going to prove that f extends to $f':B^2 - \bigcup_{i=1}^{p} \text{int } D_i \to U_k$ with $\bigcup_{i=1}^{p} f'(\partial D_i) \subset U_m$ for some discs $D_1,\cdots,D_p$ with mutually disjoint interiors.

It is so if $x_0 \notin f(S^1)$. Therefore, we consider the case where $x_0 \in f(S^1)$. Then we can represent S^1 as the union of arcs $L_0,\cdots,L_{2p+1}$ such that $f(L_{2i}) \subset U_{m+1}$, $x_0 \notin f(L_{2i+1})$ for $0 \leq i \leq p$ and $L_i \cap L_j$ is either empty or is a one-point set for $i \neq j$. Then the endpoints of L_{2i+1}, for any i, are mapped by f to the same component of U'_m (because they are mapped to the same component of U'_k). Consequently, we can find polygonal arcs L'_{2i+1}, $0 \leq i \leq p$, in B^2 and maps $f_i:L'_{2i+1} \to U'_m$ such that the following conditions are satisfied:

a. the endpoints of L_{2i+1} and L'_{2i+1} coincide,

b. $(\text{int } L'_{2i+1}) \cap (\text{int } L'_{2j+1}) = \emptyset$ for $i \neq j$,

c. $(\text{int } L'_{2i+1}) \cap S^1 = \emptyset$,

d. $f_i = f$ on $L_{2i+1} \cap L'_{2i+1}$.

Now, for each $i \leq p$, take the 2-disc D_i in B^2 bounded by $L_{2i+1} \cup L'_{2i+1}$ and let $g_i:\partial D_i \to U'_{k+1}$ be defined by $g_i = f$ on L_{2i+1} and $g_i = f_i$ on L'_{2i+1}. Then, for each i, there exist discs $D_{i,1},\cdots,D_{i,k_i}$ with mutually disjoint interiors and an extension $g'_i:D_i - \bigcup_{j=1}^{k_i} \text{int } D_{i,j} \to U'_k$ of g_i such that $g'_i(\partial D_{i,j}) \subset U'_m$ for each j. Now, it is obvious that there exist 2-discs $B_1,\cdots,B_r$ in B^2 with mutually disjoint interiors and an extension $f':B^2 - \bigcup_{i=1}^{r} \text{int } B_i \to U_k$ of f such that $f'(\partial B_i) \subset U_m$ for each i.

By Lemma 3.2, $X \in LC^1$ and we may assume that each U_k is connected.

Now, for any $k \geq 1$ consider the homology exact sequence

$$H_{p+1}(U_k,U'_k) \to H_p(U'_k) \to H_p(U_k) \to H_p(U_k,U'_k) \to H_{p-1}(U'_k)$$

Since $(H_p(U_k,U'_k))_{k=1}^{\infty}$ is stable for each p, we infer from Lemmata 2.1 and 2.2 that $(H_p(U_k))_{k=1}^{\infty}$ is stable for $p < n$ and satisfies the Mittag-Leffler condition for $p = n$.

For each k take a weak homotopy equivalence

$$f_k:(X_k,x_k) \to (U_k,x_0)$$

where (X_k,x_k) is a pointed CW complex (see [S], Theorem 1 on p. 412). Then, for each k there is a map $p_{k,k+1}:(X_{k+1},x_{k+1}) \to (X_k,x_k)$ such that $f_k \cdot p_{k,k+1} \simeq i_{k,k+1} \cdot f_{k+1}$, where $i_{k,k+1}$ is the inclusion from (U_{k+1},x_0) to (U_k,x_0) (see [S], Theorem 22 on p. 404).

Let $\underline{X} = (X_k,p_{k,k+1})$. Since each f_k induces isomorphisms of homology and homotopy groups (see [S], Theorem 25 on p. 406), we infer that pro-$\pi_1(\underline{X})$ is trivial, pro-$H_p(\underline{X})$ is stable for $p < n$ and satisfies the Mittag-Leffler condition for $p = n$. Then Lemma 3.1 in $[D_5]$ says that pro-$\pi_p(\underline{X})$ is stable for $p < n$ and satisfies the Mittag-Leffler condition for $p = n$. Hence $(\pi_p(U_k,x_0))_{k=1}^{\infty}$ is stable for $p < n$ and satisfies the Mittag-Leffler condition for $p = n$. Consequently, for each k there exists $m > k$ such that for each nieghborhood U of x_0 in X and for any map $f:\partial B^p \to U_m$ $(p \leq n+1)$ there exists a PL-ball $B_1^p \subset B^p$ and an extension $f':B^p - \text{int}\, B_1^p \to U_k$ of f with $f'(\partial B_1^p) \subset U$.

By Lemma 3.2, $X \in LC^n$.

An immediate consequence of Theorem 4.3 and Theorem 7.1 in [Hu] (p. 168) is

4.4 <u>Corollary</u>. Let X be a finite-dimensional metrizable space such that $X - \{x_0\} \in ANR$ for some non-isolated point $x_0 \in X$. Then $X \in ANR$ iff $e(X,x_0)$ is nearly 1-movable and its homology pro-groups are stable.

<u>Convention</u>. If $\mathcal{U}$ and $\mathcal{V}$ are collections of open sets in a

space X, then $L(\mathcal{V},\mathcal{U})$ is the statement: any partial realization of a simplicial complex K in $\mathcal{V}$ extends to a full realization of K in $\mathcal{U}$ (see [D-S]).

4.5. Lemma. Let X be a compactum which is locally n-connected for each $n \geq 0$ and $X - x_0 \in ANR$ for some $x_0 \in X$. Then $X \in ANR$ if for each neighborhood U of x_0 there exists a neighborhood V of x_0 in U such that the inclusion $i(V,U)$ induces a trivial shape morphism.

Proof. Suppose $\mathcal{U}$ is an open covering of X. Take $W \in \mathcal{U}$ containing x_0 and let $V \subset W$ be a neighborhood of x_0 such that $i(V,W)$ induces a trivial shape morphism. Choose the star refinement $\mathcal{U}_1$ of $\{V\} \cup \{U - \{x_0\}: U \in \mathcal{U}\}$. Take an element $U_1 \in \mathcal{U}_1$ containing x_0 and a star refinement $\mathcal{U}_2$ of $\{U_1\} \cup \{U - \{x_0\}: U \in \mathcal{U}_1\}$. By the assumption there exists an open covering $\mathcal{U}_3$ of $X' = X - \{x_0\}$ such that $L(\mathcal{U}_3,\mathcal{U}_2')$ holds, where $\mathcal{U}_2' = \{U - \{x_0\}: U \in \mathcal{U}_2\}$.

Suppose $f: L \to X$ is a partial realization of a finite simplicial complex K in $\{U_2\} \cup \mathcal{U}_3$, where $U_2 \in \mathcal{U}_2$ contains x_0. Take all vertices $\{v_i\}_{i \in J}$ of K which are mapped to U_2 by f. Consider subcomplexes $K_1 = K - \bigcup_{i \in J} \operatorname{st} v_i$ and $L_1 = L \cap K_1$. Then $f/L_1: L_1 \to X'$ is a partial realization of K_1 in $\mathcal{U}_3$ and it extends to a full realization $f': K_1 \to X'$ of K_1 in $\mathcal{U}_2'$.

Notice that

$$f(L_1 \cap \operatorname{cl}(\bigcup_{i \in J} \operatorname{st} v_i)) \subset U_2 \cup \bigcup \{U \in \mathcal{U}_3: U \cap U_2 \neq \emptyset\} \subset U_1$$

and therefore $f'(K_1 \cap \operatorname{cl}(\bigcup_{i \in J} \operatorname{st} v_i)) \subset \bigcup \{U \in \mathcal{U}_1: U \cap U_1 \neq \emptyset\} \subset V$.

Using Claim 1 from the proof of Lemma 3.1 it is easy to show that $f'/(K_1 \cap \operatorname{cl}(\bigcup_{i \in J} \operatorname{st} v_i))$ is null-homotopic in W. Therefore, we can extend f' to $f'': K \to X$ such that $f''/L = f$ and $f''(\operatorname{cl}(\bigcup_{i \in J} \operatorname{st} v_i)) \subset W$. Then, f'' is a full realization of K in $\mathcal{U}$ extending f. Thus

$X \in ANR$ (see [Hu], p. 122).

Remark. Theorem 4.5 is proved in [K-L] under the stronger assumption that X is locally contractible.

4.6. Theorem. Let X be a compactum such that $e(X,x_0)$ is movable and $X - \{x_0\} \in ANR$ for some non-isolated point $x_0 \in X$. Then $X \in ANR$ iff $\text{pro-}H_k(e(X,x_0))$ is stable for $k \geq 0$ and almost all homology groups of X are trivial.

Proof. In view of Theorem 4.3, it suffices to prove $X \in ANR$ provided $\text{pro-}H_k(e(X,x_0))$ is stable for $k \geq 0$ and $H_k(X) = 0$ for $k \geq N$.

By Theorem 4.2, $X \in LC^0$.

Special case. Assume $X - \{x_0\}$ is a polyhedron (i.e., it is homeomorphic to the body of a simplicial complex). Then, there exists a decreasing sequence $\{P_n\}_{n=1}^{\infty}$ of subpolyhedra of $X - \{x_0\}$ such that $cl(X - P_n)$ is compact for each n, $P_{n+1} \subset \text{int } P_n$ for $n \geq 1$, each space $P_n' = P_n \cup \{x_0\}$ is connected and $(P_n)_{n=1}^{\infty}$ is a cofinal subsystem of $e(X,x_0)$. Since $e(X,x_0)$ is movable, it is nearly 1-movable and by Theorem 4.3, the space P_n' is locally k-connected for all k and all $n \geq 1$. Without loss of generality, we may assume (in view of movability of $e(X,x_0)$) that for each $m > n$ there exists a map $g_{m,n}: P_n \to P_m$ homotopic in P_{n-1} to the inclusion $i(P_n,P_{n-1})$ from P_n to P_{n-1}.

Claim 1. Each pointed continuum (P_m',x_0) is movable.

Proof of Claim 1. Observe that $P_m' = \varprojlim(P_m/P_n, q_{n,n+1})$, where $q_{n,n+1}: P_m/P_{n+1} \to P_m/P_n$ is the natural projection. Suppose

$k > n > m+1$. By homotopy extension theorem for ANR's (see [Hu], Theorem 2.2 on p. 117) there is a homotopy $H:P_m \times I \to P_m$ such that $H(x,0) = x$ for $x \in P_m$, $H(x,t) = x$ for $x \in cl(P_m - P_{n-1})$ and $0 \leq t \leq 1$, $H(P_n \times I) \subset P_{n-1}$ and $H(x,1) = g_{k,n}(x)$ for $x \in P_n$. Then the homotopy H induces a map $f:P_m/P_n \to P_m/P_k$ such that $q_{n-1,k}\, g \simeq q_{n-1,n}$. Thus, (P'_m, x_0) is movable.

Claim 2. For each m all homology groups of P'_m are finitely generated and almost all of them are trivial.

Proof of Claim 2. By Lemma 3.1, for each k, the group $H_k(P'_m)$ is isomorphic to pro-$H_k(P'_m)$ in pro-*Gr*. Since pro-$H_k(P'_m)$ consists of finitely generated groups, $H_k(P'_m)$ is finitely generated. Since $A = cl(X - P'_m)$ and $B = P'_m \cap A$ are compact polyhedra, there exists $M \geq 1$ such that $H_k(B) = H_k(X) = 0$ for $k \geq M$. From the Mayer-Vietoris sequence (see [S], pp. 186-189)

$$\to H_k(B) \to H_k(A) \oplus H_k(P'_m) \to H_k(X) \to$$

we get $H_k(P'_m) = 0$ for $k \geq M$.

Fix $k \geq 1$. Since $P'_k \in LC^1$, there exists $m > k$ such that any loop in P'_m is null-homotopic in P'_k. Since $P'_m \in LC^n$ for all $n \geq 1$, by Theorem 8.7 in $[D_1]$ (p. 41) the homotopy pro-group pro-$\pi_n(P'_m, x_0)$ is isomorphic to $\pi_n(P'_m, x_0)$ for all $n \geq 1$. In particular, $\pi_1(P'_m, x_0)$ is finitely generated because pro-$\pi_1(P'_m, x_0)$ consists of finitely presented groups. Hence, there is a wedge (S, s_0) of a finite number of circles and a map $g:(S, s_0) \to (P'_m, x_0)$ inducing an epimorphism of fundamental groups. Then the matching space $R = C(S) \cup_g P'_m$ is 1-connected, where $C(S)$ is a cone over S, and it is easy to prove that it is movable (see also $[D_2]$). Now we can extend the inclusion $i(P'_m, P'_k)$ to $h:R \to P'_k$ (because $\pi_1(i(P'_m, P'_k))$ is trivial).

Using Claim 2, it is obvious that all homology groups of R are finitely generated and almost all of them are trivial. Hence, (see

[S], exercise 5 on p. 420) there exists a weak homotopy equivalence $f:K \to R$, where K is a finite CW complex. Since $R \in LC^n$ for each n, the map f induces isomorphisms of all homotopy pro-groups (for LC^∞-spaces there is no distinction between homotopy groups and homotopy pro-groups - see [D_1], Theorem 8.7 on p. 41) and by Theorem 8.2.4 in [D-S] (p. 108) f is a shape equivalence. So take a map $f':R \to K$ which is a shape inverse to f. Since K is locally contractible, there exists P'_p such that f'/P'_p is null-homotopic. Hence $i(P'_p,R)$ induces a trivial shape morphism and consequently $i(P'_n,P'_k) = h\cdot i(P'_p,R)$ induces a trivial shape morphism.

By Theorem 4.5, we conclude $X \in ANR$.

<u>General case</u>. Take the space $X \times Q/\{x_0\}\times Q$, where Q is the Hilbert cube. Since $(X - \{x_0\}) \times Q$ is a Q-manifold (see [C], Theorem 44.1 on p. 106), there exists a polyhedron P such that $P \times Q$ and $(X - \{x_0\}) \times Q$ are homeomorphic (see [C], Theorem 37.2 on p. 83). Let ωP be the one-point compactification of P and $p_0 \in \omega P - P$. Then $e(X,x_0)$ and $e(\omega P,p_0)$ are isomorphic in pro-HT (this is easy) and $X \times Q/\{x_0\}\times Q$ is homeomorphic to $\omega P\times Q/\{p_0\}\times Q$ (because both spaces are one-point compactifications of homeomorphic spaces).

Observe that almost all homology groups of $X\times Q/\{x_0\}\times Q$ are trivial because the projection $q:X \times Q \to X\times Q/\{x_0\}\times Q$ is a shape equivalence (see [D-S], Corollary 4.4.4 on p. 59) and by Lemma 3.1, there is no distinction between homology groups and homology pro-groups of these spaces (by Theorem 4.3, they are locally k-connected for all k). Hence, almost all homology groups of ωP are trivial (ωP is a retract of $\omega P\times Q/\{p_0\}\times Q$) and by the special case $\omega P \in ANR$. Then, $\omega P\times Q/\{p_0\}\times Q \in ANR$ (see [D-S], Corollary 9.4.3 on p. 119) and $X\times Q/\{x_0\}\times Q \in ANR$.

Thus $X \in ANR$, as a retract of $X\times Q/\{x_0\}\times Q$.

<u>Remark</u>. Compare Theorem 4.6 with the main result of [W-W].

4.7. Theorem. Let X be a compactum such that $e(X,x_0)$ is movable and $X - \{x_0\} \in ANR$ for some non-isolated point $x_0 \in X$. If all homology pro-groups of $e(X,x_0)$ are stable and almost all of them are trivial, then $X \in ANR$.

Proof. In view of Theorem 4.6, it suffices to prove that almost all homology groups of X are trivial.

First suppose $X - \{x_0\}$ is a polyhedron and take a decreasing sequence $\{P_n\}_{n=1}^{\infty}$ of subpolyhedra of $X' = X - \{x_0\}$ such that $cl(X - P_n)$ is compact for each n and $(P_n)_{n=1}^{\infty}$ is a cofinal subsystem of $e(X,x_0)$. Since $e(X,x_0)$ is movable, we may assume that for each $m \geqslant n$ there is a map $g_{m,n}:P_n \to P_m$ such that $g_{m,n}$ is homotopic in P_{n-1} to the inclusion $i(P_n,P_{n-1})$.

Suppose pro-$H_k(e(X,x_0))$ is trivial for $k \geqslant N$. Then, for each k there exists $m \geqslant 2$ such that $H_k(i(P_m,P_1))$ is trivial. Since $i(P_m,P_1)\cdot g_{2,m} \simeq i(P_2,P_1)$, the homomorphism $H_k(i(P_2,P_1))$ is trivial for each k. Let $A = cl(X' - P_2)$, $B = A \cap P_2$ and $M = \max(\dim A,N)$. Using the Mayer-Vietoris sequence

$$\to H_k(B) \to H_k(A) \oplus H_k(P_2) \to H_k(X') \to H_{k-1}(B) \to$$

we get $H_k(X') = 0$ for $k \geqslant M+2$ because $H_k(i(P_2,X'))$ is trivial for all k and $H_k(B) = H_k(A) = 0$ for $k \geqslant M+1$.

For each $m > n$ let $q_{n,m}:(X'/P_m) \to (X'/P_n)$ be the natural projection. Then $X = \varprojlim(X'/P_n,q_{n,m})$. Since the projection $q_n:(X',P_n) \to (X'/P_n,\{P_n\})$ induces isomorphisms of the homology groups (see [Do], Corollary 4.4 in Chapter V), we have a functorial exact sequence for each n and $k \geqslant 2$

$$H_k(P_n) \to H_k(X') \to H_k(X'/P_n) \to H_{k-1}(P_n)$$

and by Lemma 2.3, they give rise to the following exact sequence:

$$H_k(X') \to \check{H}_k(X) \to \check{H}_{k-1}(e(X,x_0))$$

where $\check{H}_k(e(X,x_0)) = \underleftarrow{\lim}\ \text{pro-}H_k(e(X,x_0))$. Hence $\check{H}_k(X) = 0$ for $k > M+1$ and by Lemma 3.1 (recall that $X \in LC^\infty$ by Theorem 4.3), $H_k(X) = 0$ for $k > M+1$.

Thus, $X \in ANR$ by Theorem 4.6.

If $X - \{x_0\}$ is an arbitrary ANR, then, as in the proof of Theorem 4.6, there exists a polyhedron P such that $(X - \{x_0\}) \times Q$ is homeomorphic to $P \times Q$. Then $e(\omega P, p_0)$ is isomorphic in pro-*HT* to $e(X,x_0)$, where ωP is the one-point compactification of P and $p_0 \in \omega P - P$. By the first part of the proof, we have $\omega P \in ANR$ and therefore $X \in ANR$ as in the proof of Theorem 4.6.

Remark. Observe that the triviality of almost all homology pro-groups of $e(X,x_0)$ is not necessary for X to be an ANR. Indeed, if X is the space of orbits of the standard based-free action of Z_2 on the Hilbert cube Q (see [W-W]), then $e(X,x_0)$ is stable and is isomorphic to $K(Z_2,1)$, where x_0 is the orbit of the fixed point of the action. Therefore, infinitely many homology pro-groups of $e(X,x_0)$ are non-trivial and $X \in AR$ (see [W-W]).

5. LC^n-divisors and ANR-divisors.

5.1. Definition. A compactum X is an LC^n-divisor (ANR-divisor) provided for each compact space $Y \in LC^n$ $(Y \in ANR)$ containing X, the quotient space Y/X is locally k-connected for $k \leq n$ $(Y/X \in ANR)$.

Remark. The notion of an ANR-divisor was introduced by D.M. Hyman [Hy]. The notion of an LC^n-divisor was introduced by the author in [D_3].

Observe that every compactum is an LC^0-divisor.

5.2. Theorem. For a compactum X, the following conditions are equivalent $(n \geq 1)$:

a. X is an LC^n-divisor,

b. $Y_0/X \in LC^n$ for some metrizable space $Y_0 \in LC^n$ containing X,

c. X is nearly 1-movable, pro-$H_k(X)$ is stable for $k < n$ and satisfies the Mittag-Leffler condition for $k = n$,

d. $Y/X \in LC^n$ for each metrizable space $Y \in LC^n$ containing X.

Proof. Suppose X is contained in a metrizable space $Y \in LC^n$ and take a sequence $\{U_p\}_{p=1}^{\infty}$ of open neighborhoods of X in Y such that $cl(U_{p+1}) \subset U_p$ for each p and any neighborhood U of X contains U_p for some p.

Consider open subsets $W_p = U_p \times [0,1/p] - X \times \{0\}$ of $Y \times I - X \times \{0\}$.

Then, for each p, the diagram

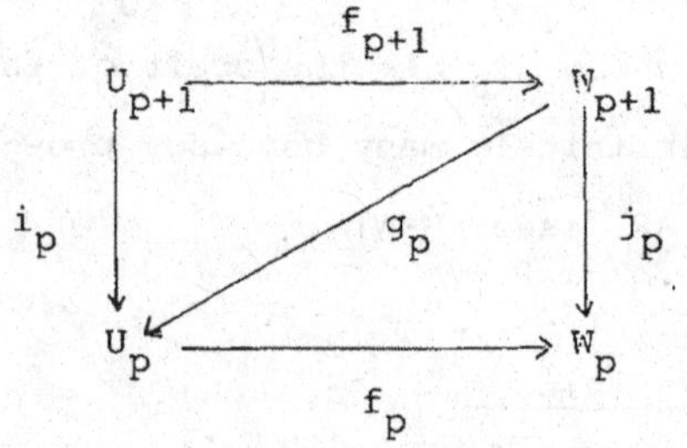

commutes up to homotopy, where i_p and j_p are inclusions, $f_m: U_m \to W_m$ is defined by $f_m(x) = (x,1/m)$ for $x \in U_m$ and $g_m: W_{m+1} \to U_m$ is defined by $g_m(x,t) = x$ for $(x,t) \in W_{m+1}$. Therefore, $(H_k(U_p))_{p=1}^{\infty}$ is isomorphic to pro-$H_k(e(Y \times I/X \times \{0\}, x_0'))$, where x_0' is the base point of $Y \times I/X \times \{0\}$ which is a metrizable space (see [E], Theorem 4.4.15 on p. 355). This is because $(W_m)_{m=1}^{\infty}$ is a cofinal subsystem of $e(Y \times I/X \times \{0\}, x_0')$.

By Lemma 3.1, pro-$H_k(X)$ is isomorphic to $(H_k(U_p))_{p=1}^{\infty}$ for $k \leq n$.

Thus, pro-$H_k(X)$ is isomorphic to pro-$H_k(e(Y \times I/X \times \{0\}, x_0'))$ for $k \leq n$.

Observe that $Y/X \in LC^n$ implies $Y \times I/X \times \{0\} \in LC^n$. Therefore, Condition b implies Condition c (in view of Theorem 4.3).

If X is nearly 1-movable, $\text{pro-}H_k(X)$ is stable for $k < n$ and satisfies the Mittag-Leffler condition for $k = n$, then $Y \times I/X \times \{0\} \in LC^n$ by Theorem 4.4. Consequently, $Y/X \in LC^n$ as a retract of $Y \times I/X \times \{0\}$. Thus, Condition c implies Condition d which completes the proof of Theorem 5.2.

Remark. Theorem 5.2 is a generalization of the main result of $[D_3]$.

The problem of characterizing ANR-divisors in an algebraic way is still open. However, we can characterize some classes of ANR-divisors.

5.3. Theorem. A compactum X whose deformation dimension is finite is an ANR-divisor iff its homology pro-groups are stable and X is nearly 1-movable.

Proof. If X is an ANR-divisor, then it is an LC^n-divisor for each n (see Theorem 5.2). Therefore, X is nearly 1-movable and its homology pro-groups are stable.

Suppose def-dim X is finite and $\text{pro-}H_n(X)$ is stable for each n. By Theorem 5.3.5 in [D-S] (p. 70), there exists a finite-dimensional compactum Y with $\text{Sh } Y = \text{Sh } X$. Embed Y in a k-dimensional ball B^k for some k. Then $\dim(B^k/Y) \leqslant k$ and $B^k/Y \in LC^k$ (because Y is nearly 1-movable and its homology pro-groups are stable). Hence, $B^k/Y \in ANR$ (see [Hu], Theorem 7.1 on p. 168). By Lemma 9.4.4 in [D-S] (p. 120), Y is an ANR-divisor and by Corollary 9.4.3 in [D-S] (p. 118), X is an ANR-divisor.

Remark. Theorem 5.3 was proved in $[D_3]$.

5.4. Theorem. A movable compactum X is an ANR-divisor iff its Čech homology groups are finitely generated and almost all of of them are trivial.

Proof. Embed X as a Z-set in the Hilbert cube Q (see [C], Theorem 11.2 on p. 14). By Corollary 4.4.2 in [D-S] (p. 57) $Sh(Q/X) = Sh(Q \cup C(X))$ and by Corollary 4.4.4 in [D-S] (p. 59), $sh(Q \cup C(X)) = Sh(Q \cup C(X)/Q)$. Since $Q \cup C(X)/Q$ is homeomorphic to the suspension ΣX of X, we have $Sh(Q/X) = Sh(\Sigma X)$.

It is well-known that $H_k(K) = H_{k+1}(\Sigma K)$ for $k \geq 1$ and for any space K (one gets it easily by applying the Mayer-Vietoris sequence). By functoriality of the Mayer-Vietoris sequence, we get $\check{H}_k(X) = \check{H}_{k+1}(\Sigma X)$ for $k \geq 1$. Thus $\check{H}_{k+1}(Q/X) = \check{H}_k(X)$ for $k \geq 1$.

Consequently, if X is an ANR-divisor, then $Q/X \in$ ANR and the Čech homology groups of X are finitely generated and almost all of them are trivial.

Suppose X is movable, $\check{H}_k(X)$ is finitely generated for each $k \geq 0$ and $\check{H}_k(X) = 0$ for $k \geq N$.

Let $\{U_p\}_{p=1}^{\infty}$ be a decreasing sequence of open neighborhoods of X in Q such that any neighborhood U of X in Q contains U_p for some p.

By Corollary 6.1.9 in [D-S] (p. 81) pro-$H_k(X)$ is stable for each k and is trivial for $k \geq N$. Therefore, $(H_k(U_p))_{p=1}^{\infty}$ is stable for each k and is trivial for $k \geq N$ (see Lemma 3.1). Let x_0 be the base point of Q/X. Then $e(Q/X, x_0)$ is isomorphic in pro-HT to $(U_p - X)_{p=1}^{\infty}$. Since X is a Z-set in Q, both $(U_p - X)_{p=1}^{\infty}$ and $(U_p)_{p=1}^{\infty}$ are isomorphic in pro-HT (see Corollary 3.5.5 in [D-S], p. 38). Thus, $e(Q/X, x_0)$ is movable, its homology pro-groups are stable and almost all of them are trivial.

By Theorem 4.7, $Q/X \in$ ANR and by Lemma 9.4.2 in [D-S] (p. 119) X is an ANR-divisor.

Remark. Theorem 5.4 was proved by A. Kadlof [K] under the additional assumption that pro-$\pi_1(X)$ is stable.

REFERENCES

[B-K] A.D. Bousfield and D.M. Kan, Homotopy limits, completions and localizations, Lecture Notes in Math. 304, Springer-Verlag, 1972.

[C] T.A. Chapman, Lectures on Hilbert cube manifolds, Lectures on CMBS, N. 28.

[Do] A. Dold, Lectures on algebraic topology, Springer-Verlag, Berlin, 1972.

[D_1] J. Dydak, The Whitehead and Smale theorems in shape theory, Dissertationes Mathematicae 156 (1979), 1-51.

[D_2] J. Dydak, On unions of movable compacta, Bull, Ac. Pol. Sci. 26 (1976), 57-60.

[D_3] J. Dydak, On LC^n-divisors, Topology Proceedings 3 (1978), 319-333.

[$D_{4,5}$] J. Dydak, On algebraic properties of continua, I and II, Bull. Ac. Pol. Sci. 27 (1979), 717-721 and 723 - 729.

[D-S] J. Dydak and J. Segal, Shape theory: An introduction, Lecture Notes in Math. 688, Springer-Verlag (1978), 1-150.

[E] R. Engelking, General topology, Monografie Matematyczne 60, Polish Scientific Publishers, Warszawa, 1977.

[G] R. Geoghegan, Open problems in infinite-dimensional topology, Topology Proc. 4 (1979), 287-338.

[Hu] S.T. Hu, Theory of retracts, Wayne State University Press, 1965.

[Hy] D.M. Hyman, ANR-divisors and absolute neighborhood contractibility, Fund. Math. 62 (1968), 61-73.

[K] A. Kadlof, Remark on ANR-divisors, preprint.

[K-L] A.H. Kruse and P.W. Liebnitz, An application of a family homotopy extension theorem to ANR-spaces, Pacific J. Math. 16

(1966), 331-336.

[M] D.R. McMillan, Jr., One-dimensional shape properties and three-manifolds, in Studies in topology, Academic Press (1975), 367-381.

[S] E. Spanier, Algebraic topology, McGraw Hill, New York, 1966.

[W-W] J.E. West and R.Y.-T. Wong, Based-free actions of finite groups on Hilbert cubes with absolute retract orbit spaces are conjugate, Proceedings of the Georgia Geometric Topology Conference, Athens, Ga., 1977.

A SIMPLE-HOMOTOPY APPROACH TO THE FINITENESS OBSTRUCTION

Steve Ferry*

§0. INTRODUCTION

The purpose of this paper is to develop Wall's finiteness obstruction ([Wa_1], [Wa_2]) from an extremely geometrical point of view. There is an analogy, which will be made precise, to the situation regarding the theory of Whitehead torsion where there are two treatments in somewhat differing styles. The first treatment is that of Whitehead [W] which quickly reduces statements in PL topology to statements above chain complexes and proceeds on this basis. The second treatment is that of M. Cohen [Co_1] in which the Whitehead group is defined geometrically. All of the *formal* properties of the theory (functoriality, sum theorem, product theorem, etc.) are easily derivable from this geometric definition. Of course, the same reduction to chain complexes is necessary in order to calculate Whitehead groups and show that the theory is non-vacuous.

Cohen's approach has proven to be very influential. Siebenmann's infinite simple homotopy theory [S] and Hatcher's higher simple homotopy theory [H] are developed along these geometric lines. It is also possible to develop a controlled simple homotopy theory (in the sense of Chapman-Connell-Hollingswoth-Quinn) along these lines. These last two cases are noteworthy because the corresponding reduction to algebra is either difficult to carry out or reduces to heretofore unknown algebraic territory.

The approach of Wall's original papers on the finiteness obstruction [Wa_1,Wa_2] is philisophically very similar to that of Whitehead. We will redevelop this theory along Cohen's more geometrical lines.

One payoff is that there is a well-understood procedure for passing from the study of Wall's finiteness obstruction to the study of stable PL or Q-manifold missing-boundary problems. See [Ch-S] or the section entitled "Splitting Theorem" in almost any paper by Chapman. Thus, a geometric version of the Chapman-Connell-Hollingsworth-Quinn controlled simple homotopy theory should be usable to develop a Q-manifold version of Quinn's controlled end theory [Q].

The techniques used in this paper are derived from Mather's influential [Ma] and from "Siebenmann's variation on West's proof that compact ANR's have finite type," an unpublished manuscript of R.D. Edwards. These techniques have been exploited relentlessly during the past few years by T.A. Chapman and the author ([Ch_1],[Ch_2],[F]). The present paper is, to the author's knowledge, the first time that this approach has been used to set up a nontrivial obstruction theory. As stated above, the main value of this approach is that it is extremely formal and therefore is adaptable to a variety of situations.

---

* Partially supported by NSF grants and the A.P. Sloan Foundation.

§1. SIMPLE HOMOTOPY PRELIMINARIES

We will assume that the reader is *completely* familiar with §4-6 of [Co_1]. For our purposes, we will need to develop this geometric theory in slightly greater generality.

DEFINITION 1.1. Let X be a topological space. A *finite relative cell complex* (over X) is a pair (L,X) where $L = L_n \supset L_{n-1} \supset \ldots \supset L_o = X$ and L_{i+1} is obtained from L_i by attaching a cell of some dimension. Note that cells are attached in no particular order and with possibly noncellular attaching maps.

DEFINITION 1.2. If (K_1,X) and (K_2,X) are finite relative cell complexes, we say that K_1 collapses to K_2 by an *elementary collapse* if $K_1 = K_2 \cup e^{n-1} \cup e^n$ and the attaching map for e^n is a homeomorphism over (when restricted to the inverse image of) the interior of e^{n-1}. A finite sequence of operations each of which is an elementary expansion or elementary collapse is called a *formal deformation.*

DEFINITION 1.3. $Wh(X) = \{(K,X) \mid K \hookrightarrow X$ is a homotopy equivalence and (K,X) is a finite relative cell complex$\}/\sim$ where $(K_1,X) \sim (K_2,X)$ iff K_1 formally deforms to K_2 rel X. We will denote the equivalence class of (K_1,X) by $[K_1,X]$ or $\tau(K_1,X)$.

DEFINITION 1.4. If $f : X \to Y$ we define $f_* : Wh(X) \to Wh(Y)$ by $f_*([K,X]) = [K \cup_f Y,Y]$.

One easily checks that Wh(X) is an abelian group. The only alteration of the proof on p. 21 of [Co] is to use reduced mapping cylinders M_D in which the copy of $X \times I$ has been collapsed to a single copy of X.

If $f : (K,X) \to (L,X)$ is a homotopy equivalence of finite relative cell complexes which is the identity on X, we define $\tau(f) = f_*([M_X(f),K]) \in Wh(L)$ where $M_X(f)$ is the reduced mapping cylinder.

If $K \supset L$ are finite simplicial complexes with L a strong deformation retract of K, then $[K,L] = 0$ in Wh(L) if and only if there exist a finite simplicial complex $\overline{K}$ and PL maps $c : \overline{K} \to K$, $d : \overline{K} \to L$ with contractible point-inverses such that the diagram

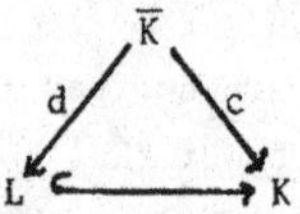

homotopy commutes. "Only if" follows from the fact that K and L have PL homeomorphic regular neighborhoods in high-dimensional Euclidean spaces. "If" is a result of M. Cohen [Co_2]. The reader is urged to attempt to prove this result for himself using the Sum Theorem (Prop. 1.6) for Whitehead torsion. The "if" implication is not used in this paper.

PROPOSITION 1.6. ([Co_1,p.76]). Suppose that $K = K_1 \cup K_2$, $K_0 = K_1 \cap K_2$, $L = L_1 \cup L_2$, $L_0 = L_1 \cap L_2$ and that $f : K \to L$ is a map which restricts to homotopy equivalences $f_\alpha : K_\alpha \to L_\alpha$, $\alpha = 0,1,2$. Let $j_\alpha : L_\alpha \to L$ and $i_\alpha : K_\alpha \to K$ be the inclusions. Then f is a homotopy equivalence and $\tau(f) = j_{1*}\tau(f_1) + j_{2*}\tau(f_2) - j_{0*}\tau(f_0)$. If f is an inclusion, $\tau(L,K) = i_{1*}\tau(L_1,K_1) + i_{2*}\tau(L_2,K_2) - i_{0*}\tau(L_0,K_0)$.

PROOF. We prove the second statement. The first follows easily. Let $r_t : L_0 \to L_0$ be a homotopy rel K_0 from id to a retraction $r_1 : L_0 \to K_0$. By the simple homotopy extension theorem, L_1 deforms rel K_0 to $L_1 \cup_{r_1} L_0$. Similarly, L_2 and L deform rel K_0 to $L_2 \cup_{r_1} L_0$ and $L \cup_{r_1} L_0$. But one easily sees that

$$(L \cup_{r_1} L_0) \underset{K_0}{\cup} L_0 = (L_1 \cup_{r_1} L_0) \underset{K_0}{\cup} (L_2 \cup_{r_1} L_0) .$$

This completes the proof of Proposition 1.6. ∎

REMARK: Proposition 1.6 could be proven for finite relative complexes, but this result is not needed in the sequel, so we omit it. We will need the following formula for the torsion of a composition.

PROPOSITION 1.7. If $f : K \to L$ and $g : L \to M$ are homotopy equivalences between finite simplicial complexes, then $\tau(g \circ f) = \tau(g) + g_*\tau(f)$.

PROOF. This follows immediately from Fact 2 on p. 22 of [Co_1]. Again, a version for finite relative complexes is possible but unnecessary.

We will need several easy lemmas about mapping cylinders in the sequel.

LEMMA MC1. If $f : X \to Y$ is a map, then the mapping cylinder M(f) is homotopy equivalent to Y. ∎

LEMMA MC2. If $f : X \to Y$ and $g : Y \to Z$ are maps then the mapping cylinder $M(g \circ f)$ is homotopy equivalent to the union along Y of M(f) and M(g). This homotopy equivalence is the identity on $X \cup Z$. ∎

LEMMA MC3. If $f,g : X \to Y$ are homotopic maps, then M(f) is homotopy equivalent to M(g) via a homotopy equivalence which is the identity on $X \cup Y$. ∎

See [Co_1,§5] or [F] for proofs.

§2. WALL'S FINITENESS OBSTRUCTION

Let X be a topological space and let $d : K^n \to X$ be a domination with right inverse u. Here, K^n is a finite n-dimensional CW-complex.

PROPOSITION 2.1. (Mather [M]). X is homotopy equivalent to an (n+1)-dimensional CW-complex.

PROOF. Let $\alpha = u \circ d : K \to K$ and let $T(\alpha)$ be the space obtained from the mapping cylinder $M(\alpha)$ by identifying the top and the bottom of $M(\alpha)$ using the identity map. Following Mather, we show that X is homotopy equivalent to a cyclic cover of $T(\alpha)$.

The argument is most easily understood via a picture (Figure 1). $T(\alpha)$ is seen to be homotopy equivalent to the intermediate space Y. This uses MC2 and the fact that $\alpha = u \circ d$. Y is then seen to be homotopy equivalent to $X \times S^1$ using MC2 and the fact that $d \circ u \simeq id$.

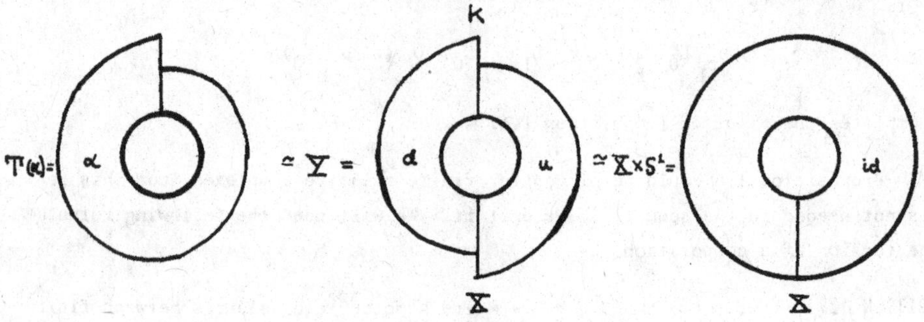

Figure 1

X is homotopy equivalent to the infinite cyclic cover of $X \times S^1$ obtained by unwrapping the S^1 factor. X is therefore homotopy equivalent to the covering space $I(\alpha)$ of $T(\alpha)$ illustrated in Figure 2.

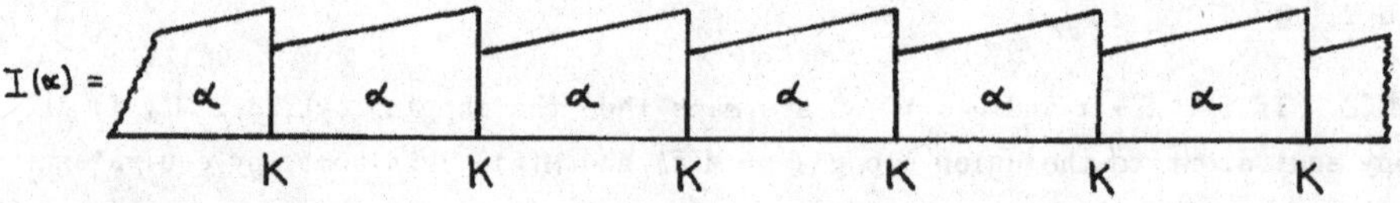

Figure 2

This completes the proof of Proposition 2.1. ∎

DEFINITION 2.2. Let $T : X \times S^1 \to X \times S^1$ be the homeomorphism $(\chi,\theta) \to (\chi,-\theta)$. Let $\phi : T(\alpha) \to X \times S^1$ be the homotopy equivalence defined in Proposition 2.1. Then we define $\sigma(X) = \phi_* \tau(\phi^{-1} T \phi) \in Wh(X \times S^1)$, where ϕ^{-1} is a homotopy inverse to ϕ.

Let X be a finitely dominated topological space. Then

THEOREM 2.3. $\sigma(X)$ is well-defined and $\sigma(X) = 0$ if and only if X is homotopy equivalent to some finite CW-complex.

PROOF. Let $d_1 : K_1 \to X$ and $d_2 : K_2 \to X$ be finite dominations with right inverses u_1 and u_2, respectively.

STEP 1. We wish to show that the elements of $Wh(X \times S^1)$ obtained by the process of Proposition 2.1 are the same.

Case I. Suppose that $K_1 \supset K_2$, $d_2 = d_1|K_2$, and $u_1 = u_2$.

Let $\alpha_1 = u_1 \circ d_1$ and let $\alpha_2 = u_2 \circ d_2$. In this case there is a collapse $g : T(\alpha_1) \to T(\alpha_2)$ and the diagram

$$\begin{array}{ccc} T(\alpha_1) & \xrightarrow{g} & T(\alpha_2) \\ & {\scriptstyle \phi_1} \searrow \quad \swarrow {\scriptstyle \phi_2} & \\ & X \times S' & \end{array}$$

commutes up to homotopy. Then

$$\phi_{1*}\tau(\phi_1^{-1}T\phi_1) = \phi_{2*}g_*\tau(g^{-1}\phi_2^{-1}T\phi_2 g)$$

$$= \phi_{2*}g_*[0 + g_*^{-1}\tau[\phi_2^{-1}T\phi_2] + 0]$$

$$= \phi_{2*}\tau[\phi_2^{-1}T\phi_2] .$$

This calculation uses the fact that $\tau(g) = 0$ and the composition formula Proposition 1.7. This completes the proof of Case I.

Case II. The general case.

Given $d_1 : K_1 \to X$ and $d_2 : K_2 \to X$ form $K_3 = M(u_2 \circ d_1)$. Define $d_3 : K_3 \to X$ by $d_3 = d_2 \circ c$, where $c : M(u_2 \circ d_1) \to K_2$ is the mapping cylinder collapse. Define $u_3 : X \to K_3$ to be the composition of u_2 with the inclusion map. Now, $d_3|K_1 \sim d_2 \circ u_2 \circ d_1 \sim d_1$ and the composition of u_1 with the inclusion map is homotopic to $u_2 \circ d_1 \circ u_1 = u_2 = u_3$. Thus we are reduced to Case I and $\sigma(X)$ is well-defined.

STEP 2. $\sigma(X) = 0 \Rightarrow X$ is homotopy equivalent to some finite complex.

If $\sigma(X) = 0$, then $\tau(\phi^{-1}T\phi) = 0$ and there exist a finite polyhedron Z and CE-PL maps $c_1 : Z \to T(\alpha)$ and $c_2 : Z \to T(\alpha)$ so that $c_2 \sim \phi^{-1}T\phi c_1$. Passing to infinite cyclic covers, we have a diagram:

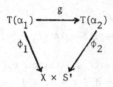

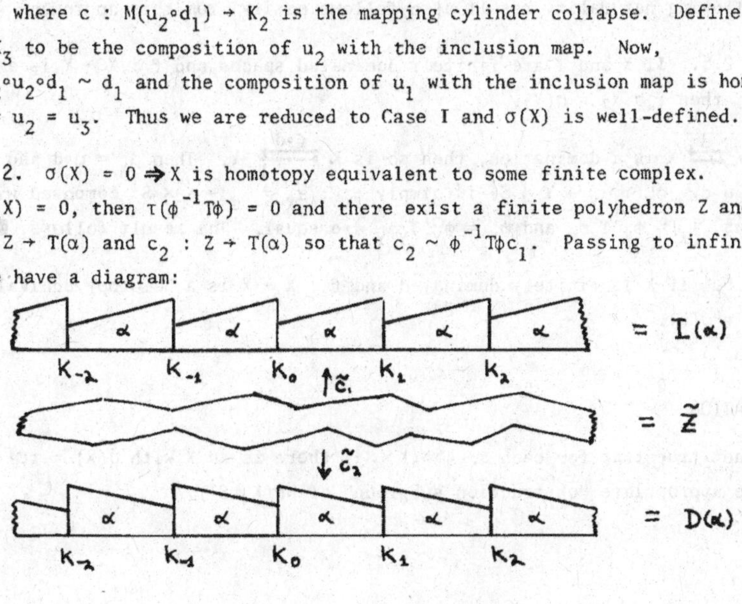

where the map $\tilde{c}_2 \circ (\tilde{c}_1)^{-1} : I(\alpha) \to I(\alpha)$ reverses the ends. We refer to this reversed copy of $I(\alpha)$ as $D(\alpha)$. If we choose N large enough, the region of $\tilde{Z}$ trapped between $(\tilde{c}_1)^{-1}(K_N)$ ($K_1, K_2, \ldots$ are all copies of K) and $(\tilde{c}_2)^{-1}(K_{-N})$ is a strong deformation retract of $\tilde{Z}$ and therefore has the homotopy type of $\tilde{Z}$, that is to say, the homotopy type of X.

STEP III. If X is homotopy equivalent to a finite complex, then $\sigma(X) = 0$.

In this case, let $d : K \to X$ be a homotopy equivalence and let $u : X \to K$ be d^{-1}. Then $\alpha \sim id : K \to K$ and $T(\alpha) \approx K \times S^1$. In this case the torsion of $\phi^{-1} T \phi$ is clearly zero. ∎

We will need the following proposition later.

PROPOSITION 2.3. If $K \underset{u}{\overset{a}{\rightleftarrows}} X$ is a finite domination and $\sigma(X) = 0$, then there exist a finite complex $\overline{K} \supset K$ and an extension $\overline{d} : \overline{K} \to X$ of d such that $u \circ \overline{d} \sim id$.

PROOF. This follows easily from the proof that $\sigma(X)$ is well-defined. ∎

(Sum Theorem)

PROPOSITION 2.4. If $X = X_1 \cup X_2$ with $X_0 = X_1 \cap X_2$ and each X_i is a finitely dominated CW-complex (ANR), then $\sigma(X) = i_{1*}\sigma(X_1) + i_{2*}\sigma(X_2) - i_{0*}\sigma(X_0)$.

PROOF. An ANR X is finitely dominated if and only if there is a homotopy $h_t : X \to X$ with $h_0 = id$ and $\overline{h_1(X)}$ compact. Using the homotopy extension theorem and the finite domination of X_i, $i = 0,1,2$, one easily constructs such an h_t which respects X_i, $i = 0,1,2$. The result then follows immediately from the Sum Theorem for Whitehead torsion. ∎

The following naturality result also follows easily from this approach.

PROPOSITION 2.5. If X and Y are finitely dominated spaces and $f : X \to Y$ is a homotopy equivalence, then $f_* \sigma(X) = \sigma(Y)$.

PROOF. If $K \underset{u}{\overset{d}{\rightleftarrows}} X$ is a domination, then so is $K \underset{u \circ f^{-1}}{\overset{f \circ d}{\rightleftarrows}} Y$. Then $\alpha_X = u \circ d$ and $\alpha_Y = u \circ f^{-1} \circ f \circ d \simeq u \circ d$. $\phi_Y : \alpha_Y \to Y \times S^1$ is simply $\phi_X : \alpha_X \simeq \alpha_Y \to X \times S^1$ composed with $f \times id$. Thus, $\psi_X = \phi_X^{-1} T_X \phi_X$ and $\psi_X = \phi_Y^{-1} T_Y \phi_Y$ are equal. The result follows. ∎

COROLLARY 2.6. If X is finitely dominated and $f : X \to X$ is a homotopy equivalence, then $f_* \sigma(X) = \sigma(X)$.

§3. REALIZATION

It is not true that for each $\tau \in Wh(K \times S^1)$ there is an X with $\sigma(X) = \tau$. We will identify the appropriate "obstruction subgroup" of $Wh(K \times S^1)$.

DEFINITION 3.1. Let $p_1 : S^1 \to S^1$ be a finite covering map. This induces a finite covering $p = p_1 \times id : X \times S^1 \to X \times S^1$. If $[L, X\times S^1] \in Wh(X\times S^1)$, let $\tilde{L}$ be the covering of L induced by p using the pullback diagram below:

$$\begin{array}{ccc} \tilde{L} & \longrightarrow & L \\ \downarrow & & \downarrow \text{retraction} \\ X \times S^1 & \xrightarrow[p \times id]{} & X \times S^1 \end{array}$$

The strong deformation retraction $L \searrow X \times S^1$ lifts to a strong deformation retraction $\tilde{L} \searrow X \times S^1$ and we obtain an element $[\tilde{L}, X\times S^1] \in Wh(X\times S^1)$. We call this element $p^*[L, X\times S^1]$. The map $p^* : Wh(X\times S^1) \to Wh(X\times S^1)$ is a homomorphism called the *transfer*. Let $Wh_0(X\times S^1)$ be the subgroup of $Wh(X\times S^1)$ consisting of elements invariant under such transfer maps.

THEOREM 3.2. (a) If X is finitely dominated then $\sigma(X) \in Wh_0(X\times S^1)$.
(b) If $\tau \in Wh_0(K\times S^1)$ and $\tau \neq 0$, then there is a finitely dominated X with $\sigma(X) \neq 0$.

PROOF. (a) Consider the double cover of ψ pictured below:

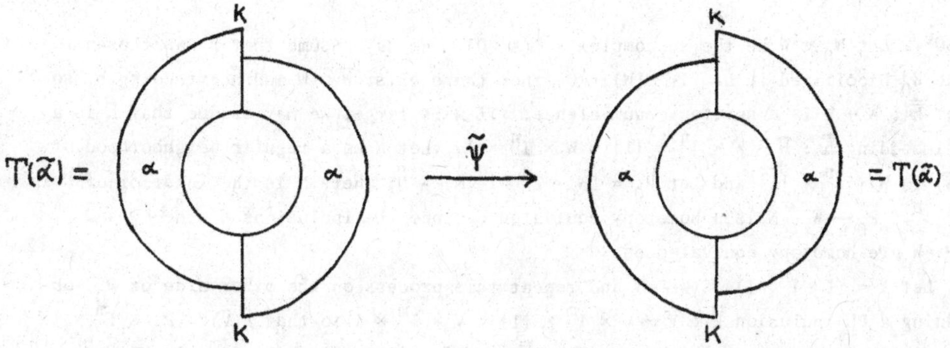

We have drawn the second picture of $T(\tilde{\alpha})$ backwards to emphasize the fact that $\tilde{\psi}$ reverses orientation over S^1. Using the fact that $\alpha\circ\alpha \sim \alpha$, we can use MC2 to define a simple homotopy equivalence $S : T(\tilde{\alpha}) \to T(\alpha)$ such that the diagram below commutes:

$$\begin{array}{ccccc} T(\tilde{\alpha}) & \xrightarrow{\tilde{\psi}} & T(\tilde{\alpha}) & \xrightarrow{\tilde{\phi}} & \\ \downarrow s & & \downarrow s & & X \times S^1 \\ T(\alpha) & \xrightarrow{\psi} & T(\alpha) & \xrightarrow{\phi} & \end{array}$$

$$p^*(\sigma(X)) = \tilde{\phi}_*\tau(\tilde{\psi}) = \tilde{\phi}_*\tau(s^{-1}\psi s) = \phi_* s_* \tau(s^{-1}\psi s)$$

$$= \phi_* s_*[\tau(s^{-1}) + s_*^{-1}\tau(\psi) + s_*^{-1}\psi_*\tau(s)]$$

$$= \phi_* s_* s_*^{-1}\tau(\psi) = \phi_*\tau(\psi)$$

This completes the proof of part (a).

PROOF. (b) Let $\tau \in Wh_0(K\times S')$, $\tau \neq 0$. Let $(L, K\times S')$ be a representative for τ and let $r : L \to K \times S'$ be a PL retraction. Lifting to the cyclic cover, we have a PL retraction $\tilde{r} : \tilde{L} \to K \times \mathbf{R}'$. Let $W = \tilde{r}^{-1}(K\times [0,\infty))$. W is a polyhedron. We need:

LEMMA 3.3. W is finitely dominated.

PROOF. Let $\tilde{r}_t$, $0 \le t \le 1$, be a homotopy from $\tilde{r}$ to id rel $K \times R^1$. This homotopy is proper. Thus, there is a compact set $C \subset W$ such that $\tilde{r}_t(W - C) \subset W$ for all t, $0 \le t \le 1$. Let $\phi : W \to [0,1]$ be a function with $\phi(c) = 0$ such that $\phi^{-1}[0,1)$ is compact. Then $h_t(\chi) = \tilde{r}_{t\cdot\phi(\chi)}(\chi)$ deforms W into the union of $K \times [0,\infty)$ with a finite complex. It is easy to modify h_t to $\bar{h}_t$ which deforms W into a finite subcomplex L of W relative to a neighborhood of $\tilde{r}^{-1}[K\times 0]$. Setting $d = i : L \hookrightarrow W$ and $u = \bar{h}_1 : W \to L$, we see that W is finitely dominated.

The next step in the proof is the following lemma:

LEMMA 3.4. If $\sigma(W) = 0$ there exist an n and a finite bicollared polyhedron $P \subset L \times I^n$ which separates the ends such that $P \hookrightarrow L$ is a homotopy equivalence.

PROOF. Let $W_0 \subset W$ be the subcomplex $\tilde{r}^{-1}(K\times 0)$. We may assume that r was chosen to make W_0 bicollared in $\tilde{L}$. If $\sigma(W) = 0$, then there exist $\bar{W} \supset W$ and $\bar{h}$ extending $\bar{h}_1$ so that $\bar{\bar{h}} : \bar{W} \to W$ is a homotopy equivalence. If n is large, we may assume that $\bar{\bar{h}}$ is a PL imbedding $\bar{\bar{h}} : \bar{W} \to W \times I^n \times \{1\} \subset W \times I^n \times I$. Let N be a regular neighborhood of $\bar{\bar{h}}(\bar{W})$ in $W \times I^n \times \{1\}$ and let $P_0 = (W_0 \times I^n \times I \cup N) - \mathring{N}$, where $\mathring{N}$ is the interior of N in $W \times I^n$. $P_0 \hookrightarrow W - \mathring{N}$ is a homotopy equivalence since the inclusions $W - \mathring{N} \hookrightarrow W$ and $P_0 \hookrightarrow W$ are homotopy equivalences.

Let $V = (\tilde{L}\times I^n \times I) - (W - \mathring{N})$ and repeat this process on the other side of P_0, obtaining a PL inclusion $\bar{\bar{k}} : \bar{V} \to V \times I^m \times \{1\} \subset V \times I^m \times I$ so that $\bar{\bar{k}}(\bar{V}) \supset P_0 \times I^m \times I$. Now, let M be a regular neighborhood of $\bar{\bar{k}}(\bar{V})$ in V and let $P = (P_0 \times I^m \times I \cup M) - \mathring{M}$. As before, P is a strong deformation retract of the other side, $(W\times I^{n+m+1} \times I - V\times I^m \times I) \cup M$, since $P \cup M$ has already been shown to be a strong deformation retract of this space and P is a strong deformation retract of $P \cup M$ since M is a regular neighborhood of a set in $V \times I^m \times \{1\}$. ∎

We continue with the proof of Theorem 3.2. Since τ is invariant under transfer, we may assume that the projection $\tilde{L} \times I^{n+m} \to L \times I^{n+m}$ imbeds a bicollar neighborhood $P \times I$ of P.

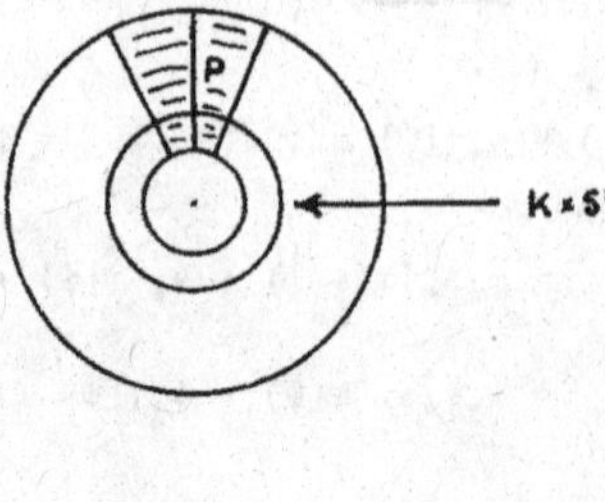

The complement of $\mathring{\widehat{P\times I}}$ in $L \times I^{n+m}$ is homeomorphic to the space between two translated copies of P in $\tilde{L} \times I^{n+m}$. It is therefore homotopy equivalent to K and we can construct a strong deformation retraction $r_t : L \times I^{n+m} \searrow K \times S'$ which takes $P \times I$ into itself and $(L \times I^{n+m} - P \times I)$ into itself. The sum theorem for Whitehead torsion [Co_1,p.76] shows that $\tau \in \mathrm{im}(Wh(K) \to Wh(K \times S^1))$. But it is geometrically clear that if $p : S^1 \to S^1$ is a double cover, then $p^*\tau = 2\tau$. The invariance of τ under transfer then shows that $\tau = 0$. ∎

COROLLARY (of the proof). If $\tau \in Wh(K \times S^1)$ and $p : S^1 \to S^1$ is the double cover, then $p^*\tau = \tau$ implies that $\sigma(W) = 0$, where W is defined as in the theorem above. Thus, if there is a nonzero element $\tau \in Wh(K \times S^1)$ which is invariant under passage to the double cover, then there is a nonzero element $\tau \in Wh(K \times S^1)$ which is invariant under passage to all finite covers. ∎

REMARK: This corollary is also clear from the Bass-Heller-Swann decomposition of $Wh(K \times S^1)$.

REFERENCES

[Ch-S] T.A. Chapman and L.C. Siebenmann, Finding a boundary for a Hilbert cube manifold, Acta Mathematica 137 (1976), 171-208.

[Ch_1] T.A. Chapman, Homotopy conditions which detect simple homotopy equivalences, Pacific J. Math. 80 (1979), 13-45.

[Ch_2] __________, Whitehead torsion and the Wall finiteness obstruction, to appear.

[Co_1] M. Cohen, *A Course in Simple-Homotopy Theory*, Springer-Verlag, Berlin and New York.

[Co_2] __________, Simplicial structures and transverse cellularity, Annals of Math. 85 (1967), 218-245.

[H] A. Hatcher, Higher simple homotopy theory, Annals of Math. 102 (1975), 101-139.

[F] S. Ferry, Homotopy, simple homotopy, and compacta, Topology 19 (1980), 101-110.

[M] M. Mather, Counting homotopy types of topological manifolds, Topology 4 (1965), 93-94.

[Q] F. Quinn, Ends of maps, I, Annals of Math. 110 (1979), 275-331.

[S] L.C. Siebenmann, Infinite simple homotopy types, Indag. Math. 32 (1970), 479-495.

[Wa_1] C.T.C. Wall, Finiteness conditions for CW-complexes, Annals of Math. 106 (1977), 1-18.

[Wa_2] __________, Finiteness conditions for CW-complexes II, Proc. Roy. Soc. Ser. A., 295 (1966), 129-139.

[W] J.H.C. Whitehead, Simple homotopy types, Amer. J. Math. 72 (1952), 1-57.

University of Kentucky
Department of Mathematics
Lexington, Kentucky 40506

GENERALIZED THREE-MANIFOLDS

R. C. Lacher*

Naturally occuring as upper semicontinuous cell-like decomposition spaces of 3-manifolds and topological factors of (3+k)-manifolds, generalized 3-manifolds underly much of 3-dimensional topology. Recent successes in high dimensions, resulting in remarkably simply characterizations of both n-manifolds and generalized n-manifolds, $n \geq 5$, have stimulated an upsurge in interest in the geometric topology of generalized 3-manifolds. This article reviews recent work in this field and suggests directions for further study.

1. *Preliminaries*.

I will begin with a listing of basic facts and concepts, all of which are dimension-independent. Readers encountering these ideas for the first time should read the first 9 sections of [19] for motivation, historical perspective, and a discussion of the more specialized aspects of the high-dimensional ($n \geq 5$) cases. All spaces are assumed to be Hausdorff. Euclidean n-space is denoted by $\mathbb{R}^n$; B^n is the closed unit ball in $\mathbb{R}^n$; and S^n is the one-point compactification of $\mathbb{R}^n$.

A *euclidean neighborhood retract* (abbreviated ENR) is a retract of a neighborhood in euclidean space, or, equivalently, a finite-dimensional locally compact separable metric ANR.

A G-*homology* n-*manifold* (without boundary) is a space X such that $\check{H}^i(X, X-\{x\};G) \simeq \check{H}^{n-i}(\{x\};G)$ for all $x \in X$ and all integers i, where $\check{H}^*$ denotes Cech cohomology. Notice that if X is locally contractible then this condition is equivalent to $H_i(X, X-\{x\};G) \simeq \check{H}^{n-i}(\{x\};G)$ for all $x \in X$ and all i, where H_* is singular homology.

A G-*homology* n-*manifold* (with boundary) is a space X such that $\check{H}^i(X, X-\{x\};G)$ is isomorphic to either $\check{H}^{n-i}(\{x\};G)$ or 0 for all $x \in X$. (Again, for locally contractible spaces the group $\check{H}^i(X, X-\{x\};G)$ may be replaced by $H_i(X, X-\{x\};G)$.) The *boundary* of such X is the subset ∂X consisting of points $x \in X$ with $\check{H}^i(X, X-\{x\};G) = 0$ for all i; and the *interior* is $\mathrm{int}X = X - \partial X$. It should be clear from context whether I am assuming $\partial X = \emptyset$ in a particular discussion.

A *generalized* n-*manifold* (with boundary) is an ENR that is a $\mathbb{Z}$-homology n-manifold (with boundary), where $\mathbb{Z}$ is the ring of integers.

A *singular point* of a G-homology n-manifold X (with or without boundary) is a point of X that has no closed neighborhood homeomorphic to B^n. The set of singular points of X is denoted by S(X) and called the singular set of X; the complement X-S(X) is called the manifold set and denoted by M(X). An n-manifold (with boundary) is a generalized n-manifold (with boundary) that has no singular points.

A closed subset Z of a generalized n-manifold X is said to be 1-LCC (1-locally co-connected) in X if X is 1-LC mod Z (cf. [18], p.504). When $\dim X \leq n-3$, 1-LCC is a kind of local tameness condition on $Z \subset X$.

A resolution of a generalized manifold X without boundary is a pair (M,f) where M is a topological manifold without boundary and f is a proper cell-like mapping of M onto X. Results from the theory of cell-like mappings [18] imply that M and X must have the same dimension and further that the only finite-dimensional spaces that are proper cell-like images of n-manifolds are generalized n-manifolds. A resolution (M,f) of X is called conservative if $f^{-1}(x)$ is a single point for all $x \in M(X)$.

Orientability and duality. Suppose X is a locally contractible G-homology n-manifold without boundary, where G is either $\mathbb{Z}$ or $\mathbb{Z}_p$, the ring of integers modulo the prime p. (These assumptions could be weakened somewhat; please refer to Bredon's work for best-possible results.) Thus X is locally contractible and $H_i(X, X-\{x\};G) \simeq \check{H}^{n-i}(\{x\};G)$ for all $x \in X$. The last condition merely says $H_i(X,X-\{x\};G)$ is G for i=n and 0 for $i \neq n$, but it emphasizes that this is "Alexander duality for points", or "infinitesimal orientability". A result of G.E. Bredon [3] is that, in the presence of some form of local connectedness, infinitesimal orientability implies local orientability: for each $x \in X$ there exists a neighbourhood U of x in X such that $H_n(X,X-U;G) \to H_n(X,X-y;G)$ is an isomorphism for all $y \in U$. This breakthrough leads to a complete theory of orientability and duality that parallels the usual theory for manifolds.

Geometric topology of generalized manifolds. The two primary types of generalized manifolds (in the narrow sense of my definition above) are closed, cell-like images of manifolds and manifold factors. In fact, there are no examples known not to come from both sources. The two primary questions about a generalized n-manifold X (without boundary) arising from these observations are: Is X the proper, cell-like image of an n-manifold (i.e., does X have a resolution), and is $X \times R^k$ an (n+k)-manifold for some k? The recent successes in this study in high dimensions, combined with dramatic results on shrinking of cell-like decompositions of high-dimensional manifolds, provide direction for research into generalized 3-manifolds. (See [9], [13], [19], [24].) This article focuses on aspects of current research related specifically to singularities and resolutions of generalized 3-manifolds and their possible role in characterizations of the 3-manifold property.

2. Examples.

So far, there have been two primary sources of specimens of generalized 3-manifolds: cell-like upper semicontinuous (abbreviated CU) decompositions of 3-manifolds, and end-point compactifications of non-compact 3-manifolds. Examples may be divided into two catogeries: "hard", meaning truly extant, and "soft", meaning based on assumed failure of the Poincaré conjecture.

A distinguishing feature of dimension three in the study of generalized manifolds is that it is the smallest dimension admitting singular (i.e., non manifold) examples (see [27]). Consequently, a generalized 3-manifold cannot have a polyhedral singularity.

Suppose, in fact, that X is a generalized 3-manifold with boundary and that $p \in X$ has a neighborhood N in X such that N is homeomorphic to the join $A*B$, where A and B are non-void compacta. Then A and B are generalized manifolds (possibly with boundary) of dimension a and b, respectively, and $a+b = 2$; consequently A and B are manifolds with boundary. Analysis of the homology groups of A in dimensions 0 and 1 shows that A is either an a-sphere or an a-cell, $0 \leq a \leq 2$, and a similar analysis holds for B. Therefore $A*B$ is a 3-manifold (with boundary), and $p \in M(X)$.

<u>Decomposition spaces</u>. The study of cell-like (or point-like, which is the same as cellular) decompositions of 3-manifolds began in earnest with R.H. Bing's work on the sum of two Alexander horned spheres [1] and has flourished ever since, providing many interesting examples of generalized 3-manifolds. Without attempting a review of this body of literature, a few comments based on it are essential. (See [10] for a good place to start working into the subject of CU decompositions.)

The first comment is philosophical: It may be difficult to decide whether a generalized 3-manifold has any singular points at all. (This difficulty was well-established in high dimensions in, for example, the "double suspension problem"; see [8].) A particular case in point is Bing's "dogbone space": at one time, Bing had shown this CU decomposition space X of $\mathbb{R}^3$ was not topologically $\mathbb{R}^3$, but he did not know a topological property that distinguished X from R^3 (see [2], Section 6).

Consider examples of the form $X = S^3/K$, where K is a single cell-like subset of S^3. M.Brown [6] showed that K is cellular if and only if $S(X) = \emptyset$. If K is an arc with non-simply connected complement, then X has one singular point p and $X-\{p\}$ is not simply connected. (In fact, the fundamental group of $X-\{p\}$ is not finitely generated; see [14].) On the other hand, if K is the continuum of J.H.C. Whitehead [26], then X has one singular point p and $X-\{p\}$ is contractible. In any case, p must be "wild" in X: If $\{p\} \subset X$ is 1-LCC then $K \subset M^3$ satisfies the cellularity criterion of D.R. McMillan, Jr. [20] and hence $M^3/K = X$ is a manifold. (A much more general result is discussed in Section 4.)

Consider examples of the form $X = S^3/G$ where G is a CU decomposition of S^3 and the closure of the image of the non-degenerate elements of G is a 0-dimensional set in X. Assuming X is not a manifold, dim $S(X) = 0$. In this case, points of S(X) may be either tame or wild in X [2], but S(X) itself must be wild. (See Section 4.)

Finally, for some really bad news, consider some properties of the "ghastly" generalized 3-manifold of R.J. Daverman and J.J. Walsh [11]. This example is a CU decomposition space X of $\mathbb{R}^3$ (or S^3) such that (1) every mapping $B^2 \to X$ that is one-one on ∂B^2 contains a non-void open set in its image, and (2) no embedding $\{x\} \subset X$ is 1-LCC. Thus X contains no Dehn disks and every point of X is wild. In particular, $S(X) = X$.

Question. Do there exist a compact, connected, simply connected generalized 3-manifold X, without boundary, and a point $p \in X$, such that the fundamental group of $X - \{p\}$ is non-trivial but finitely generated?

CU decomposition spaces of manifolds display many properties from sublime ("almost E^3", as in the dogbone space) to ridiculous ("ghastly"), but they all have one property in common: each is a generalized manifold with a resolution. For potentially unresolvable examples we must turn to another source.

Compactifications. If X is a compact generalized manifold with S(X) 0-dimensional then X is the end-point compactification of M(X). In particular, if p is an isolated point of S(X) and N is a compact neighborhood of p such that $N \cap S(X) = \{p\}$ and ∂N is a compact 2-manifold, then $M(N) = N - \{p\}$ "carries" the singularity of X at p in the sense that the end-point (or one-point) compactification of M(N) is $N = M(N) \cup \{p\}$. Thus any isolated singularity of X is determined by a (nonsingular) 3-manifold with compact boundary.

An elementary "soft" example is obtained by removing the interior of each of the 3-cells $B\left(\frac{1}{n}, \frac{1}{4n}\right)$, n=1, 2, ..., in S^3 and sewing fake 3-cells in their place. The result is a compact generalized 3-manifold with one singular point (formerly the origin in S^3). This singularity is "carried" by an open 3-manifold $M = \Sigma_1 \# \Sigma_2 \# \ldots$, the locally finite connected sum of exotic homotopy 3-spheres Σ_j. In Section 3 I discuss the fact that no compact generalized 3-manifold containing infinitely many pairwise disjoint fake cubes can have a resolution.

Far more interesting and subtle are the "soft" examples of M.Brin [4] and Brin and McMillan [5]. These are compact generalized 3-manifolds with irreducible manifold set (in particular, containing no fake cubes) and admitting no resolution.

The Brin-McMillan construction begins with an exotic homotopy 3-sphere Σ^3 written as a union $N_1 \cup N_2$ of cubes with handles, where $N_1 \cap N_2 = \partial N_1 \cap \partial N_2$. McMillan shows in [22] that there exists a homeomorphism h of Σ^3 onto itself that is isotopic to the identity on Σ^3 and that takes N_1 into its interior in such a way that the inclusion $h(N_1) \subset \text{int}N_1$ is null-homotopic. As a consequence, the continuum $K = \bigcap_{j=1}^{\infty} h^j(N_j)$ is cell-like. (The notation h^j means h iterated j times.) The complement $U = \Sigma^3 - K$ is an acyclic open 3-manifold. Note that $\hat{U}$, the end-point compactification of U, is a generalized 3-manifold with resolution $\Sigma^3 \to \Sigma^3/K$. Brin and McMillan use an infinite cyclic cover to "sum" copies of U together, as follows.

Find a solid torus T and a 2-cell D in Σ^3-K such that $D \cap T = \partial D \cap \partial T$ is a curve on T homotopic to the core of T (so that T $\cup$ (neighborhood of D) = 3-cell).. Let A be an arc in Σ^3 with $A \cap K$ and $A \cap T$ each consisting of an endpoint of A and $A \cap D = \emptyset$. Assume T, D, A are polyhedral. Let p: $V \to (\Sigma^3\text{-int } T)$ be a universal covering. Since Σ^3-int T has the homotopy type of S^1, p is an infinite cyclic covering. V can be realized as the pasting together of infinitely many copies of "Σ^3-int T split along D" end to end, together with an action of Z on V generated by a "shift" mapping t: $V \to V$; in this setting p: $V \to (\Sigma^3\text{-int } T)$ is the quotient of V by $\mathbb{Z}$.

Using t and the fact that "Σ^3-int T split along D" is contractible, it is easily seen that the end-point compactification $\hat{V}$ of V is a generalized 3-manifold with $\partial\hat{V}$ a 2-sphere. Let V* be the generalized 3-manifold without boundary obtained by attaching a 3-cell Q to $\hat{V}$ along $\partial\hat{V}$, and let $K^* = Q \cup p^{-1}(K \cup A)$. Then K* is cell-like, so $V^*/_{K^*}$ is a generalized 3-manifold without boundary. But this quotient is the end-point compactification of $M = p^{-1}(\Sigma^3 - (K \cup A \cup T))$, a 3-manifold with one end. $\hat{M}$ is the Brin-McMillan example based on Σ^3. They show:

(1) M is an increasing union of cubes with handles,
(2) M does not embed in any compact 3-manifold, and
(3) $\hat{M}$ has no resolution.

The fact ($\exists$resolution $\to$ $\exists$conservative resolution, discussed in Section 3) is enough to show that (2) $\to$ (3). A proof of (2) uses the fact that M contains infinitely many pairwise disjoint copies of Σ^3-K-A which is homeomorphic to Σ^3-K, together with the fact that Σ^3-K is an acyclic open manifold two copies of which cover Σ^3 (by pushing N_1, hence K, into its complement). See Corollary 1 of [23].

The examples of Brin [4] are also end-point compactifications of open 3-manifolds M, but M is constructed directly as an increasing union of compact manifolds. Brin's M may be constructed with any 0-dimensional compactum as end points of M or singular set of $\hat{M}$. The end structure of the universal cover V of a polyhedral homotopy S^1 is straightforward to analyze; neighborhoods of an end deform to the end (keeping the endpoint fixed), so $\hat{V}$ is an ENR. This covering structure is the key to the relative simplicity of the Brin-McMillan argument. Brin's construction requires direct proof that $\hat{M}$ is a generalized manifold.

3. Acyclic images

This section reviews some results and consequences (both fact and conjecture) of a paper by J. Bryant and myself [7]. These results can be loosely described in three statements:

(1) A generalized 3-manifold with a resolution admits a conservative resolution;

(2) A generalized 3-manifold with a $\mathbb{Z}_2$-resolution admits a resolution; and

(3) A locally contractible space with a $\mathbb{Z}_2$-resolution is a generalized 3-manifold.

All of these results can be stated in one theorem.

Definitions. Let $f:X\to Y$ be a mapping (i.e., continuous function). f is proper if f(A) is closed in Y for every closed set A in X and $f^{-1}(y)$ is compact for every point y in Y. f is monotone if $f^{-1}(y)$ is compact and connected for every $y\in Y$. f is G-acyclic if $\check{H}^i(f^{-1}(y);G) \simeq \check{H}^i(\{y\};G)$ for all $i\in\mathbb{Z}$ and $y\in Y$.

THEOREM 1. Suppose M^3 is a 3-manifold without boundary and f is a proper $\mathbb{Z}_2$-acyclic mapping of M^3 onto the locally contractible space X. Then X is a generalized 3-manifold with a conservative resolution.

Because it uses a number of deep results about 3-manifolds, including McMillan's neighborhood theorems [21] and the results of G. Kozlowski and Walsh [15] on finite-dimensionality of cell-like images, the proof given in [7] appears to rely inextricably on the manifold nature of M^3. The third conclusion (3), that any $\mathbb{Z}_2$-acyclic image of a 3-manifold must be a generalized 3-manifold, seems more natural in the category of generalized 3-manifolds:

Question. Suppose X is a generalized 3-manifold and f is a proper, $\mathbb{Z}_2$-acyclic mapping of X onto Y. Suppose also that Y is locally contractible. Must Y be a generalized 3-manifold?

A related question concerns the "$\mathbb{Z}_2$ implies $\mathbb{Z}$" aspect of the theorem:

Question. Suppose X is an ENR and a $\mathbb{Z}_2$-homology 3-manifold. Must X be a $\mathbb{Z}$-homology 3-manifold?

The acyclic images theorem implies strict geometric constraints on the nature of M(X), if X has a resolution. For example, the following is a Kneser-type finiteness theorem, due to T. Knoblauch [17] in the case where X is a 3-manifold.

COROLLARY 1.1 Suppose X is a compact generalized 3-manifold with a resolution. Then there exists a positive integer k such that, whenever $K_1, \ldots, K_{k+1}$ are pairwise disjoint compact subsets of M(X), some neighborhood of some K_i embeds in $\mathbb{R}^3$.

Proof. Let $f: M^3 \to X$ be a conservative resolution, and let k be the Knoblauch number for M^3.

COROLLARY 1.2 Suppose X is a space containing a compact subset X_0 that contains arbitrarily large (finite) numbers of pairwise disjoint fake 3-cells. Then X is not the image of a proper $\mathbb{Z}_2$-acyclic mapping on a 3-manifold without boundary. Theorem 1 can be modified as follows (see [7]):

THEOREM 2. Suppose M^3 is an orientable 3-manifold without boundary and f is a proper monotone mapping of M^3 onto the locally contractible $\mathbb{Z}_2$-homology 3-manifold X. Suppose further that there is a 0-dimensional subset Z of X such that $\check{H}^1(f^{-1}(x);\mathbb{Z}_2) = 0$ for all $x \in X-Z$. Then X is a generalized manifold with a (conservative) resolution.

So far we have made no assumption on the nature of S(X). The following result of Brin and McMillan [5] delineates the zero-dimensional singular set case as natural and interesting in itself.

COROLLARY 2.1 Let X be a compact generalized 3-manifold without boundary and with 0-dimensional singular set. The following statements (1)-(4) are equivalent:

(1) X has a resolution;

(2) M(X) embeds in a compact 3-manifold;

(3) S(X) has a neighborhood N in X such that $N \cap M(X)$ embeds in a compact (orientable) 3-manifold;

(4) S(X) has a neighborhood N in X such that $N \cap M(X)$ embeds in $\mathbb{R}^3$.

Proof. The implication (1)→(2) follows from the existence of conservative resolutions (and does not depend on 0-dimensionality of S(X)). The implication (4)→(3) is obvious.

For the implication (2)→(4), suppose $M(X) \subset M^3$, a compact 3-manifold.

We may take M^3 to have no boundary. Let U be an open neighborhood of S(X) in X that is orientable. Then $U_1 = U \cap M(X)$ is an orientable open set in the closed manifold M_1^3=orientable double cover of M^3. A result of Knoblauch [16] implies that U_1-K embeds in $\mathbb{R}^3$ for some compact $K \subset U_1$. Let N = U-K.

For the implication (3)→(1), suppose N is a compact neighborhood of S(X) such that $N \cap M(X)$ embeds in the compact (orientable) 3-manifold W^3 via the mapping h. By taking N and subsequently W^3 small enough, we may assume that W^3 is orientable (if it were not to start with) and that $\partial W^3 = h(\partial N)$. Then N is homeomorphic to the end-point compactification of $W - h(N \cap M(X))$ which is homeomorphic to W/G, the non-degenerate elements of G being components of $W - h(N \cap M(X))$. Theorem 2 now implies X has a resolution.

4. Zero-dimensional singular sets

Consider a generalized 3-manifold X without boundary and with $\dim S(X) \leq 0$. Let $p \in X$. Then p has arbitrarily small compact orientable generalized manifold neighborhoods N with ∂N a (compact, orientable) 2-manifold and $\partial N \cap S(X) = \emptyset$; if p has arbitrarily small such N with the genus of ∂N less than or equal to n, we say X has genus $\leq$ n at p. If X has genus $\leq$ n at p but does not have genus $\leq$ n-1 at P, we say X _has genus_ n _at_ p. If X does not have genus $\leq$ n at p for any integer n, we say X _has genus_ ∞ at p. Let g(X,p) denote the genus of X at p.

A generalized 3-manifold X is said to _satisfy condition_ KF (for Kneser Finiteness) if, for each compact subset X_0 of X, there exists an integer $k = k(X_0)$ such that X_0 contains at most k pairwise disjoint fake 3-cells. According to Section 3, any resolvable generalized 3-manifold satisfies condition KF. The next result is proved in [7] but goes back to C.H. Edwards, Jr. [12].

THEOREM 3. Suppose X is a generalized 3-manifold without boundary satisfying KF and Z is a closed 0-dimensional 1-LCC subset of X. If $S(X) \subset Z$ then $S(X) = \emptyset$ and Z is tame in X.

COROLLARY 3.1 Suppose X is a generalized 3-manifold without boundary satisfying KF and $\dim S(X) \leq 0$. Then $g(X,p) = 0$ for all $p \in X$ if and only if $S(X) = \emptyset$.

Question. If X satisfies KF and $\dim S(X) \leq 0$, is $M(X) = \{p \in X \mid g(X,p) = 0\}$?

If $X = S^3$/(Whitehead continuum) then S(X) is a single point p and $g(X,p) = 1$. It is not difficult to modify the Whitehead construction to produce cell-like compacta $K_n \subset S^3$ such that $X_n = S^3/K_n$ has exactly one singular point p_n and $g(X_n,p_n) = n$, $n = 1, 2, \ldots$, or $n = \infty$.

Brin's examples, constructed from exotic homotopy spheres, show that, for any compact 0-dimensional topological type T, there exists a compact generalized 3-manifold X_T with $M(X_T)$ irreducible (thus satisfying KF), with $S(X_T) \approx T$, with $g(X_T,p) = 1$ for all $p \in S(X_T)$, and with X having no resolution. Thus even assuming KF and $g(X,p) \leq 1$ for all p, resolution is stymied by the Poincaré conjecture. Brin went on to show the equivalence of the Poincaré conjecture with the conjecture that generalized 3-manifolds X with KF, 0-dimensional (or discrete) singular set, and $g(X,p) \leq 1$ for all p have a resolutions [4]. Brin, McMillan, and T.L. Thickstun have pushed this equivalence to the entire 0-dimensional case (see [5] and [25]):

THEOREM 4. Let M^3 be the class of all compact generalized 3-manifolds without boundary, M^3_{KF} the subclass of those $X \in M^3$ satisfying KF, and M^3_0 the subclass consisting of those $X \in M^3$ with $X-S(X)$ irreducible. Let C be M^3, M^3_{KF}, or M^3_0. The following three statements are equivalent:

(1) There are no exotic homotopy 3-spheres;

(2) If $X \in C$, $S(X) = \{p\}$, and $g(X,p) = 1$ then X has a resolution;

(3) If $X \in C$ and $\dim S(X) = 0$ then X has a resolution.

Because of this entanglement with the Poincaré conjecture, it would appear worthwhile to find conditions on $X \in M^3_{KF}$, or the germ of X at $p \in S(X)$, that imply X has a resolution. For example, if $\dim S(X) \leq 0$ and $S(X)$ is 1-LCC in X, or $g(X,p) = 0$ for all $p \in S(X)$, then X has a resolution by Theorem 3.

Problem. Find a property P of pairs (X,Z), where X is a generalized 3-manifold without boundary and Z is a closed 0-dimensional subset of X containing S(X), that is equivalent to (or at least implies) X having a resolution.

A related problem is that of finding a suitable definition of DDP. DDP should be a property of generalized 3-manifolds X and should be equivalent to $S(X) = \emptyset$ (assuming X satisfies KF). The nearer DDP is to being a "homotopical" or "map approximation" property, the better. For example, we could let DDP mean "locally homeomorphic to $\mathbb{R}^3$", and DDP would be equivalent to $S(X) = \emptyset$, but DDP is not "homotopical". The condition "$S(X) \subset Z$ where Z is a closed 1-LCC subset of X" satisfies these criteria, but it is still not suitable since many potential singular sets might be wild in X. For the 0-dimensional case this suitability question can be made precise using decompositions.

Problem. Find a property Q of pairs (X,Z), where X is a generalized 3-manifold without boundary and Z is a closed 0-dimensional subset of X containing S(X), satisfying the following constraints.

(1) If X satisfies KF and (X,Z) satisfies Q then X has a resolution.

(2) If G is a closed 0-dimensional upper semicontinuous cell-like decomposition of the 3-manifold M^3 and if $(M^3/G,G)$ satisfies Q then G is shrinkable.

References

1. R.H. Bing, A homeomorphism between the 3-sphere and the sum of two Alexander horned spheres, Ann. of Maths. (2) 56 (1952), 354-362.
2. ________, Decompositions of E^3, Topology of 3-Manifolds (M.K. Fort, ed.), Prentice-Hall, Englewood Cliffs, N.J., 1962, pp. 5-20.
3. G.E. Bredon, Wilder manifolds are locally orientable, Proc. Nat. Acad. Sci. U.S.A. 63 (1969), 1079-1081.
4. M. Brin, Generalized 3-manifolds whose non-manifold set has neighborhoods bounded by tori, Trans. Amer. Math. Soc. (forthcoming).
5. M. Brin and D.R. McMillan, Jr., Generalized three-manifolds with zero-dimensional non-manifold set (to appear).
6. M. Brown, A proof of the generalized Schoenflies theorem, Bull. Amer. Math. Soc., 66 (1960), 74-76.
7. J.L. Bryant and R.C. Lacher, Resolving acyclic images of three-manifolds, Math. Proc. Camb. Phil. Soc. 88 (1980), 311-319.
8. J.W. Cannon, Shrinking cell-like decompositions of manifolds: co-dimension three, Ann. of Math. (2) 110 (1979), 83-112.
9. ________, The characterization of topological manifolds of dimension $n \geq 5$, Proc. Int. Cong. Math. (Helsinki, 1978), Academia Scientiarum Fennica, 1980.
10. R.J. Daverman, Decomposition Spaces (lectures given in Austin, Texas, Summer 1980) (to appear).
11. R.J. Daverman and J.J. Walsh, A ghastly generalized n-manifold (to appear).
12. C.H. Edwards, Jr., Open 3-manifolds which are simply connected at infinity, Proc. Amer. Math. Soc. 14 (1963), 391-395.
13. R.D. Edwards, The topology of manifolds and cell-like maps. Proc. Int. Cong. Math. (Helsinki, 1978), Academia Scientiarum Fennica, 1980.
14. D.E. Galewski, J.G. Hollingsworth, and D.R. McMillan, Jr., On the fundamental group and homotopy type of open 3-manifolds, General Topology and Appl. 2 (1972), 299-313.
15. G. Kozlowski and J.J. Walsh, The cell-like mapping problem, Bull. Amer. Math. Soc. (2) 2 (1980), 315-316.
16. T. Knoblauch, Imbedding deleted 3-manifold neighborhoods in E^3. Illinois J. Math. 18 (1974), 598-601.
17. ________, Imbedding compact 3-manifolds in E^3, Proc. Amer. Math. Soc. 48 (1975), 447-453.
18. R.C. Lacher, Cell-like mappings and their generalizations, Bull. Amer. Math. Soc. 83 (1977), 495-552.

19. R.C. Lacher, Resolutions of generalized manifolds, Proc. Steklov Maths. Inst. 154 (forthcoming).

20. D.R. McMillan, Jr., Strong homotopy equivalence of 3-manifolds, Bull. Amer. Math. Soc. 73 (1967), 718-722.

21. ______________, Acyclicity in 3-manifolds, Bull. Amer. Math. Soc. 76 (1970), 942-964.

22. ______________, Heegaard splittings of homology 3-spheres and homotopy 3-spheres (to appear).

23. D.R.McMillan, Jr., and T.L. Thickstun, Open three-manifolds and the Poincaré conjecture, Topology 19 (1980), 313-320.

24. F. Quinn, Ends of maps, I, Ann. of Math. (2) 110 (1979), 275-331.

25. T.L. Thickstun, Open acyclic 3-manifolds, a loop theorem, and the Poincaré conjecture (to appear).

26. J.H.C. Whitehead, A certain open manifold whose group is unity, Quart. J. Maths. (Oxford series) 6 (1935), 268-279.

27. R.L. Wilder, Generalized Manifolds, Amer. Math. Soc. Colloq. Publ. 32 (American Mathematical Society, Providence, R.I., 1949).

Florida State University
Tallahassee, Florida 32306
U.S.A.

---

*Supported in part by NSF Grant MCS78-00405.

Some properties of deformation dimension

by

Sławomir Nowak and Stanisław Spież (Warszawa)

Introduction. In this note we study properties of the deformation dimension of topological spaces. The notion of deformation dimension was formally introduced by J. Dydak (see [D]). For compact metric spaces this notion is equivalent to the notion of the Borsuk fundamental dimension (see [B]).

In the section 1 of this note we give some general properties of the deformation dimension of topological spaces. In the section 2 a cohomological characterization of the deformation dimension of topological spaces is described (for compact metric spaces it was done by S. Nowak [N]). This characterization does not cover the case of deformation dimension 2 (see [S_1], here section 3). The cohomological characterization is useful in the study of the deformation dimension of the Cartesian product of topological spaces. In section 4 there are given results relating to the deformation dimension of the Cartesian product of topological spaces with this special deformation dimension 2.

By HP we denote the full subcategory of the homotopy category HT whose objects are polytopes. The procategory of a category C is denoted by pro-C. If a morphism of pro-HT $\underline{q} = \{q_\beta\}_{\beta\in\Sigma} : X \to \underline{Y} = \{Y_\alpha, q_\alpha^\beta, \Sigma\}$ satisfies the continuity condition (see [D-S], p. 22) then we say that the system $\underline{Y}$ is associated with X. The Čech system $\check{C}(X) = \{K(\mathcal{U}_\alpha), \rho_\alpha^\beta, \Sigma\}$ of X, where $\mathcal{U}_\alpha$ is an open numerable locally finite cover of X is associated with X.

1. Deformation dimension.

Deformation dimension of a topological space X (of a pointed topological space (X,x)) is the smallest integer n such that any map $f: X \to Y$ (respectively $f:(X,x) \to (Y,y)$) where Y is a CW complex and $y \in Y^{(o)}$, is homotopic (resp. homotopic in the pointed sense) to a map whose values lie in the n-skeleton

$Y^{(n)}$ of Y.

The deformation dimension of X (or (X,x)) is denoted by def-dim X (resp. def-dim (X,x)).

The following statements hold (see [D] and [D - S])

(1.1) If X is a topological space and $x \in X$, then def-dim X = def-dim (X,x)

(1.2) If $Sh(X) \leq Sh(Y)$, then def-dim $X \leq$ def-dim Y.

(1.3) For every topological space X we have $\dim X \geq Sd(X) = \min\{\dim Y : Sh(Y) \geq Sh(X)\} \geq$ def-dim X.

(1.4) Let $\{X_\alpha, [p_\alpha^\beta], \Sigma\} = \underline{X} \in$ pro-HP be associated with X. Then the following conditions are equivalent

(a) def-dim $X \leq n$

(b) for every $\alpha \in \Sigma$ there is $\beta \in \Sigma$ such that $\beta \geq \alpha$ and $p_\alpha^\beta : X_\beta \to X_\alpha$ is homotopic to $q : X_\beta \to X_\alpha$, where $q(X_\beta) \subset X_\alpha^{(n)}$.

(c) there exists an $\underline{Y} = \{Y_\alpha, [q_\alpha^\beta], \Sigma'\} \in$ Ob pro-HP isomorphic to $\underline{X}$ such that $Y_\alpha^{(n)} = Y_\alpha$ for every $\alpha \in \Sigma'$.

For every topological space X we denote by $H_n(X;G)$ and $H^n(X;G)$ (respectively) the n-dimensional Čech homology and cohomology groups of X with coefficients in G based on locally finite numerable open coverings of X.

Using (1.4) one can easily prove that

(1.5) $H_n(X;G) = 0 = H^n(X;G)$ for every $n >$ def-dim X and every group G.

For every compactum X the deformation dimension def-dim X is equal to the fundamental dimension $Fd(X)$. Therefore def-dim $X \leq n$ iff def-dim $X_0 \leq n$ for each component X_0 of a compactum X. This statement is not true for the case when X is not a compactum.

(1.6) Example. Let $A_n = \{(x,y) \in E^2 : x^2 + y^2 = (1 + \frac{1}{n})^2\} \cap \{(x,y) \in E^2 : 1 \geq x\}$ and $X = \bigcup_{n=1}^{\infty} A_n \cup \{(1,0)\}$.

Every component of X is a contractible compactum and therefore def-dim $X_0 = 0$ for every component X_0 of X. Since $H_1(X;Z) \neq 0$, we obtain that def-dim X = dim $X = 1$.

Let us recall that a topological space X is strongly paracompact (see [E], p. 404) if X is a Hausdorff space and every open cover of X has a star-finite open refinement (a family of sets R is star-finite iff every set $A \in R$ meets only finitely members of R). Every Lindelöf space is strongly paracompact ([E], p. 406). Therefore every compact topological space and every separable metric space are strongly paracompact.

Let $\mathcal{U}|_A = \{U \in \mathcal{U} : A \cap U \neq \emptyset\}$ for every open cover $\mathcal{U}$ of X and every subset A of X.

One can prove that if A is a closed subset of a strongly paracompact space X, then the Čech system $\check{C}(A)$ of A and the system $\check{C}(X)|_A = \{K(\mathcal{U}_\alpha|_A), [\nu_\alpha^\beta], \Sigma\}$ are isomorphic, where $\check{C}(X) = \{K(\mathcal{U}_\alpha), [\mu_\alpha^\beta], \Sigma\}$ is a Čech system of X and $\nu_\alpha^\beta : K(\mathcal{U}_\beta|_A) \to K(\mathcal{U}_\alpha|_A)$ is the restriction of μ_α^β to $K(\mathcal{U}_\beta|_A)$ and $K(\mathcal{U}_\alpha|_A)$ for $\alpha, \beta \in \Sigma$ and $\beta \geq \alpha$.

Applying the last fact one can prove that the following theorem holds:

(1.7) THEOREM. Let X_1 and X_2 be closed subsets of a strongly paracompact space $X = X_1 \cup X_2$. Then def-dim $X \leq \max($def-dim X_1, def-dim X_2, def-dim$(X_1 \cap X_2) + 1)$.

2. <u>Cohomology and deformation dimension</u>. Let X be a topological space and suppose that L and K are local systems of abelian groups on X. We say that L and K are homotopy equivalent iff there exists a local system H of abelian groups on $X \times [0,1] = Y$ such that L and K are induced respectively by maps $p: X \to Y$ and $q: X \to Y$ defined by formulas

$$p(x) = (x,0) \quad \text{and} \quad q(x) = (x,1) \quad \text{for every } x \in X.$$

The homotopy equivalence of local systems on X is an equivalence relation. If L and K belong to the same equivalence class, then the groups $H_n(X;L)$ and $H_n(X;K)$ (or $H^n(X;L)$ and $H^n(X;K)$) are canonically isomorphic.

If $f_1, f_2 : X \to Y$ are homotopic and L_i is induced by f_i and a local system K on Y for $i = 1,2$, then L_1 and L_2 are homotopically equivalent.

Therefore one can talk about cohomology with coefficients in the class of homotopically equivalent local systems and about homomorphisms between cohomology groups with coefficients in the classes of local systems which are induced by the homotopy class of a map.

In particular we obtain the following

(2.1) PROPOSITION. Let L and K be local systems on polytopes X and Y such that L is induced by K and a map $f:X \to Y$ which is homotopic to a map whose values lie in the n-skeleton $Y^{(n)}$ of Y. Then the homomorphism $(f)^*:H^m(Y;K) \to H^m(X;L)$ is trivial for $m > n$.

In view of all of these facts we can define Čech cohomology groups with coefficients in local systems of groups.

Let $\underline{X} = \{X_\alpha, [p_\alpha^\beta], \Sigma\}$ Ob pro-HP be associated with a topological space X and assume that for every $\alpha \in \Sigma$ we have a local system L_α on X_α. If for all $\alpha, \beta \in \Sigma$ satisfying the inequality $\alpha \leq \beta$ the local system L_β, and the local system induced by L_α and $p_\alpha^\beta : X_\beta \to X_\alpha$ are homotopically equivalent, then the pair $\underline{L} = (\underline{X}, \{L_\alpha\}_{\alpha\in\Sigma})$ is called Čech local system of groups on X and $H^n(X;\underline{L}) = \text{lim dir}\ \{H^n(X_\alpha, L_\alpha), (p_\alpha^\beta)^*, \Sigma\}$ is called a Čech cohomology group of X with coefficients in $\underline{L}$.

Let $(\underline{X}, \{L_\alpha\}_{\alpha\in\Sigma}) = \underline{L}$ and $\underline{K} = (\underline{Y}, \{K_\beta\}_{\beta\in\Sigma'})$ be Čech local systems of abelian groups on X and Y and suppose that
Mor pro-HP $\ni \underline{f} = \{f_\beta\}_{\beta\in\Sigma'} : \underline{X} \to \underline{Y}$.

If for every $\beta \in \Sigma'$, there is $\alpha \in \Sigma$, $\alpha \geq f(\beta)$, such that L_α is homotopically equivalent to the local system induced by K_β and the map $f_\beta \circ p^\alpha_{f(\beta)} : X_\alpha \to Y_\beta$; then one can observe that $\underline{f}$ induces a homomorphism $(\underline{f})^* : H^n(Y;\underline{K}) \to H^n(X;\underline{L})$ and that the assignation $(\underline{f})^*$ to $\underline{f}$ has functorial properties.

Suppose that X is a topological space. We denote by $c[X]$ the maximum (finite or infinite) of all integers n such that there is a Čech local system $\underline{L}$ on X with $H^n(X;\underline{L}) \neq 0$.

A topological space X is said to be shape connected iff there is $\underline{X} = \{X_\alpha, [p^\beta_\alpha], \Sigma\} \in$ pro-HP associated with X such that for every $\alpha \in \Sigma$ there exists $\beta \in \Sigma, \beta \geq \alpha$ such that p^β_α is homotopic to a map $q: X_\beta \to X_\alpha$ with $q(X^{(1)}_\beta) \subset X^{(0)}_\alpha$.

If X is 1-shape connected, then $c[X]$ equals the maximum of all integers n such that there exists an abelian group G with $H^n(X;G) \neq 0$.

We shall now generalize the cohomological characterizations of the fundamental dimension (see [N]) to the case of the deformation dimension.

(2.2) THEOREM. Let X be a topological space with def-dim $X < \infty$. Then $c[X] \leq$ def-dim $X \leq \max(2, c[X])$. If $c[X] = 0$, then def-dim $X = 0$.

Similarly as in the case of compacta we obtain the inequality def-dim $X \leq \max(2, c[X])$ as an application of the theory of obstructions to deformations which was given by S.T. Hu and which was presented in $[Hu_1]$ and $[Hu_2]$ (Exercise E, Chapter VI of $[Hu_1]$ or $[Hu_2]$ p. 203).

The proof of the second part of Theorem (2.2) which states that $c[X] = 0$ implies def-dim $X = 0$ one obtains by a generalization and an easy modification of the proof of this statement for the compact case (see [K-S]).

In the next section we describe a continuum X with def-dim $X = 2$ and $c[X] = 1$.

Let X be movable and $\{X_\alpha, p^\beta_\alpha, \Sigma\} \in$ pro-HP be associated with X. Then for any $\alpha \in \Sigma$ there is $\alpha' \geq \alpha$ such that for each $\beta \geq \alpha$ there is $r: X_\alpha \to X_\alpha$ with $p^\beta_\alpha r \simeq p^{\alpha'}_\alpha$. If $c[X] \leq m \geq 3$, then using the theory of obstructions to deformations (the same as before) one can prove that for every n there exists a homotopy $\varphi_n : X_{\alpha'} \times [0,1] \to X_\alpha$ joining $p_\alpha^{\alpha'}$ with a map whose values belong to $X_\alpha^{(m)}$. We may assume that φ_{n-1} and φ_n agree on $X_{\alpha'}^{(n-2)}$. Applying this fact we can prove that $p_\alpha^{\alpha'}$ is homotopic to a map q with $q(x) \in X_\alpha^{(m)}$ for $x \in X_{\alpha'}$.

This allows us to prove

(2.3) THEOREM. If X is a movable topological space and def-dim $X = \infty$ then $c[X] = \infty$.

3. <u>An example of a metric continuum X with def-dim $X = 2$ and $c[X] = 1$.</u> Let r_j be a word (not necessary reduced) in symbols $g_1, g_2, \ldots, g_k$, for $j = 1,2,\ldots,m$. The cellular model of $(g_1, g_2, \ldots, g_k : r_1, r_2, \ldots, r_m)$ is a CW-complex with a single 0-cell, one 1-cell e_i^1 for each g_i and one 2-cell $e_{r_j}^2$ for each r_j is attached to the 1-skeleton according to the instruction provided by the word r_j; the 1-cells are oriented at first. Let φ_{r_j} denote the respective characteristic map from the 2-ball B with the counter clockwise orientation $B = \{x \in E^2 : |x| \le 1\}$.

Let Q and Q' be the cellular models of $(a,b;\ r,a)$ and $(a,b;\ r',s')$ respectively, where r and s are words (in the symbols a and b) not equivalent to the trivial word and $r' = rsr^{-1}s^{-1}$ and $s' = r^2s^2r^{-2}s^{-2}$. The 1-skeletons of Q and Q' are equal to the bouquet of the two circles e_a and e_b.

For $i = 1,2$, let A_i be the sum of $4i$ rays in B from the point 0. We assume that for $i = 1$ (resp. $1 = 2$) the ends of this these rays divide the boundary $\dot{B}$ of B for the segments $L_{i.j}$, where $j = 1,2,\ldots,4i$ such that the restriction of the map $\varphi_{r'}$ (resp. $\varphi_{s'}$) to $L_{i,j}$ describe the word r if $j = 1$ (resp. $j = 1$ or 2), s if $j = 2$ (resp. $j = 3$ or 4), r^{-1} if $j = 3$ (resp. $j = 5$ or 6), s^{-1} if $j = 4$ (resp. $j = 7,8$). The quotient space $B_i = B/A_i$ is the bouquet $\bigvee_{j=1}^{4i} B_{i,j}$ of $4i$ 2-balls $B_{i,j}$ (the ball $B_{i,j}$ corresponds to $L_{i,j}$) with base point b_i. Denote by p_i the quotient map from B to B_i.

Let $f:Q' \to Q$ be a map which satisfies the following conditions:

(i) f is the identity of the 1-skeleton of Q'

(ii) $f\circ\varphi_{r'} = f_1 p_1$ for a map $f_1:B_1 \to Q$ such that

$$f_1\circ h_{1,j} = \begin{cases} \varphi_r & \text{when } j=1,3 \\ \varphi_s & \text{when } j=2,4 \end{cases}$$

where $h_{1,j}:B \to B_{1,j}$ is a homomorphism for $j=1,2,3,4$.

(iii) $f\ \varphi_{s'} = f_2\ p_2$ for a map $f_2:B_2 \to Q$ such that

$$f_2\circ h_{2,j} = \begin{cases} \varphi_r & \text{when } j=1,2,5,6 \\ \varphi_s & \text{when } j=3,4,7,8 \end{cases}$$

where $h_{2,j}:B \to B_{2,j}$ is a homeomorphism for $j=1,\ldots,8$.

One can prove (see [S_1]) that

(3.1) For any local system L of abelian groups on Q the map $f:Q' \to Q$ induces the trivial homomorphism $(f)^*:H^2(Q;L) \to H^2(Q':L_f)$, where L_f is the local system induced by L and f,

and that the following statement holds:

(3.2) Let $f:(P,x) \to (Q,y)$ be a map of connected finite CW-complexes which induces epimorphism $f_\#:\pi_1(P,x) \to \pi_1(Q,y)$. Suppose that the rank of the group $\pi_1(P,x)$ is equal to the rank of the group $\pi_1(Q,y)$. If f is deformable to the 1-skeleton $Q^{(1)}$ of Q (i.e., there is a map $f':P \to Q^{(1)}$ such that f is homotopic to $i\circ f'$, where $i:Q^{(1)} \to Q$ is the inclusion) then the group $\pi_1(P,x)$ is free.

Let $r_1 = a^2$ and $s_1 = b^2$. Suppose that we have defined the words r_i and s_i for i $1,2,\ldots,n$. Let $r_{n+1} = r_n s_n r_n^{-1} s_n^{-1}$ and $s_n = r_n^2 s_n^2 r_n^{-2} s_n^{-2}$. Observe that r_n and s_n are not equivalent to the trivial word for each natural n. Let Q_n be a cellular model of $(a,b;\ r_n,s_n)$ and the map $p_n^{n+1}:Q_{n+1} \to Q_n$ be defined as the map $f:Q' \to Q$ for $n=1,2,\ldots$. The group $\pi_1(Q_n:x_n)$ is not a free group and has the rank equal 2 and $(p_1^n)_\#:\pi_1(Q_n,x_n) \to \pi_1(Q_1,x_1)$ is epimorphism for each $n=1,2,\ldots$ (x_n is the point of 0-skeleton of Q_n). Thus by (3.2) the map $p_1^n:Q_n \to Q_1$ is not deformable to the 1-skeleton

$Q_1^{(1)}$ of Q_1, so the continuum $X = \lim \operatorname{inv} \{Q_n, p_n^m\}$ has the deformation dimension def-dim $X = 2$. By (3.1) for any local system of abelian groups L_n on Q_n the homomorphism $(p_n^{n+1})^*: H^2(Q_n; L_n) \to H^2(Q_{n+1}; L'_n)$ is trivial, where L'_n is the local system of coefficients induced by p_n^{n+1} and L_n. Thus $H^2(X; \underline{L}) = 0$ for any Cech local system of coefficients $\underline{L}$ on X and we have the following statement:

(3.3) There exists a continuum X with def-dim $X = 2$ and with $c[X] = 1$.

(3.4) REMARK. J. Stallings [Sta] and R. Swan [Sw] have proved Eilenberg-Ganea Conjecture: Groups of cohomological dimension one are free. Thus, if K is a connected 2-dimensional CW-complex (not necessarily finite) with $c[K] = 1$, then K has the homotopy type of 1-dimensional CW-complex.

4. The deformation dimension of the Cartesian product of topological spaces with the deformation dimension 2. In this section we will consider connected topological spaces, however this restriction is not essential. We say that a topological space Y belongs to the class F if

$$\text{def-dim } Y = \max \{n : H^n(Y;G) \neq 0 \text{ for every nontrivial abelian group } G\}.$$

If X is a topological space with $2 \neq$ def-dim $X < \infty$, then def-dim $X = c[X]$. By the Künneth formula (compare [N]) it follows that if def-dim $X \neq 2$ then

$$\text{def-dim } (X \times Y) = \text{def-dim } X + \text{def-dim } Y$$

for every $Y \in F$. We know that there is a metric continuum X with def-dim $X = 2$ such that

$$\text{def-dim } (X \times Y) < \text{def-dim } X + \text{def-dim } Y$$

for every topological space Y with $0 <$ def-dim $Y < \infty$.

In this section, we present some results (see $[S_2]$) relating to the deformation dimension of the Cartesian product of connected topological spaces with the deformation dimension 2.

If X is a connected topological space with $\text{def-dim}\, X = 2$ and $\text{pro-}\pi_1(X,x)$ isomorphic to the trivial group then $c[X] = 2$ and

$$\text{def-dim}\,(X \times Y) = \text{def-dim}\, X + \text{def-dim}\, Y$$

for every $Y \in F$.

Let $\underline{G} = \{G_\alpha, p_\alpha^\beta, \Sigma\}$ be an inverse system of groups. We say that $\underline{G}$ contains elements of infinite order (briefly $\text{In}\ \underline{G} \neq 0$) if there is an index $\alpha \in \Sigma$ such that for every $\beta \in \Sigma$, $\alpha \leq \beta$, the image $p_\alpha^\beta(G_\beta)$ contains elements of infinite order. We say that $\underline{G}$ has a torsion (briefly $\text{Tor}\ \underline{G} \neq 0$) if there is an index $\alpha \in \Sigma$ such that for every $\beta \in \Sigma$, $\alpha \leq \beta$, the homomorphism p_α^β maps some torsion element of G_β onto a nontrivial element. Observe that if $\text{Tor}\ \underline{G} = 0$ and $\text{In}\ \underline{G} = 0$, then $\underline{G}$ is isomorphic to the trivial group.

One can prove the following

(4.1) THEOREM. Let X be a connected topological space with $\text{def-dim}\, X \leq 2$ and $\text{Tor}(\text{pro-}\pi_1(X,x)) \neq 0$. Then $\text{def-dim}\, X = c[X] = 2$ and $\text{def-dim}\,(X \times Y) = \text{def-dim}\, X + \text{def-dim}\, Y$ for every $Y \in F$.

We can also prove the following

(4.2) THEOREM. If Y is a connected topological space with $\text{In}(\text{pro-}\pi_1(Y,y)) \neq 0$ then $\text{def-dim}\,(X \times Y) \geq \text{def-dim}\,(X \times S^1)$ for every topological space X.

From Theorem (4.2) it follows

(4.3) COROLLARY. If a connected topological space $Y \subset E^3$ has a nontrivial shape then $\text{def-dim}\,(X \times Y) \geq \text{def-dim}\,(X \times S^1)$ for every topological space X.

(4.4) COROLLARY. Let Y_i be a connected topological space with $\text{In}(\text{pro-}\pi_1(Y_i)) \neq 0$ for each $i = 1,2,\dots,k$. Then $\text{def-dim}\,(Y_1 \times Y_2 \times \dots \times Y_k) \geq k$.

(4.5) COROLLARY. Let Y_i be a connected topological space with $\text{def-dim}\, Y_i = 1$, then $\text{def-dim}\,(Y_1 \times Y_2 \times \dots \times Y_k) = k$.

Let Y be a connected topological space with $\text{def-dim}\, Y = 2$ and $c[Y] = 1$. Then $\text{pro-}\pi_1(Y,y)$ is not isomorphic to the trivial group and by Theorem (4.1) Tor $(\text{pro-}\pi_1(Y,Y)) = 0$. Thus In $(\text{pro-}\pi_1(Y,Y)) \neq 0$. By Theorem (4.2), $\text{def-dim}\,(X \times Y) \geq \text{def-dim}\,(X \times S^1)$ for every topological space X. So we obtain the following

(4.6) COROLLARY. If Y is a connected topological space with $\text{def-dim}\, Y = 2$ and $c[Y] = 1$, then $\text{def-dim}\,(X \times Y) = \text{def-dim}\,(X \times S^1)$ for every topological space X with $\text{def-dim}\, X > 0$.

We also obtain

(4.7) COROLLARY. Let Y_i be a connected topological space with $\text{def-dim}\, Y_i = 2$ and $c[Y_i] = 1$ for each $i = 1,2,\ldots,k$, $k \geq 2$. Then $\text{def-dim}\,(Y_1 \times Y_2 \times \ldots \times Y_k) = k$.

A topological space X is 2-shape connected if the Čech system $\check{C}(X) = \{K(\mathcal{U}_\alpha), [\mu_\alpha^\beta], \Sigma\}$ of X is 2-connected i.e. for any $\alpha \in \Sigma$ there is $\beta \in \Sigma$, $\alpha \leq \beta$, such that for any map $f: S^2 \to K(\mathcal{U}_\beta)$ the composition $\mu_\alpha^\beta \circ f$ is homotopically trivial. If X is not shape 2-connected and $\text{def-dim}\, X = 2$, then $c[X] = 2$ (see [N]). We can also prove the following

(4.8) THEOREM. If X_i is not shape 2-connected for each $i = 1,2,\ldots,k$, then $\text{def-dim}\,(X_1 \times X_2 \times \ldots \times X_k) = 2k$.

(4.9) THEOREM. Let X be a movable topological space with $\text{def-dim}\, X = 2$. If X is not shape 2-connected then $\text{def-dim}\,(X \times Y) = \text{def-dim}\,(S^2 \times Y)$ for every topological space Y.

Let $f: P \to Q$ be a map of the pseudoprojective plane of order m into a 2-dimensional CW-complex Q, which induces a nontrivial homomorphism $f_\#: \pi_1(P,p) \to \pi_1(Q,q)$. If n^2 does not divide m, where n is the order of $\text{im}\, f_\#$, then the homomorphism $f_{\#,2}: \pi_2(P,p) \to \pi_2(Q,q)$ is not trivial. It follows

(4.10) PROPOSITION. A movable connected topological space X with Tor $(\text{pro-}\pi_1(X,x)) \neq 0$ and $\text{def-dim}\, X = 2$ is not shape 2-connected.

A consequence of the above proposition is the following

(4.11) COROLLARY. If X is a shape 2-connected movable connected topological space with $\text{def-dim}\, X = 2$, then In $(\text{pro-}\pi_1(X,x)) \neq 0$.

By Theorems (4.2), (4.9) and Corollary (4.11) we obtain

(4.12) COROLLARY. Let X_i be a movable connected topological space with $1 \lneq \text{def-dim } X_i \leq 2$ for each $i = 1,2,\ldots,k$. Then $\text{def-dim } (X_1 \times X_2 \times \ldots \times X_k) \geq k$.

For every integer $n \geq 3$ there is a family $\{X_i\}_{i=1}^{\infty}$ of finite polyhedra with $\text{def-dim } X_i = n$ (for every positive integer i); such that $\text{def-dim } (X_1 \times \ldots \times X_k) = n$ for every k. The following example (4.14) will show that the assumption of movability in the corollary (4.12) is essential.

Let n be a positive integer and let $\underline{\pi}$ be an inverse sequence of groups (abelian if $n > 1$). We can say that a metric continuum (X,x) is a space of shape type $(\underline{\pi},n)$ iff $\text{pro-}\pi_n(X,x)$ is isomorphic to $\underline{\pi}$ and $\text{pro-}\pi_k(X,x)$ is isomorphic to the trivial group for every $k \neq n$. One can prove, using the Whitehead theorem in shape theory, the following

(4.13) PROPOSITION. If (X,x) and (Y,y) are pointed metric continua of shape type $(\underline{\pi},n)$ with finite deformation dimension, then $Sh(X,x) = Sh(Y,y)$.

(4.14) EXAMPLE. For any integer $k \geq 2$, let $Y_n(k)$ be the pseudo-projective plane of order $k^{(2^n)}$ and let $q_n^{n+1}: Y_{n+1} \to Y_n$ be a map which induces an epimorphism of 1-homotopy groups and the trivial homomorphism of 2-homotopy groups. Let $Y(k)$ be the inverse limit of the inverse sequence $\{Y_n(k), q_n^{n+1}\}$. Observe that $Y(k)$ is a metric continuum of shape type $(\underline{\pi},1)$. If k and k' are integers relatively prime $(k > 1,\ k' > 1)$, then one can check that $\text{pro-}\pi_1(Y(k) \times Y(k'))$ and $\text{pro-}\pi_1(Y(k \cdot k'))$ are isomorphic. Since $Y(k) \times Y(k')$ and $Y(k \cdot k')$ are continua of shape type $(\underline{\pi},1)$ with finite deformation dimension, by the proposition (4.13) we have $Sh(Y(k) \times Y(k')) = Sh(Y(k \cdot k'))$. Thus if $k_1,k_2,\ldots,k_n$ are different primes, $Y(k_1) \times Y(k_2) \times \ldots \times Y(k_n)$ has the same shape as $Y(k_1 \cdot k_2 \cdot \ldots \cdot k_n)$ and so $\text{def-dim } (Y(k_1) \times Y(k_2) \times \ldots \times Y(k_n)) = 2$.

Problems

The following problems are open:

1. Is it true that $Fd(X) = \text{def-dim}\, X$ for every topological space?
2. Is it true that $c[X] = 2$ for every movable space X (or movable compactum X) with $\text{def-dim}\, X = 2$?
3. Let X_i be a movable compactum with $\text{def-dim}\, X_i = 2$ for each $i = 1,2,\ldots,k$. Is it true that $\text{def-dim}\,(X_1 \times X_2 \times \ldots \times X_k) = 2k$?
4. Is every compactum X (every compactum X with $\dim X < \infty$) CE-equivalent to a compactum Y with $Fd(X) = \dim Y$?
5. Let X be a compactum of a finite deformation dimension. Is there a map $f: X \to Y$ being a shape equivalence such that the dimension of Y is finite?

References

[B] K. Borsuk, Theory of Shape, Warszawa 1975.

[D] J. Dydak, The Whitehead and Smale theorems in shape theory, Dissertationes Mathematicae 156(1979), 1-51.

[D-S] J. Dydak and J. Segal, Shape theory, an introduction, Lecture Notes in Mathematics 688, Springer-Verlag 1978.

[E] R. Engelking, General Topology, Warszawa 1977.

[Hu_1] Sze-Tsen Hu, Homotopy Theory, Academic Press, New York and London 1959.

[Hu_2] Sze-Tsen Hu, Theory of Retracts, Wayne State University Press, 1965.

[K-S] A. Kadlof and S. Spież, Remark on the fundamental dimension of cartesian product of metric sompacta, Bull. Acad. Polon. Ser. Math. (to appear).

[N] S. Nowak, Algebraic theory of fundamental dimension, Dissertationes Mathematicae 187 (1980).

[Spa) E. H. Spanier, Algebraic Topology, New York 1966.

[S_1] S. Spież, An example of a continuum X with $Fd(X \times S^1) = Fd(X) = 2$, Bull. Acad. Polon. Ser. Math. (to appear).

[S_2] S. Spież, On the fundamental dimension of the Cartesian product of compacta with the fundamental dimension 2, Fund. Math. (to appear).

[Sta] J. R. Stallings, On torsion-free groups with infinitely many ends, Ann. Math. 88(1968), pp. 312-344.

[Sw] R. Swan, Groups of cohomological dimension "one", Journal of Algebra 12(1969), pp. 585-601.

Institute of Mathematics, University of Warsaw.

DIMENSION, COHOMOLOGICAL DIMENSION, AND CELL-LIKE MAPPINGS

John J. Walsh

1. INTRODUCTION

These notes explore the relationship between the theory of (covering) dimension and the theory of integral cohomological dimension. The purpose is to develop a "self-contained" description of each theory that easily establishes their equivalence for finite dimensional spaces and illuminates the potential differences for infinite dimensional spaces. The motivating application of such a development is to understand the equivalence of the two problems.

(P_1) Does there exist a cell-like map between compacta that raises dimension?

(P_2) Does there exist an infinite dimensional compactum that has finite cohomological dimension?

One direction of the equivalence is a consequence of the Vietoris-Begle Mapping Theorem [Be] (see Appendix B) while the other direction was established more recently by R. D. Edwards [Ed]. See Section 6.

For the most part, these notes deal with known, even "classical," results. The unified "geometric" approach permits easy access to cohomological dimension and, most importantly, an insightful view of its relationship with covering dimension and cell-like maps. The author is indebted to R. D. Edwards for several lucid conversations out of which these materials developed.

2. PRELIMINARIES

With the exception of the classifying spaces described next and used throughout, spaces are separable and metric.

We shall use K_n to denote an Eilenberg-MacLane space satisfying

$$\pi_k(K_n, *) \cong \begin{cases} \pi_k(S^n, *) & k \leq n \\ 0 & k \geq n+1 \, . \end{cases}$$

On the one hand, a standard and useful model for K_n is a CW-complex obtained from the n-sphere S^n by attaching cells having dimensions $\geq n + 2$, for the skeleta of such a model satisfy $K_n^{(n)} = K_n^{(n+1)} = S^n$. On the other hand, we prefer not to limit ourselves to a single pre-chosen CW-structure for K_n. In all cases, K_n is to be a CW-complex, generally not locally finite. "Different" K_n's are homotopy equivalent and, being CW-complexes, are absolute neighborhood extensors for metric spaces.

The important feature is that the K_n's classify Čech cohomology with integral coefficients; precisely, $\check{H}^n(X; Z) \cong [X, K_n]$, the brackets denoting homotopy classes of maps and the isomorphism being natural. This fact is never needed in these notes but it "justifies" the definition of cohomological dimension presented in the next section.

An indispensable tool is the general form of Borsuk's homotopy extension theorem. Following its statement are two technical results needed later. The first is a "standard" fact about CW-complexes. The second can be established by imposing additional constraints during a "standard" proof of the Homotopy Extension Theorem.

Homotopy Extension Theorem. Let A be a closed subset of a space X, let L be an absolute neighborhood extensor, and let $H:(A \times I) \cup (X \times \{0\}) \to L$. Then H extends to a homotopy $\tilde{H}:X \times I \to L$.

Proposition 2.1. A CW-complex L has the property that for each skeleton $L^{(q)}$ there is an open set $U \supset L^{(q)}$ and a homotopy $\{h_t:L \to L | 0 \le t \le 1\}$ satisfying h_0 = identity and each h_t = identity on $L^{(q)}$, $h_t(L^{(k)}) \subset L^{(k)}$ for each skeleton, and $h_1(U) \subset L^{(q)}$.

Proposition 2.2. Let A be a closed subset of a space X, let U be an open subset of an absolute neighborhood extensor L, and let $H:(A \times I) \cup (X \times \{0\}) \to L$ with $H(A \times I) \subset U$. Then there is an extension $\tilde{H}:X \times I \to L$ such that $\tilde{H}(x, t) = H(x, 0)$ whenever $H(x, 0) \in L - U$ and $\tilde{H}(x, t) \in U$ whenever $H(x, 0) \in U$.

3. DIMENSION AND COHOMOLOGICAL DIMENSION: A COMPARISON

We choose not to use one of the standard inductive or cover definitions of dimension but rather to use a characterization as our "working" definition. Dimension Theory by W. Hurewicz and H. Wallman remains an excellent source for a development of the theory for separable metric spaces; the following equivalent formulation occurs as Theorem VI4 in [HW; p. 83].

Definition of dimension: a space X has dimension $\le n$, written $\dim X \le n$, provided each map $\alpha:A \to S^n$, from a closed subset into the n-sphere, extends to a map $\tilde{\alpha}:X \to S^n$.

The choice of characterization is dictated by our intended comparison.

Definition of cohomological dimension: a space X has cohomological dimension $\le n$, written $\text{c-dim}\, X \le n$, provided each map $\alpha:A \to K_n$, from a closed subset into an Eilenberg-MacLane space $K_n = K(Z, n)$, extends to a map $\tilde{\alpha}:X \to K_n$.

Cohomological dimension can be defined with respect to any coefficient group G by using an Eilenberg-MacLane space $K(G, n)$ and there is a rich literature justifying the generality. For our purposes, we need only integral coefficients and, consequently, restrict the term "cohomological dimension" and the symbol "c-dim" to this setting. (Other symbols occurring in the literature are "$\dim_Z$" and "D(; G)".)

The definitions are completed by setting

$\dim X = n$ provided $\dim X \le n$ and $\dim X > n - 1$;

$\text{c-dim}\, X = n$ provided $\text{c-dim}\, X \le n$ and $\text{c-dim}\, X > n - 1$;

dim $X = \infty$ provided dim $X \geq n$ for each n ; and
c-dim $X = \infty$ provided c-dim $X \geq n$ for each n .

The remainder of the section is devoted to developing the basics of both theories of dimension and to establishing their equivalence for spaces with finite dimension.

(D_1) If $Z \subset X$ and dim $X \leq n$, then dim $Z \leq n$.
(CD_1) If $Z \subset X$ and c-dim $X \leq n$, then c-dim $Z \leq n$.

Outline of the proof of (D_1) and (CD_1) . Both are evidently true for closed subsets $Z \subset X$.

An open subset $Z \subset X$ is the union of a locally finite countable collection of closed sets $\{Z_1, Z_2, \ldots\}$. An extension of a map α defined on a relatively closed subset $A \subset Z$ is obtained by recursively extending an extension α_k defined on $A \cup Z_1 \cup \ldots \cup Z_k$ to α_{k+1} defined on $A \cup Z_1 \cup \cdots \cup Z_k \cup Z_{k+1}$ and by using the α_k 's to determine an extension $\tilde{\alpha}$ defined on Z .

For an arbitrary subset $Z \subset X$, we need the content of the following

Exercise 1. A map $\alpha : A \to P$, from a subset $A \subset X$ to an absolute neighborhood extensor P , is homotopic to a map $\alpha' : A \to P$ which extends to a map β defined on an open subset $U \subset X$ containing A .

Following an appeal to the Homotopy Extension Theorem, we assume that the map α defined on a relatively closed subset $A \subset Z$ extends to a map β defined on an open set $U \supset A$. Choose disjoint open sets V and W with $A \subset V$ and $Z - U \subset W$. The intersection $\overline{V} \cap U$ is a relatively closed subset of the open set $U \cup W$ and we have already given an argument assuring that the restriction of β to $\overline{V} \cap U$ extends to a map $\tilde{\beta}$ defined on $U \cup W$. The latter map restricts to yield the desired extension of α to Z .

(D_2) If dim $X \leq n$, then dim $X \leq n + 1$.
(CD_2) If c-dim $X \leq n$, then c-dim $X \leq n + 1$.

Proof of (D_2) . An appeal to the Homotopy Extension Theorem establishes that it suffices to verify the stronger assertion that every map $\alpha : A \to S^{n+1}$, defined on a closed subset $A \subset X$, is null-homotopic. Let B be the (n+1)-cell which forms the "closed upper hemisphere" of S^{n+1} . Since $\mathrm{Fr}B \cong S^n$ and dim $X \leq n$, the restriction of α to $\alpha^{-1}(\mathrm{Fr}B)$ extends to a map $\beta : \alpha^{-1}(B) \to \mathrm{Fr}B$. The maps β and the restriction of α are homotopic as maps into B and the homotopy can be chosen to fix the points of $\alpha^{-1}(\mathrm{Fr}B)$. Consequently, α is null-homotopic being homotopic to a map whose image is contained in the contractible subset $S^n - \mathrm{Int}(B)$.

The proof of (CD_2) is delayed until we have established the characterization of cohomological dimension appearing in the next section. A proof using path fibrations suggested by S. Ferry is presented in Appendix A; a useful by-product is that it uses the relationship between K_{n+1} and K_n analogous to the (n+1)-sphere being the suspension of the n-sphere.

For the remainder of the section, we assume that the CW-complex K_n has a structure determined by attaching cells of dimension $\geq n + 2$ to the n-sphere S^n. Then $K_n^{(n)} = K_n^{(n+1)} = S^n$ and we write $K_n^{(n+i+1)} = K_n^{(n+i)} \cup (\cup e^{n+i+1})$ suppressing the characteristic maps and not caring whether there are finitely or infinitely many cells of a given dimension.

<u>Proposition</u> 3.1. If $\dim X \leq n + 1$, then each map $\alpha: X \to K_n$ is homotopic to a map $\beta: X \to S^n \subset K_n$ with $\beta = \alpha$ on $\alpha^{-1}(S^n)$.

<u>Proof</u>. The important observation is that for $i \geq 1$ and for a map $\gamma: A \to K_n^{(n+i)}$ defined on a subset $A \subset X$, since $\dim A < n + i$, there is a homotopy $\{\gamma_t: A \to K_n^{(n+i)}\}$ with $\gamma_0 = \gamma$, $\gamma_t = \gamma$ on $\gamma^{-1}(K_n^{(n+i-1)})$, and $\gamma_1(A) \subset K_n^{(n+i-1)}$; the homotopy is an amalgamation of homotopies defined on each individual (n+i)-cell of $K_n^{(n+i)}$. Continuing with a "downward" induction, a homotopy $\{\tau_t: A \to K_n^{(n+i)}\}$ is produced such that $\tau_0 = \gamma$, $\tau_t = \gamma$ on $\gamma^{-1}(S^n)$, and $\tau_1(A) \subset S^n = K_n^{(n+1)}$. This establishes the proposition for compact spaces while the general case requires further refinements.

Set $X^{(n+i)} = \alpha^{-1}(K_n^{(n+i)})$ for $i \geq 2$. The strategy is to successively "pull" each $X^{(n+i)}$ into $K_n^{(n+1)} = S^n$.

As a first step, specify an open set $U \supset K_n^{(n+2)}$ and a homotopy $\{h_t: K_n \to K_n\}$ as in Proposition 2.1. Let $\{\alpha_t: X^{(n+2)} \to K_n^{(n+2)}\}$ be a homotopy with $\alpha_0 = \alpha$ on $X^{(n+2)}$, $\alpha_t = \alpha$ on $\alpha^{-1}(S^n)$, and $\alpha_1(X^{(n+2)}) \subset S^n$. Using Proposition 2.2, we obtain an extension $\{\tilde{\alpha}_t: X \to K_n\}$ with $\tilde{\alpha}_0 = \alpha$, $\tilde{\alpha}_t = \alpha$ on $\alpha^{-1}(K_n - U)$, and $\tilde{\alpha}_t(x) \in U$ whenever $\alpha_t(x) \neq \alpha(x)$. The composed homotopy $\{A_t = h_t \circ \alpha_t: X \to K_n\}$ has the properties that $A_0 = \alpha$, $A_t = A_0$ on $A_0^{-1}(S^n)$, $A_1(\alpha^{-1}(U)) \subset S^n$, and $A_1^{-1}(K_n^{(n+i)}) \supset X^{(n+i)}$ for each i, as A_1 agrees with $\alpha \circ h_1$ on $X - \alpha^{-1}(U)$ and h_1 is a skeleta preserving map.

Repeating this argument using A_1 produces a homotopy from A_1 to a map A_2 with each stage of the homotopy agreeing with A_1 on $A_1^{-1}(S^n)$, with $A_2(V) \subset S^n$ for some open set V containing $A_1^{-1}(K_n^{(n+3)}) \supset X^{(n+3)}$, and with $A_2^{-1}(K_n^{(n+i)}) \supset X^{(n+i)}$ for each i.

Continuing inductively produces a homotopy from α to the map $A_\infty = \lim A_i$ which satisfies $A_\infty(X) \subset S^n$.

Exercise 2. For a space X having dimension $\leq n$, every map $\alpha: X \to K_{n+1}$ is null-homotopic.

Exercise 3. For a complex L, simplicial or CW, a map $\alpha: L \to K_{n+1}$ is null-homotopic if and only if the restriction to $L^{(n+1)}$ is null-homotopic.

We are prepared to establish the fundamental relationship between the two approaches to dimension theory.

Theorem 3.2.

(a) For any space, c-dim $X \leq$ dim X .

(b) For any space with dim $X < \infty$, c-dim X = dim X .

Proof. The first conclusion is justified by observing that if dim $X \leq n$, then Property (D_1) and Proposition 3.1 yield that any map $\alpha : A \to K_n$ defined on a closed subset $A \subset X$ is homotopic to a map $\beta : A \to S^n$ and by using the Homotopy Extension Theorem to conclude that, since β extends to a map $\tilde{\beta} : X \to S^n$, α extends to a map $\tilde{\alpha} : X \to K_n$.

In order to verify the second conclusion we set dim $X = n + 1$ and observe that if it were the case that c-dim $X \leq n$, then each map $\alpha : A \to S^n$ defined on a closed subset $A \subset X$ not only would extend to a map $\tilde{\alpha} : A \to K_n$ but also, in view of Proposition 3.1, would extend to a map $\tilde{\alpha} : X \to S^n$ implying that dim $X \leq n$.

The situation not covered by this result remains an unresolved issue.

A Classical Problem. Is there an infinite dimensional space which has finite cohomological dimension?

Perhaps it is more interesting to ask the question for compacta but apparently the general case remains an open problem.

We conclude the section by using the observation that the 1-sphere S^1 is a $K(Z, 1)$ to establish the

Corollary 3.3. For any space, dim $X \leq 1$ if and only if c-dim $X \leq 1$.

4. DIMENSION AND COHOMOLOGICAL DIMENSION: INVERSE LIMITS

We are prepared to characterize the dimension and cohomological dimension of compacta in terms of properties of inverse sequences determining the compacta. The characterizations lead to a proof of (CD_2) and, in the next section, are used to provide a characterization of cohomological dimension which serves to illuminate "the problem" central to the classical question concerned with the equivalence of the two theories.

For the statements of the next theorems and for the remainder of the section $\{P_q, f_q\}$ denotes an inverse sequence of compact polyhedra with a compactum X as its limit.

Theorem 4.1. The compactum X has dimension $\leq n$ if and only if for each integer k and each $\varepsilon > 0$ there is an integer $j > k$, a triangulation L_k of P_k , and a map $g_{jk} : P_j \to L_k^{(n)}$ which is ε-close to f_{jk} .

Theorem 4.2. (R. D. Edwards) The compactum X has cohomological dimension $\leq n$ if and only if for each integer k and each $\varepsilon > 0$ there is a triangulation L_k of P_k and an integer $j > k$ such that for any triangulation L_j of P_j there is a map

$g_{jk}:L_j^{(n+1)} \to L_k^{(n)}$ which is ε-close to the restriction of f_{jk}.

Proof of Theorem 4.1. We establish the reverse implication first. Let $\{P_q^A, f_q^A\}$ be an inverse sequence consisting of subpolyhedra and restrictions and having limit a closed subset $A \subset X$; for convenience we require that $f_q^A(P_q^A) \subset \mathrm{Int}(P_{q-1}^A)$. For each map $\alpha:A \to S^n$, there is an integer k and a map $\alpha_k:P_q^A \to S^n$ such that α is homotopic to $\alpha_k \circ f_k^A$. Choose $\varepsilon > 0$ so that the ε-neighborhood of $f_{ik}(P_i^A)$ is contained in P_k^A for $i > k$ and so that ε-close maps into P_k^A are homotopic as maps into P_k^A. Let L_k, $j > k$, and $g_{jk}:P_j \to L_k^{(n)}$ be as hypothesized for the choice of ε. Since S^n is (n-1)-connected, α_k extends to a map $\tilde{\alpha}_k:P_k^A \cup L_k^{(n)} \to S^n$. Since g_{jk} restricts to a map homotopic to f_{jk}^A, the composition $\tilde{\alpha}_k \circ g_{jk} \circ f_{\infty j}$ restricts to a map homotopic to $\alpha_k \circ f_{jk}^A \circ f_{\infty j}^A = \alpha_k \circ f_{\infty k}^A$ and, consequently, to a map homotopic to α. Once again an appeal to the Homotopy Extension Theorem yields an extension of α.

Conversely, if $\dim X \leq n$, then choose any triangulation L_k of P_k whose simplices have diameters $< \varepsilon/2$. The argument given in the first paragraph of the proof of Proposition 3.1 produces a homotopy $\{h_t:X \to P_k\}$ such that $h_0 = f_{\infty k}$, $h_1(X) \subset L_k^{(n)}$, and $h_t(f_{\infty k}^{-1}(\sigma)) \subset \sigma$ for each $\sigma \in L_k$. Since $L_k^{(n)}$ is an ANE, there are maps $g_{ik}:P_i \to L_k^{(n)}$ (for large values of i) such that $\lim_{i\to\infty} g_{ik} \circ f_{\infty i} = h_1$. A choice of j sufficiently large assures that $\tau \cap \rho \neq \emptyset$ whenever $g_{jk}(x) \in \rho \in L_k$ and $f_{jk}(x) \in \tau \in L_k$ and we conclude that f_{jk} and g_{jk} are ε-close.

Proof of Theorem 4.2. Proceed as in the proof of Theorem 4.1 up to and including the choice of $\varepsilon > 0$ and let $j > k$, L_k, L_j, and $g_{jk}:L_j^{(n+1)} \to L_j^{(n)}$ be as hypothesized for the choice of ε. Since the restrictions of g_{jk} and f_{jk} to $L_j^{(n+1)} \cap P_j^A$ are homotopic, the Homotopy Extension Theorem yields an extension $\tilde{g}_{jk}:P_j^A \cup L_j^{(n+1)} \to P_k^A \cup L_k^{(n)}$ with the restriction $\tilde{g}_{jk}:P_j^A \to P_k^A$ homotopic to f_{jk}^A. Since K_n is (n-1)-connected, α_k extends to a map $\tilde{\alpha}_k:P_k^A \cup L_k^{(n)} \to K_n$. Since the homotopy groups of K_n are trivial in dimensions $\geq n+1$, the composition $\tilde{\alpha}_k \circ \tilde{g}_{jk}:P_j^A \cup L_j^{(n+1)} \to K_n$ extends to a map $h:P_j \to K_n$. The composition $h \circ f_{\infty j}:X \to K_n$ restricts to the map $\alpha_k \circ \tilde{g}_{jk} \circ f_{\infty j}^A$ which is homotopic to $\alpha_k \circ f_{jk}^A \circ f_{\infty j}^A = \alpha_k \circ f_{\infty k}^A$ and, consequently, is homotopic to α. We conclude that α has an extension $\tilde{\alpha}:X \to K_n$.

We interrupt the proof in order to describe a CW-complex obtained from a simplicial complex by replacing simplices having dimension $\geq n+1$ with carefully specified "compatible" Eilenberg-MacLane spaces, a simplex σ being replaced by $K(\pi_n(\sigma^{(n)}), n)$ where $\sigma^{(n)}$ denotes the n-skeleton of σ. The next exercises establish that the Eilenberg-MacLane spaces just identified are essentially no more complicated than a $K(Z, n)$ for $n \geq 2$.

Exercise 4. Show that the nth-homotopy group $\pi_n(\sigma^{(n)})$ of the n-skeleton of a simplex where $\dim \sigma \geq n+1$ is a finitely generated free abelian group for $n \geq 2$.

Exercise 5. Show that the cartesian product $\prod_1^r K(Z, n)$ is an Eilenberg-MacLane space of the form $K(\bigoplus_1^r Z, n)$ and, consequently, if c-dim $X \leq n$, then any map $\alpha: A \to K(\bigoplus_1^r Z, n)$ defined on a closed subset $A \subset X$ extends to a map $\tilde{\alpha}: X \to K(\bigoplus_1^r Z, n)$

For an integer $n \geq 1$ and a simplicial complex L decomposed as $L = L^{(n)} \cup \sigma_1 \cup \sigma_2 \cup \cdots \cup \sigma_s$ with the simplices ordered so that $n + 1 \leq \dim \sigma_i \leq \dim \sigma_{i+1}$, we construct a CW-complex $\hat{L} = L^{(n)} \cup K(\sigma_1) \cup K(\sigma_2) \cup \cdots \cup K(\sigma_s)$ satisfying:

(a) $\hat{L}^{(n)} = \hat{L}^{(n+1)} = L^{(n)}$ and $L^{(n)} \cap K(\sigma_i) = \sigma_i^{(n)}$;

(b) $K(\sigma_i)$ is an Eilenberg-MacLane space $K(\pi_n(\sigma_i^{(n)}), n)$ and is a subcomplex of $\hat{L}$;

(c) $K(\sigma_i) \cap K(\sigma_j) = \sigma_i \cap \sigma_j$ whenever $\dim \sigma_i \cap \sigma_j \leq n$ and $K(\sigma_i) \cap K(\sigma_j) = K(\sigma_i \cap \sigma_j)$ whenever $\dim \sigma_i \cap \sigma_j \geq n + 1$.

The construction of $\hat{L}$ is by induction on s starting with $\hat{L} = L^{(n)}$ for $s = 0$. Assume the construction of a CW-complex $L^{(n)} \cup K(\sigma_1) \cup \cdots \cup K(\sigma_k)$ satisfying properties (a)-(c) and denote the subcomplex $\sigma_{k+1}^{(n)} \cup (\cup\{K(\sigma_i): \sigma_i$ is a proper face of $\sigma_{k+1}\})$ by $K(\dot{\sigma}_{k+1})$, $\dot{\sigma}_{k+1}$ being the simplicial boundary of σ_{k+1}. Property (a) assures that the subcomplex $K(\dot{\sigma}_{k+1})$ is obtained from $\sigma_{k+1}^{(n)}$ by attaching cells having dimensions $\geq n + 2$. Therefore, the inclusion $\sigma_{k+1}^{(n)} \to K(\dot{\sigma}_{k+1})$ induces isomorphisms of homotopy groups in dimensions $\leq n$, permitting the construction of $K(\sigma_{k+1})$ by attaching cells having dimensions $\geq n + 2$ to $K(\dot{\sigma}_{k+1})$. The reader can easily verify that properties (a)-(c) continue to hold for $L^{(n)} \cup K(\sigma_1) \cup \cdots \cup K(\sigma_k) \cup K(\sigma_{k+1})$.

The construction just described is particularly simple when $n = 1$ for the choices $K(\sigma_i) = \sigma_i^{(1)}$ and $\hat{L} = L^{(1)}$ are adequate; suprisingly, this is the one case we will not be able to use.

<u>The Completion of The Proof of Theorem</u> 4.2. Since c-dim $X \leq 1$ implies that $\dim X \leq 1$ (see Corollary 3.3), the converse follows from Theorem 4.1 when $n = 1$.

Now we require $n \geq 2$, an assumption needed below. Choose a triangulation L_k of P_k whose simplices have diameters $< \varepsilon/2$. Let $\hat{L}_k = L_k^{(n)} \cup K(\sigma_1) \cup \cdots \cup K(\sigma_s)$ be an associated CW-complex of the type just described. Observe that, since $n \geq 2$, Exercises 4 and 5 establish that each map $\alpha: A \to K(\sigma_i)$ defined on a closed subset $A \subset X$ extends to a map $\tilde{\alpha}: X \to K(\sigma_i)$. Therefore, a map $h: X \to \hat{L}_k$ satisfying $h = f_{\infty k}$ on $f_{\infty k}^{-1}(L_k^{(n)})$ and $h(f_{\infty k}^{-1}(\sigma_i)) \subset K(\sigma_i)$ for each i can be constructed by successively extending the restriction of $f_{\infty k}$ to $f_{\infty k}^{-1}(L_k^{(n)})$ over $f_{\infty k}^{-1}(\sigma_i)$ for $i = 1,2,\ldots,s$.

Since $\hat{L}_k$ is an absolute neighborhood extensor, there are maps $h_{ik}: P_i \to \hat{L}_k$ (for large values of i) such that $\lim_{i\to\infty} h_{ik} \circ f_{\infty i} = h$. Choosing a sufficiently large value for j, we conclude that f_{jk} and h_{jk} are "close" provided the measurement is made in L_k; precisely,

(1) if $f_{jk}(x) \in \tau \in L_k$ and $h_{jk}(x) \in \rho \in L_k^{(n)}$, then $\rho \cap \tau \neq \emptyset$; or

(2) if $f_{jk}(x) \in \tau \in L_k$ and $h_{jk}(x) \in K(\sigma_i)$, then $\sigma_i \cap \tau \neq \emptyset$.
For any triangulation L_j of P_j, the restriction of h_{jk} to $L_j^{(n+1)}$ can be adjusted over $K(\sigma_i)$ successively for $i = s, s-1, \ldots, 1$ to produce a map $g_{jk}:L_j^{(n+1)} \to \hat{L}_k^{(n+1)} = L_k^{(n)}$, agreeing with h_{jk} on $h_{jk}^{-1}(L_k^{(n)})$, such that $g_{jk}(x) \in K(\sigma_i)$ whenever $h_{jk}(x) \in K(\sigma_i)$. These constraints combine with (1) and (2) to imply that g_{jk} is ε-close to the restriction of f_{jk}.

The first application of this characterization is a proof of Property (CD_2): if c-dim $X \leq n$, then c-dim $X \leq n + 1$; an alternate proof is given in Appendix A.

<u>Proof of</u> (CD_2). We establish that for a space with c-dim $X \leq n$ every map $\alpha:X \to K_{n+1}$ is null-homotopic. Specify an inverse sequence $\{P_i, f_i\}$ of polyhedra with limit X, a map $\alpha:X \to K_{n+1}$, and maps $\alpha_i:P_i \to K_{n+1}$ such that $\lim_{i\to\infty} f_{\infty i}\circ\alpha_i = \alpha$. Choose k so that for $j \geq k$ the maps $\alpha_k \circ f_{kj}$ and α_j are homotopic and the maps $\alpha_j \circ f_{\infty j}$ and α are homotopic; the last homotopy reduces the problem to showing that some α_j is null-homotopic for $j > k$.

Name an $\varepsilon > 0$ such that ε-close maps into P_k are homotopic and let L_k, $j > k$, and $g_{jk}:L_j^{(n+1)} \to L_k^{(n)}$ be as in the conclusion of Theorem 4.2 for this choice of ε and some triangulation L_j of P_j. Since g_{jk} and the restriction of f_{jk} are homotopic, the Homotopy Extension Theorem yields an extension $\tilde{g}_{jk}:L_j \to P_k$ homotopic to f_{jk}. The restriction of α_k to $L_k^{(n)}$ is null-homotopic (see Exercise 2) yielding the conclusion that $\alpha_k \circ g_{jk}$ is null-homotopic. In turn, we conclude that $\alpha_k \circ \tilde{g}_{jk}$ is null-homotopic (see Exercise 3). Since f_{jk} and $\tilde{g}_{jk}$ are homotopic, $\alpha_k \circ f_{jk}$ is null-homotopic forcing α_j to be null-homotopic.

5. A GENERALIZATION

We present a generalization of the characterization of cohomological dimension stated in Theorem 4.2. It represents an exploitation of the inductive consequence of (CD_2) that c-dim $X \leq n$ implies c-dim $X \leq n + i$ for $i \geq 1$ and it sets forth the stability problem central to the question of equivalence of dimension and cohomological dimension.

Once again we specify an inverse sequence $\{P_q, f_q\}$ of compact polyhedra having limit a compactum X.

<u>Theorem</u> 5.1. (R. D. Edwards) The compactum X has cohomological dimension $\leq n$ if and only if, given an integer $i \geq 1$, for each integer k and each $\varepsilon > 0$ there is a triangulation L_k of P_k and an integer $j > k$ such that for any triangulation L_j of P_j there is a map $g_{jk}:L_j^{(n+i)} \to L_k^{(n)}$ which is ε-close to the restriction of f_{jk}.

<u>Proof</u>. Theorem 4.2 establishes the result for each integer n and $i = 1$ and we proceed inductively, assuming the result for each integer n and the integer $i - 1$.

Let $s > k$ and L_k be as in Theorem 4.2 for the choice $\varepsilon/2$. Let $\hat{\varepsilon}$ be such that, for any pair of points $x,y \in P_s$ $\hat{\varepsilon}$-close, $f_{sk}(x)$ and $f_{sk}(y)$ are $\varepsilon/2$-close.

Let $j > j_s$ and L_{js} be as in Theorem 5.1 for $n+1$, $i-1$, and $\hat{\varepsilon}$; we are invoking both induction and the content of (CD_2) that c-dim $X \leq n$ implies c-dim $X \leq n+1$.

We are assured a map $g_{sk}: L_s^{(n+1)} \to L_k^{(n)}$ $\varepsilon/2$-close to the restriction of f_{sk} and a map $g_{js}: L_j^{(n+i)} \to L_s^{(n+1)}$ $\hat{\varepsilon}$-close to the restriction of f_{js} for any choice of triangulation L_j of P_j.

The proof is completed by setting $g_{jk} = g_{sk} \circ g_{js}: L_j^{(n+i)} \to L_k^{(n)}$ and assigning to the reader the easy task of verifying that g_{jk} and the restriction of f_{jk} are ε-close.

Remark. The adjunction space

$$L_j \underset{g_{jk}}{\cup} L_k^{(n)}$$

has a "natural" structure of a finite CW-complex having no cells in dimensions $n+1, n+2, \ldots, n+i$. A consequence (for example see [MS]) is that a compactum X having cohomological dimension $\leq n$ is the limit of a sequence $\{L_q, f_q\}$ satisfying:

(1) each L_q is a finite CW-complex;

(2) for each q, there is an integer $i(q)$ such that $L_q^{(n)} = L_q^{(n+1)} = \cdots = L_q^{(n+i(q))}$ and $\lim_{q\to\infty} i(q) = \infty$; and

(3) for each q, there is an $\varepsilon(q) > 0$ such that $\sup\{\operatorname{diam} f_{\infty q}^{-1}(e): e \text{ is a cell of } L_q\} < \varepsilon(q)$ and $\lim_{q\to\infty} \varepsilon(q) = 0$.

Adopting the "usual" approach to Čech cohomology, it becomes transparent that $\check{H}^r(X, A) = 0$ for $r \geq n+1$ and A a closed subset of X and, consequently, the inclusion induced homomorphism $\check{H}^n(X) \to \check{H}^n(A)$ is surjective.

6. COHOMOLOGICAL DIMENSION AND CELL-LIKE MAPPINGS

We are ready to establish the equivalence of the problems (P_1) and (P_2) stated in the introduction. One direction is a consequence of the classical result of Vietoris while the other is a consequence of a relatively recent result due to R. D. Edwards [Ed].

Theorem. (Vietoris; R. D. Edwards) A compactum has cohomological dimension $\leq n$ if and only if it is the image of a cell-like map defined on a compactum having dimension $\leq n$.

Proof. First, suppose that $f: Z \to X$ is cell-like and $\dim Z \leq n$. If A is a closed subset of X and f_A denotes the restriction of f to $f^{-1}(A)$, then the inequality c-dim $Z \leq n$ (see Theorem 3.2) and the Vietoris Theorem (see Appendix B) yield that, in the commuting square

$$\begin{array}{ccc} [X, K_n] & \xrightarrow{i^\#} & [A, K_n] \\ \downarrow f^\# & & \downarrow f_A^\# \\ [Z, K_n] & \xrightarrow{j^\#} & [f^{-1}(A), K_n] \end{array},$$

the inclusion induced function $j^{\#}$ is surjective and the functions $f^{\#}$ and $f_A^{\#}$ are set isomorphisms. Consequently, the inclusion induced function $i^{\#}$ is surjective and c-dim $X \leq n$.

Second, assume that c-dim $X \leq n$; the remainder of the section is devoted to producing a compactum Z having dimension $\leq n$ and a cell-like map $\pi:Z \to X$. Specify an inverse sequence $\{P_q, f_q\}$ of compact polyhedra with limit X . At the heart of the construction to follow is Theorem 4.2. In order to avoid excessive notation when applying Theorem 4.2, once we have specified an $\varepsilon > 0$, we will assume without mention that the intermediate polyhedra have been dropped from the sequence and that $j = k + 1$. We use $N_\varepsilon(x)$ to denote the closed ε-neighborhood centered at the point x .

Setting Up Notation. Recursively, we specify numbers $0 < \varepsilon(q) < \delta(q)/3 < 1$, triangulations L_q of P_q , and maps $g_q:L_q^{(n+1)} \to L_{q-1}^{(n)}$ satisfying:

(a) for each $x \in P_q$, the inclusion $N_{\varepsilon(q)}(x) \subset N_{\delta(q)}(x)$ is null-homotopic;

(b) g_q and the restriction of f_q are $(\varepsilon(q-1)/3)$-close and, for each $x \in P_q$, diam $[f_q(N_{\delta(q)}(x)] < \varepsilon(q-1)/3$; and

(c) for $i > j$ and $x \in P_i$, diam $[f_{ij}(N_{\varepsilon(i)}(x)] < \varepsilon(j)/2^i$.

The choice of $0 < \delta(0) < 1$ is arbitrary and is followed by choices in order of $\varepsilon(0)$ and L_0 . The general pattern is to choose in order $\delta(q)$, $\varepsilon(q)$, and L_q and, then, to specify g_q .

The Map $\pi:Z \to X$. The restriction of g_q to $L_q^{(n)}$ is denoted $\hat{g}_q$. Set Z equal to the limit of the inverse sequence $\{L_q^{(n)}, g_q\}$. Although $\hat{g}_{\infty q} \neq f_{q+1} \circ \hat{g}_{\infty,q+1}$, M. Brown [Br] has shown that the maps $\hat{g}_{\infty q}:Z \to L_q^{(n)} \subset P_q$ "induce" a map $\pi:Z \to X$ nevertheless. Evidently, dim $Z \leq n$ but it is far from transparent that π is cell-like. In order to determine this we need to describe π in excruciating detail.

Given a point $z = (z_q) \in Z$, we associate a sequence of points in $\prod_{q=0}^{\infty} P_q$

$$x^0 = (z_0, z_1, z_2, \ldots)$$

$$x^1 = (f_1(z_1), z_1, z_2, \ldots)$$

$$x^2 = (f_{20}(z_2), f_2(z_2), z_2, z_3, \ldots)$$

$$\vdots$$

$$x^j = (f_{j0}(z_j), f_{j1}(z_j), \ldots, f_j(z_j), z_j, z_{j+1}, \ldots)$$

$$\vdots$$

and set $\pi(z) = \lim_{j\to\infty} x^j$. In order to have a "global" description of π , we define $\pi_j:Z \to \prod_{q=0}^{\infty} P_q$ by setting $\pi_j(z) = x^j$ and proceed to show that the π_j 's converge uniformly. The critical computations are that

$$(\dagger)\quad \begin{array}{l} f_{j+1}(z_{j+1}) \text{ and } z_j \text{ are } \varepsilon(j)\text{-close and, for } i > j , \\ f_{i+1,j}(z_{i+1}) \text{ and } f_{i,j}(z_i) \text{ are } (\varepsilon(j)/2^i)\text{-close} \end{array}$$

The first is a consequence of $g_{j+1}(z_{j+1}) = z_j$ and Condition (b) that f_{j+1} and g_{j+1} are $(\varepsilon(j)/3)$-close. The second is a consequence of

$f_{i+1,j}(z_{i+1}) = f_{i,j}(f_{i+1}(z_{i+1}))$, the first that $f_{i+1}(z_{i+1})$ and z_i are $\varepsilon(i)$-close, and Condition (c).

Requiring that the metrics d_q for P_q are bounded by 1 and that $d((s_q),(r_q)) = \sum_{q=0}^{\infty} [d_q(s_q, r_q)/2^q]$ is the metric for $\prod_{q=0}^{\infty} P_q$, a direct computation using (†) shows that π_j and π_{j+1} are $1/2^j(\varepsilon(0) + \varepsilon(1)/2 + \cdots + \varepsilon(j-1)/2^{j+1} + \varepsilon(j))$-close (i.e., $(3/2^j)$-close) and, therefore, that the π_j 's converge uniformly.

Letting y_j denote the jth-coordinate of $\pi((z_q))$, we have that $y_j = \lim_{i>j} f_{ij}(z_i)$ and, therefore,

$$f_j(y_j) = \lim_{i>j} f_j(f_{ij}(z_i)) = \lim_{i>j} f_{i,j-1}(z_i) = y_{j-1} ,$$

establishing that $\pi(Z) \subset X$.

<u>Describing</u> $\pi^{-1}(x)$. Let $x = (x_q) \in X$. An easy consequence of Condition (b) is that

$$\hat{g}_q(N_{\delta(q)}(x_q) \cap L_q^{(n)}) \subset N_{\varepsilon(q-1)}(x_{q-1})$$

and, consequently, both $\{N_{\delta(q)}(x_q) \cap L_q^{(n)}, \hat{g}_q\}$ and $\{N_{\varepsilon(q)}(x_q) \cap L_q^{(n)}, \hat{g}_q\}$ are inverse sequences (the $\hat{g}_q$'s being restricted). A second consequence of the above containment is that these inverse sequences have a common limit which we now show is $\pi^{-1}(x)$.

Let $z = (z_q) \in \lim \{N_{\varepsilon(q)}(x_q) \cap L_q^{(n)}, \hat{g}_q\}$. The jth-coordinate of $\pi(z)$ is equal to $\lim_{i>j} f_{ij}(z_i)$. Since $z_i \in N_{\varepsilon(i)}(x_i)$, Condition (c) implies that $f_{ij}(z_i)$ and x_j are $\varepsilon(j)/2^i$-close. Therefore, $\lim_{i>j} f_{ij}(z_i) = x_j$ and $\pi(z) = x$.

For the reverse containment, we start with $z = (z_q) \notin \lim \{N_{\delta(q)}(x_q) \cap L_q^{(n)}, \hat{g}_q\}$ and show that $\pi(z) \neq x$. Choose j such that $z_j \notin N_{\delta(j)}(x_j) \cap L_j^{(n)}$. The inequality $\varepsilon(j) < \delta(j)/3$ assures that $N_{2\varepsilon(j)}(z_j) \cap N_{\varepsilon(j)}(x_j) = \emptyset$. A consequence of (†) is that the jth-coordinate of $\pi(z)$ and z_j are $(\varepsilon(j) + \varepsilon(j)/2^{j+1} + \varepsilon(j)/2^{j+2} + \ldots)$-close ; that is, the jth-coordinate of $\pi(z)$ is contained in $N_{2\varepsilon(j)}(z_j)$ implying that $\pi(z) \neq x$.

<u>Cell-likeness</u>. We need to require more than is specified in Condition (a):

(a)' for each $x \in P_q$, there is a subcomplex U of L_q such that $N_{\varepsilon(q)}(x) \subset U \subset N_{\delta(q)}(x)$ and such that the inclusion $N_{\varepsilon(q)}(x) \subset U$ is null-homotopic. The statement of Theorem 4.2 does not yield the necessary flexibility to assure that L_q can be chosen so that (a)' holds; however, the only constraint on the choice of L_q was that its simplices have diameters $< \varepsilon(q-1)/6$. In order to incorporate Condition (a)', choose a finite collection $\mathcal{U}_q$ of subpolyhedra of P_q , whose interiors cover P_q , having diameters $< \delta(q)$. Then choose $\varepsilon(q)$ and L_q .

Let $x = (x_q) \in X$ and specify subcomplexes U_{q-1} of L_q as in Condition (a)'. It follows easily from the earlier description of $\pi^{-1}(x)$, that $\pi^{-1}(x)$ is the limit of the inverse sequence $\{U_q \cap L_q^{(n)}, \hat{g}_q\}$. Finally, we establish that $\pi^{-1}(x)$ is cell-like by showing that the restriction of $\hat{g}_q \circ \hat{g}_{q+1}$ is null-homotopic as a map from $U_{q+1} \cap L_{q+1}^{(n)}$ to $U_{q-1} \cap L_{q-1}^{(n)}$.

Since $\hat{g}_{q+1}(U_{q+1} \cap L_q^{(n)}) \subset N_{\varepsilon(q)}(x_q)$, the restriction of $\hat{g}_{q+1}$ to $U_{q+1} \cap L_{q+1}^{(n)}$ is null-homotopic as a map into $U(q)$ (see Condition (a)'). Since

$\dim U_{q+1} \cap L_{q+1}^{(n)} \leq n$, this restriction is null-homotopic as a map into $U_{(q)} \cap L_q^{(n+1)}$. Composing such a null-homotopy with g_q (finally, we use the hypothesis c-dim $X \leq n$) , yields the sought after null-homotopy for the restriction of $\hat{g}_q \circ \hat{g}_{q+1}$.

APPENDIX A

We present a proof suggested by S. Ferry that c-dim $X \leq n$ implies c-dim $X \leq n + 1$. It exploits the important relationship that the loop space $\Omega(K_{n+1})$ is a $K(Z, n)$; the latter is a consequence of the equivalence $[\Sigma X, Y] = [X, \Omega Y]$ (see [Sp; p. 42]).

For convenience, we choose K_{n+1} to be a locally finite simplicial complex and specify a base boint k_0 . The path space $P(K_{n+1}, k_0)$ consists of all paths $\omega: I \to K_{n+1}$ with $\omega(0) = k_0$ with the compact-open topology. The evaluation map $\phi: P(K_{n+1}, k_0) \to K_{n+1}$ defined by $\phi(\omega) = \omega(1)$ is a Hurewicz fibration with fibers $K(Z, n)$'s (since $\phi^{-1}(k_0) = \Omega(K_{n+1})$ is a $K(Z, n)$). See [Sp; p. 75].

Since $P(K_{n+1}, k_0)$ is contractible, in order to show that c-dim $X \leq n + 1$ it suffices to exhibit a lift $\tilde{\alpha}: A \to P(K_{n+1}, k_0)$ for each map $\alpha: A \to K_{n+1}$ defined on a closed subset $A \subset X$. The lift $\tilde{\alpha}$ is constructed inductively by repeatedly using the observation that $\phi^{-1}(C)$ is a $K(Z, n)$ for each contractible subpolyhedron C of K_{n+1} .

Choose a locally finite closed cover $\mathcal{U}$ of K_{n+1} consisting of compact, contractible subpolyhedra with the additional property that ever non-empty intersection of elements of $\mathcal{U}$ is contractible. Let $\mathcal{C}$ consists of all non-empty intersections of elements of $\mathcal{U}$ and let $\mathcal{C}$ be partially ordered by set inclusion. Set $M_0 = \{\text{maximal elements of } \mathcal{C}\}$ and, for $q > 1$, set $M_q = \{\text{maximal elements of } \mathcal{C} - \bigcup_{i<q} M_i\}$.

Suppose that c-dim $A \leq n$ and $\alpha: A \to K_{n+1}$. For purposes of induction, assume that a partial lift $\alpha_q: \alpha^{-1}(\cup\{B \in M_0 \cup \cdots \cup M_q\}) \to P(K_{n+1}, k_0)$ has been constructed. For each element $C \in M_{q+1}$, $\phi^{-1}(C)$ is a $K(Z, n)$ and, therefore, α_q can be extended to a map $\alpha_C: \alpha^{-1}(C) \to \phi^{-1}(C)$. Since $\phi \circ \alpha_C$ and the restriction of α to $\alpha^{-1}(C)$ are homotopic relative to $\alpha^{-1}(C \cap (\cup\{\ \in M_0 \cup \ldots \cup M_q\}))$, α_C can be adjusted to yield a lift extending α_q . These extended lifts piece together to produce α_{q+1} and the α_q 's piece together to produce $\tilde{\alpha}$.

APPENDIX B

We present a proof of a strong version of the Vietoris-Begle Mapping Theorem for cell-like maps between compacta. The result is valid for cell-like maps between metric spaces (that is, proper maps having cell-like point inverses) and a proof can be found in unpublished notes of G. Kozlowski; the proof presented here is a variation of proofs in [Koz]. Arguments similar to those used below can be found in greater detail in a set of notes by R. Ancel [An]; the Relative Lifting Property for cell-like maps stated below, but not proved, is a easy consequence of the techniques developed therein.

<u>Theorem</u>. Let P be an absolute neighborhood extensor for which there is an integer n such that $\pi_j(P) = 0$ for $j \geq n + 1$ and let $f:X \to Y$ be a cell-like map between compacta. Then $f^{\#}:[Y, P] \to [X, P]$ is bijective.

We briefly recall a construction from [Du] referring the reader there for additional details. Suppose A is a closed subset of a space Y and u is a locally finite cover of $Y-A$ by open sets and satisfies:

(1) each neighborhood of $a \in FrA$ contains infinitely many elements of u ; and

(2) for any neighborhood W of $a \in A$, there is a neighborhood V , $a \in V \subset W$, such that $\cup\{U \in u : U \cap V \neq \emptyset\} \subset W$.

There is a "natural" topology for $N \cup A$, where N is the nerve of u , such that the identity map on A and a canonical map from $Y-A$ to N combine to produce a map $c:Y \to N \cup A$.

The following property possessed by cell-like maps can be established using techniques found in [La; Section 3], for example.

<u>Relative Lifting Property</u>. Let $f:X \to Y$ be a cell-like map between compacta that is one to one over a closed subset $A \subset Y$ and suppose $X \subset Q$ is embedded in an ANR Q . Given $\varepsilon > 0$ and an integer n , there is a cover u of $Y-A$ and map $c:Y \to N \cup A$ as described above for which there is a map $\gamma:N^{(n)} \cup A \to Q$ such that $\gamma = f^{-1}$ on A and the diagram

$$
\begin{array}{c}
X \hookrightarrow Q \\
{}^{f}\swarrow \qquad \nwarrow^{\gamma} \\
Y \xrightarrow[c]{} N \cup A \hookleftarrow N^{(n)} \cup A
\end{array}
$$

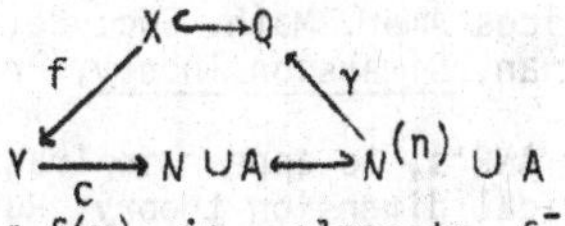

ε-commutes in the sense that $\gamma \circ c \circ f(x)$ is ε-close to $f^{-1}(x)$ whenever $c \circ f(x) \in N^{(n)} \cup A$.

The proof of the theorem consists of three steps and is based on a valuable principle exploited by G. Kozlowski in his work on hereditary shape equivalences [Koz].

First Step. Suppose that $f:X \to Y$ is a cell-like map that is one to one over a closed subset $A \subset Y$ and suppose that $\alpha:A \to P$ is a map and that $\alpha \circ f:f^{-1}(A) \to P$ extends to a map $\beta:X \to P$. We use the Relative Lifting Property to produce an extension $\tilde{\alpha}:Y \to P$ of α .

Specifically, embed X in the Hilbert cube I^{∞} and name an open subset Q containing X so that β extends to $\tilde{\beta}:Q \to P$. Specify an open cover u of $Y-A$ for which there is a map $\gamma:N^{(n)} \cup A \to Q$ with $\gamma = f^{-1}$ on A (any $\varepsilon > 0$ is sufficient). The composition $\tilde{\beta} \circ \gamma$ extends to a "polyhedral" neighborhood of A and, further, extends to $\Gamma:N \cup A \to P$ since $\pi_j(P) = 0$ for $j \geq n + 1$. The composition $\tilde{\alpha} = \Gamma \circ c$ is an extension of α .

Second Step ($f^{\#}$ is surjective). The mapping cylinder $M(f)$ of the map f is viewed as the adjunction space $X \times I \underset{f\times\{1\}}{\cup} Y$ and the induced map $F:X \times I \to M(f)$ is cell-like. We apply the result of the First Step to the map F and the closed subset $X \times \{0\} \subset M(f)$. A map $\alpha:X \to P$ is viewed as a map $\alpha:X \times \{0\} \to P$, the extension β exists trivially, and the resulting extension $\tilde{\alpha}:M(f) \to P$ composes to produce a homotopy $\tilde{\alpha}\circ F$ from α to $(\tilde{\alpha} \mid Y)\circ f$ and, consequently, $f^{\#}([\tilde{\alpha} \mid Y]) = [\alpha]$

Third Step ($f^{\#}$ is injective). The double mapping cylinder $DM(f)$ of f is viewed as the adjunction space $X \times I \underset{f\times\{0,1\}}{\cup} Y \times \{0,1\}$ and there is an induced cell-like map $F:DM(f) \to Y \times I$ (F = identity on $Y \times \{0,1\}$ and agrees with f on each level $X \times \{t\}$ for $0 < t < 1$). We apply the result of the First Step to the map F and the subset $Y \times \{0,1\}$. A pair of maps $\alpha_0,\alpha_1:Y \to P$ produce a map $\alpha:Y \times \{0,1\} \to P$ ($\alpha = \alpha_i$ on $Y \times \{i\}$) and a homotopy between $\alpha_0\circ f$ and $\alpha_1\circ f$ induces an extension $\beta:DM(f) \to P$. The extension $\tilde{\alpha}:Y \times I \to P$ is a homotopy between α_0 and α_1. Consequently, if $f^{\#}([\alpha_0]) = f^{\#}([\alpha_1])$, then $[\alpha_0] = [\alpha_1]$.

REFERENCES

[An] F. D. Ancel, Cell-like maps and the Kozlowski-Walsh Theorem—some alternative proofs, seminar notes, University of Texas, Austin, 1978.

[Be] E. G. Begle, The Vietoris mapping theorem for bicompact spaces, Ann. of Math. (2) 56(1952) 354-362.

[Br] M. Brown, Some applications of an approximation theorem for inverse limits, Proc. Amer. Math. Soc. 11(1960) 478-483.

[Du] J. Dugundji, An extension of Tietze's Theorem, Pacific J. Math. 1(1951) 353-367.

[Ed] R. D. Edwards, A theorem and a question related to cohomological dimension and cell-like maps, Notices Amer. Math. Soc. 25(1978) A-259.

[HW] W. Hurewicz and H. Wallman, Dimension Theory, Princeton University Press, Princeton, N.J., 1941.

[Koz] G. Kozlowski, Images of ANR's, to appear in Trans. Amer. Math. Soc.

[Ku] V. J. Kuźminov, Homological dimension theory, Russian Math. Surveys 23(1968) 1-45.

[La] R. C. Lacher, Cell-like mappings and their generalizations, Bull. Amer. Math. Soc. 83(1977) 495-552.

[MS] S. Mardešić and J. Segal, ε-mappings onto polyhedra, Trans. Amer. Math. Soc. 109(1963) 146-164.

[Na] K. Nagami, Dimension Theory, Academic Press, New York and London, 1970.

[Sp] E. H. Spanier, Algebraic Topology, McGraw-Hill, New York, 1966.

Department of Mathematics
University of Tennessee
Knoxville, Tennessee 37916

EMBEDDING COMPACTA UP TO SHAPE

L.S.Husch and I.Ivanšić

1. Introduction

Shape theorists have developed some techniques which allow one to do homotopy theory on larger class of spaces. It is natural to ask whether these tools are useful in studying the embedding of spaces into Euclidean spaces or, more generally, into manifolds. For example, Borsuk in his book [B] listed a number of problems relating these two theories. Since the 1976 Dubrovnik's post-graduate course "Shape theory and pro-homotopy", many results on this topic have been proved and in this paper, we shall attempt a survey. Some of these results have been published but many are still in preprint form.

We restrict ourselves to compact metric spaces X and say that X can be <u>embedded up to shape in</u> M if there exists a compactum $Y \subseteq M$ such that $\mathrm{Sh}\, X = \mathrm{Sh}\, Y$. First, let us recall the classical embedding theorem for finite dimensional metric compacta:

THEOREM 1.1. (Menger-Nöbeling [H - W]). *Every n-dimensional metric compactum* X *can be embedded into* $(2n+1)$*-dimensional Euclidean space* E^{2n+1}.

Much work in embedding theory has been an attempt to reduce the dimension of the range space. However, Flores [Fl] has constructed for each n an n-dimensional polyhedron which does not embed into E^{2n}.

Now let us recall Stalling's unpublished theorem on embedding of polyhedra up to homotopy type which has turned out to be very useful in studying the embedding up to shape problem.

THEOREM 1.2. (Stallings [St]). *Let* P *be a polyhedron of dimension* n , M *a* PL *manifold without boundary of dimension* q *and* $q-n \geq 3$. *If a map* $f:P \to M$ *is* $(2n-q+1)$*-connected, then there is a subpolyhedron* Q *in* M *of dimension* n *and a* (simple) homotopy equivalence $h:P \to Q$ such that the diagram

$$\begin{array}{ccc} P & \xrightarrow{\ f\ } & M \\ & \searrow^{h} & \uparrow j \\ & & Q \end{array} \qquad (*)$$

commutes up to homotopy, where j denotes the inclusion map.

A map f is r-connected if $\pi_i(f) = 0$ for $i \leq r$, or equivalently if f induces isomorphisms of homotopy groups in dimensions $< r$ and an epimorphism in dimension r. For $q = 2n$ and $M = E^{2n}$ every map $f:P \to E^{2n}$ is 1-connected, so by Theorem 1.2 every n-dimensional polyhedron, $n \geq 3$, can be embedded up to homotopy type in E^{2n}. 1-dimensional polyhedra are clearly embeddable up to homotopy type in E^2; also, 2-dimensional polyhedra are embeddable up to homotopy type in E^4 what is shown in [Cu]. Let us point out that for $n = 2^k$, $k \geq 1$, this dimensional restriction is the best possible; namely, in [P] it is shown that real projective space P^n can not be embedded up to homotopy type into E^{2n-1}.

Since representatives of the shape class of a compactum X may obviously have different dimensions it is natural that theorems on embedding up to shape are more appropriately stated in terms of the shape dimension (or fundamental dimension) which is defined as follow.

$$\mathrm{sd}\, X = \min\{\dim Y \mid \mathrm{Sh}\, X \leq \mathrm{Sh}\, Y\}$$

where $\leq$ denotes the shape domination. But since this minimum is always obtained, we may always assume, if necessary, that a considered representative has the dimension equal to shape dimension.

We shall use the ANR system approach to shape theory which has turned out so far to be more convenient for embedding up to shape type problems. The forthcoming book [M - S] will be a good reference for this subject. In this context we shall usually have X represented as the inverse limit of an inverse sequence of polyhedra and a shape equivalence is described in the following way:

Let $X = \varprojlim \{X_i, p_i\}$ and $Y = \varprojlim \{Y_i, q_i\}$ be inverse sequences of polyhedra; then $\mathrm{Sh}\, X = \mathrm{Sh}\, Y$ if and only if there exist subsequences $\{X_{i(j)}\}$ of $\{X_i\}$ and

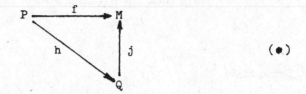

$\{Y_{k(j)}\}$ of $\{Y_k\}$ and maps

$$f_j\colon X_{i(j)} \longrightarrow Y_{k(j)}\ ,\quad g_j\colon Y_{k(j+1)} \longrightarrow X_{i(j)}$$

for each j, such that the diagram

$$\begin{array}{ccc} X_{i(j)} & \xleftarrow{\ p'_j\ } & X_{i(j+1)} \\ {\scriptstyle f_j}\Big\downarrow & \nwarrow{\scriptstyle g_j} & \Big\downarrow{\scriptstyle f_{j+1}} \\ X_{k(j)} & \xleftarrow[\ q'_j\ ]{} & Y_{k(j+1)} \end{array} \qquad (**)$$

is homotopy commutative for each j where

$$p'_j = p_{i(j)} \cdots p_{i(j+1)-1} \quad \text{and} \quad q'_j = q_{k(j)} \cdots q_{k(j+1)-1} \ .$$

Some shape theoretic notions will be defined in the text when needed. Most of the embedding results are in the pointed case, which by ignoring the base point give results in the unpointed case ; but to avoid a complicate notion we shall leave out the base point. Whenever a statement involves conditions on homotopy pro-groups and similar, the base points are assumed.

A natural question raised by these results is whether any n-dimensional continuum embeds up to shape in E^{2n} [B]? For $n = 1$, Borsuk has noted that the solenoid provides a negative answer to this question ; more generally, Borsuk (see [B], pp. 154-160) has shown that a 1-dimensional continuum embeds up to shape in E^2 if and only if it is movable.

The proof of the following embedding up to shape theorem by Ivanšić [I] contains fundamental steps which are used in all other embedding up to shape theorems which follow.

THEOREM 1.3. Let M be a q-dimensional PL manifold and let

$$M \xleftarrow{p_0} X_1 \xleftarrow{p_1} X_2 \xleftarrow{p_2} X_3 \longleftarrow \cdots$$

be an inverse sequence where all X_i, $i = 1,2, \ldots$, are polyhedra of dimension $\leq n$, and all bounding maps p_i $(i = 0,1,2,\ldots)$ are $(2n-q+1)$-connected. If $q-n \geq 3$, then the $\varprojlim \{X_i, p_i\}$ embeds up to shape into M.

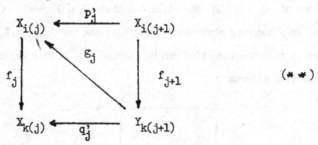

Proof. The proof uses Theorem 1.2. in the inductive step. Namely, by Theorem 1.2. there is a polyhedron $Q_1 \subset M$ and a homotopy equivalence $h_1 : X_1 \to Q_1$ such that the appropriate diagram (✱) commutes up to homotopy. If N_1 is a regular neighborhood of Q_1 in M one still has that $j_1 h_1 : X_1 \to N_1$ ($j_1 : Q_1 \hookrightarrow N_1$, the inclusion) is a homotopy equivalence, and that the map $j_1 h_1 p_2 : X_2 \to N_1$ is $(2n-q+1)$-connected, so that the induction can be carried on. In this way one obtains the homotopy commutative diagram

$$\begin{array}{ccccccc} & & X_1 & \xleftarrow{\;p_1\;} & X_2 & \leftarrow & \cdots \\ & & \downarrow & & \downarrow & & \\ M & \hookleftarrow & N_1 & \hookleftarrow & N_2 & \leftarrow & \cdots \end{array}$$

in which each vertical arrow is a homotopy equivalence, so $\mathrm{Sh}\,(\varprojlim \{X_i, p_i\}) = \mathrm{Sh}(\bigcap_i N_i)$.

Note that $p_0 : X_1 \to M$ induces a map $f : X = \varprojlim X_i \to M$ $(f_i = p_0 p_1 \cdots p_{i-1} : X_i \to M)$ or a shape morphism $\underline{f} : X \to M$, and a shape equivalence $\underline{h} : X \to Y = \bigcap_i N_i$ such that the following diagram

$$\begin{array}{ccc} X & \xrightarrow{\;\underline{f}\;} & M \\ & \underline{h} \searrow & \uparrow \underline{j} \\ & & Y \end{array}$$

commutes in the shape category, where $\underline{j}$ is the shape morphism induced by inclusion.

From the above proof is clear that the following holds.

THEOREM 1.4. *Let M be a q-dimensional PL manifold and*

$$X_1 \xleftarrow{\;p_1\;} X_2 \xleftarrow{\;p_2\;} X_3 \leftarrow \cdots .$$

an inverse sequence of polyhedra such that if whenever $f_i : X_i \to M$ is an embedding and U_i is a neighborhood of $f_i(X_i)$ then $f_i p_i : X_{i+1} \to U_i$ is homotopic in U_i to an embedding. Then $\varprojlim \{X_i, p_i\}$ embeds up to shape in M.

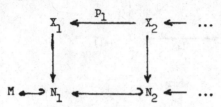

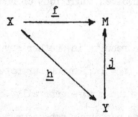

2. Borsuk's Index e(X)

Borsuk in [B] has defined the index e(X), which we call Borsuk's index, as follows:

If sd (X) $< \infty$, then e(X) is the minimum of the numbers n such that E^n contains Y with Sh X $\leq$ Sh Y . If sd X $= \infty$, then e(X) $= \infty$. It follows from the definition of e(X) that

$$\mathrm{Sh}(X) \leq \mathrm{Sh}(Y) \text{ implies } e(X) \leq e(Y) .$$

Theorem 1.1. implies that

$$e(X) \leq 2 \, \mathrm{sd}(X) + 1 .$$

The first question concerning e(X) is the following : If e(X) $\leq$ n can X be embedded up to shape into E^n ?

EXAMPLE 2.1. The negative answer to this question was given by Kadlof [K] . Let Q be polyhedron in E^3 obtained from a 3-cube I^3 with knotted hole as shown in the figure. Q has the homotopy type of a finite 2-dimensional polyhedron such that $\pi_1(Q)$ is a group with the presentation $(a,b;\ a^2 = b^3)$ and

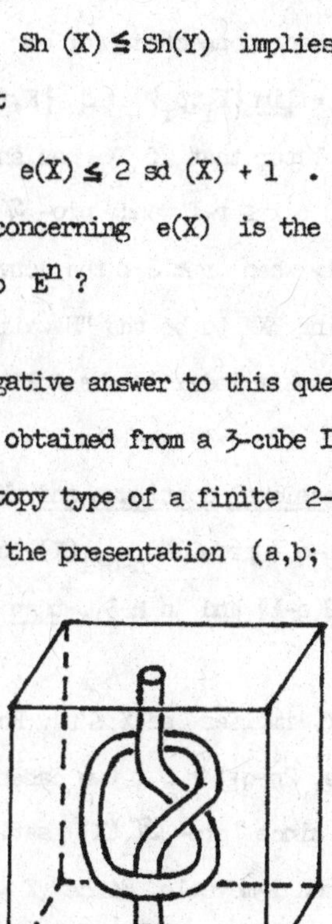

$\pi_i(Q) = 0$ for $i \geq 2$. Then $Z = Q \vee S^2 \vee S^2$ can be embedded in E^3 . But from Dunwoody [D] it follows that there are two finite 2-dimensional complexes X and Y such that $\pi_1(X) = \pi_1(Y) = G$, $\pi_2(X) = ZG$, $X \not\simeq Y$, but $X \vee S^2 \simeq Y \vee S^2$. The theory of the 2-type of complexes shows that $X \simeq Q \vee S^2$. Therefore $Y \vee S^2 \simeq Q \vee S^2 \vee S^2 = Z$ and since Y is a retract of $Y \vee S^2$ one obtains Sh (Y) $\leq$ Sh (Z). Since $Z \subset E^3$

we have $e(Y) \leq 3$. Assuming now that there is a continuum $P \subset E^3$ such that $Sh(P) = Sh(Y)$ one infers from the classification of pointed compact FANR's in E^3 (see [K] for the relevant references), that $X \simeq Y$, the contradiction.

On the other hand, Husch and Ivanšić $[H - I]_1$ have found that is possible to put some additional conditions on X under which one can obtain a positive answer to the above question. These conditions include that X is a continuum which has <u>shape finite</u> r-<u>skeleton</u> $(r \geq 1)$ i.e. there exists a finite pointed CW-complex K and a shape morphism $\underline{f}:K \to X$ which is r-shape connected (for our purposes we think of $\underline{f}$ as $\{f_i\} : \{K,id\} \to \{X_i,p_i\}$ where $X = \varprojlim \{X_i,p_i\}$ and $\{K,id\}$ is the trivial inverse sequence). It has been noticed later that if X has shape finite r-skeleton then X has stable pro- $\pi_i(X)$ for $0 \leq i \leq r-1$ and pro- $\pi_r(X)$ satisfies the Mittag-Leffler condition, and conversely when $r \geq 2$; the converse does not hold when $r = 1$, as it can be seen by taking X to be the "Hawaiian earring". The main result of $[H - I]_1$ concerning the Borsuk's index is the following theorem which was originally stated in term of the finite (2n-q+1)-skeleton.

THEOREM 2.2. <u>Let</u> $Y \subset E^q$ <u>be an n-dimensional continuum and</u> X <u>a continuum which has stable</u> pro- $\pi_i(X)$ <u>for</u> $0 \leq i \leq 2n-q$, pro- $\pi_{2n-q+1}(X)$ <u>satisfies Mittag-Leffler condition and</u> $Sh\,X \leq Sh\,Y$. <u>If</u> $3n < 2(q-1)$ <u>and</u> $n \geq 3$, <u>then</u> X <u>embed up to shape in</u> E^q .

<u>Outline of the proof</u>. $Sh\,X \leq Sh\,Y$ implies $sd\,X \leq n$. For $q \geq 2n+1$ Theorem 1.1. implies $e(X) \leq q$. For $q < 2n+1$, i.e. $2n-q+1 > 0$, two cases are considered. When $2n-q+1 = 1$ the proof is straighorward since pro- $\pi_1(X)$ satisfies the Mittag-Leffler condition, or equivalently X is pointed 1-movable. Since $X = \varprojlim \{X_i,p_i\}$ such that all p_i are 1-connected (see [Kr]). Theorem 1.3. implies that X embeds up to shape in E^q . The case $2n-q+1 \geq 2$ can be seen in the following way. First one proves that there is an inverse sequence $\{X_i,p_i\}$ of CW-complexes of dimension $\leq n$ such that $Sh\,X = Sh \varprojlim \{X_i,p_i\}$ and all bonding maps $p_i : X_{i+1} \to X_i$ are (2n-q+1)-connected (See [F]). Let $\{U_i\}$ be a sequence of PL neighborhoods of Y such that $Y = \bigcap_i U_i$. Since $Sh\,X \leq Sh\,Y$, there exist shape morphisms $\underline{f} = \{f_i\} : \{X_i,p_i\} \to \{U_i, \text{inclusions}\}$ and $\underline{g} = \{g_i\} : \{U_i, \text{inclusions}\} \to \{X_i,p_i\}$ such that $\underline{g}\,\underline{f} = \underline{id}_X$. By choosing subsequences we may assume that we have the following diagram

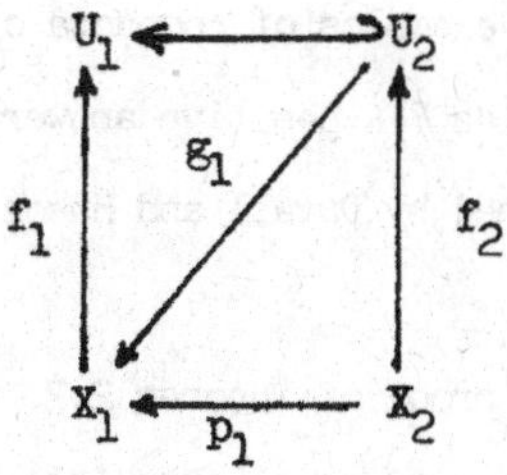

where the square and the right triangle are homotopy commutative. Now the rest of the proof consists in adding embedded handles to U_2 obtaining $V \subset E^q$ such that

$$X_2 \xrightarrow{f_2} U_2 \hookrightarrow V$$

is (2n-q+1)-connected (see $[H - I]_1$ for the argument). By Theorem 1.3. X embeds up to shape into $V \subset E^q$.

COROLLARY 2.3. Let $Y \subset E^q$ be an n-dimensional continuum and let X be a continuum which has the shape of finite complex and $\mathrm{Sh}\, X \leq \mathrm{Sh}\, Y$. If $3n < 2(q-1)$ and $n \geq 3$, then X embeds up to shape in E^q.

QUESTION. Is it true that for every X $e(X) \leq 2\, \mathrm{sd}\, (X)$ holds ?

3. Shape Embedding Index se (X)

The shape embedding index we define analogously as Borsuk's index as follows:

If $\mathrm{sd}\,(X) < \infty$, then se (X) is the minimum of the numbers n such that E^n contains Y with $\mathrm{Sh}\,(X) = \mathrm{Sh}\,(Y)$. If $\mathrm{sd}\, X = \infty$, then $\mathrm{se}\,(X) = \infty$.

From the definition, it follows that

$$e(X) \leq \mathrm{se}\,(X)\,. \qquad (1)$$

Kadlof's example, described in Section 2, tells that strict inequality holds for some X.

The following inequality follows from Theorem 1.1.

$$\mathrm{se}\,(X) \leq 2\,\mathrm{sd}\,(X) + 1\,, \qquad (2)$$

The 1-dimensional solenoid A, is an example where equality is achieved. A natural question is whether there exist examples of compacta of higher shape dimension for which the equality in (2) holds ? A positive answer for shape dimensions of the form 2^k, $k > 1$, has been obtained by Duvall and Husch $[D - H]_1$. We shall describe this example below.

It has been already noticed in the proof of Theorem 2.2. that if X is pointed 1-movable, $\mathrm{sd}\, X \leq n$, $n \geq 3$ then $\mathrm{se}\,(X) \leq 2\, \mathrm{sd}\,(X)$ (See [I]) . So in order to obtain the equality in (2) X can not be pointed 1-movable. For the sake of completeness we quote the definition of pointed r-movability in the used form :

X is pointed r-movable if given an inverse sequence $\{(X_i,x_i),p_i\}$ such that $\varprojlim \{(X_i,x_i),p_i\} = (X,x)$ and given i ,then there exists $j \geq i$ such that if $f:(K,y) \to (X_j,x_j)$ is a mapping of a pointed finite CW-complex of dimension $\leq r$, then for all $k \geq i$ there exists $g:(K,y) \to (X_k,x_k)$ such that f and g are homotopic in (X_i,x_i) when composed with the appropriate bonding maps .

EXAMPLE 3.1. Let $n = 2^k$, $k > 1$. Start with the one-sphere $A_1 = S^1$ and let $X_1 = S^1 \vee P^n$ be the wedge of S^1 and n-dimensional real projective space P^n . Consider the retraction $\beta_1 : X_1 \to A_1$ which sends P^n to the wedge point. Let $A_2 = S^1$ and let $\alpha_1 : A_2 \to A_1$ be the double cover. Then the pull-back construction for β_1 gives us X_2 , a double cover $r_1 : X_2 \to X_1$ and a retraction $\beta_2 : X_2 \to A_2$. Inductively, for $A_i = S^1$ and double cover $\alpha_i : A_{i+1} \to A_i$, and pull-back construction as in the first step we obtain the following two inverse sequences

$$\begin{array}{ccccccc} X_1 & \xleftarrow{r_1} & X_2 & \xleftarrow{r_2} & X_3 & \leftarrow & \cdots \\ \downarrow{\scriptstyle \beta_1} & & \downarrow{\scriptstyle \beta_2} & & \downarrow{\scriptstyle \beta_3} & & \\ A_1 & \xleftarrow{\alpha_1} & A_2 & \xleftarrow{\alpha_2} & A_3 & \leftarrow & \cdots \end{array}$$

and a shape morphism $\underline{\beta}$ between them. Let $X = \varprojlim \{X_i, r_i\}$ and $A = \varprojlim \{A_i, \alpha_i\}$. Then $\underline{\beta}$ is obviously a shape retraction ; since the selenoid A is not pointed 1-movable, X is also not pointed 1-movable. It is proved in $[D - H]_1$ that X is not embeddable up to shape into E^{2n} . The proof is based on the facts that in this case no differentiable embedding $f:P^n \to E^{2n}$ admits a normal field (non-zero cross section of the normal bundle) and that $A \subset X$ is not pointed 1-movable. As a consequence, we also have that sd (X) = n . In [H - D]$_2$ there are constructed some other examples of compacta for $n = 2^k$, $k > 1$, which cannot be embedded up to shape in E^{2n} . Up to now we have examples in dimensions $n = 1$ and $n = 2^k$, $k > 1$, so we still may ask the following question.

QUESTION [B]. Is it true that for every n = 1,2,3,.... there exists a space X (or a compactum) such that se (X) = 2 sd (X) + 1 ?

In the rest of Section we present results which are obtained concerning the index se (X) . Some of the proofs require considerable argument which involves the problem of embedding finite covers into vector bundles and resolving double point singularities (Whitney's trick), so at such places we shall quote the reference where the proof can be found. Some of these technique lead to embedding theorems of some n-dimensional continua into E^{2n} . These theorems may be of some interest in embedding theory in general. If a continuum X is the inverse limit of compact connected n-manifolds then we call it <u>n-manifold-like continuum</u>. The shape dimension of an n-manifold like continuum is $\leq n$. We now state the following theorem which gives sufficient conditions for se (X) $\leq$ 2 sd (X) .

THEOREM 3.2. (a) X <u>is pointed 1-movable continuum and</u> sd Y $\neq$ 2 .

(b) X <u>is an n-manifold-like continuum</u>, $X = \varprojlim \{M_i, p_i\}$, $n \geq 3$, <u>and one of the following conditions is satisfied</u> :

(i) <u>for infinitely many</u> i's , $p_{i*}\pi_1 M_i$ <u>has infinite index in</u> $\pi_1 M_{i-1}$;

(ii) <u>for</u> n <u>even</u>, $n \neq 2^k$ <u>for any</u> k , <u>each</u> p_i <u>is a covering map</u> ;

(iii) <u>for</u> n <u>even and for infinitely many</u> i's , M_i <u>is orientable</u> ;

(iv) <u>for</u> n <u>odd and</u> $\varprojlim^1 \{\pi_1(M_i)^{(2)}\} = 0$ <u>where</u> $\pi_1(M_i)^{(2)}$ <u>is the subgroup of</u> $\pi_1(M_i)$ <u>generated by elements of order two</u> ;

(v) <u>for</u> n <u>odd each</u> p_i <u>is a regular covering</u>.

<u>If any one of these conditions is satisfied then</u> se (X) $\leq$ 2 sd (X) .

<u>Comments of proofs.</u> (a) The case sd (X) ≥ 3 is described above, so only the case sd (X) ≤ 1 , deserves some comments. It follows from [T] where a classification of pointed movable continua of sd ≤ 1 is given. Since pointed 1-movable continua are pointed movable the classification applies. The case sd (X) = 2 is still unknown.

We only point out that if X is ponited 1-movable and $\text{sd}\, X = 2$ then X is the inverse limit of polyhedra of dimension ≤ 2 with bonding maps that induce epimorphisms of fundamental groups (see [Kr]).

(b) By Theorem 2.2, it suffices to consider the case when for infinitely many i's $p_{i*}\pi_1 M_i$ is a proper subgroup of $\pi_1 M_{i-1}$. Let $q_i : \tilde{M}_i \to M_i$ be the covering space corresponding to $p_{i*}\pi_1 M_{i+1}$; by covering space theory there exists a lift $g_i : M_{i+1} \to \tilde{M}_i$. If the index of $p_{i+1*}\pi_1 M_{i+1}$ in $\pi_1 M_i$ is infinite, $\tilde{M}_i$ is an open manifold and, hence, $\tilde{M}_i$ deformation retracts onto a $(n-1)$-polyhedron. It is not difficult to show that in case (i) that $\text{sd}\,(X) = n-1$ and, hence, $\text{se}\,(X) \leq 2n-1$. Therefore it suffices to consider the case when for each i, $p_{i*}\pi_1 M_i$ has finite index in $\pi_1 M_{i-1}$. Let q_i and g_i be as above ; we shall attempt to apply Theorem 1.4. Suppose M_i is embedded in E^{2n} and let U_i be a normal bundle for M_i. If q_i can be homotoped to an embedding q_i' in U_i, then $q_i' g_i$ is a 1-connected map of M_{i+1} into a sufficiently small neighborhood of $q_i'(\tilde{M}_i)$ in U_i. But the Whitney trick allows us to homotope $q_i' g_i$ to an embedding in U_i. Hence, one is led to the problem of embedding finite covers into bundles. As mentioned above, the arguments are quite involved and the reader is referred to $[D - H]_2$ for details.

Let us state one more theorem concerning the shape embedding index which tells us when $\text{se}\,(X)$ can be smaller than $2\,\text{sd}\,(X)$. We say that a continuum X is <u>r-shape connected</u> if its homotopy pro-groups vanish in dimensions $\leq r$, i.e. pro-$\pi_i(X) \cong 0$ for $i \leq r$. That X has <u>stable shape</u> means that it has the shape of a polyhedron P.

THEOREM 3.3. (a) *If X is an r-shape connected continuum and $\text{sd}\,(X) - r \geq 3$, then* $\text{se}\,(X) \leq 2\,\text{sd}\,(X) + 1 - r$;

(b) *If X is an r-shape connected continuum which is $(r+1)$-pointed movable and $\text{sd}\,(X) \geq 3$, then* $\text{se}\,(X) \leq 2\,\text{sd}(X) - r$.

<u>Comments of proofs</u>. (a) First one shows that $X = \varprojlim \{X_i, p_i\}$ where all X_i are r-connected polyhedra (see [I]). Since every mapping of X_1 into Euclidean space E^q is r-connected one may choose q to be $2\,\text{sd}\,(X) + 1 - r$ and see that Theorem 1.3. applies.

(b) This result was proved in $[H - I]_2$ and is a corollary of a generalization of Theorem 1.2. in the shape category, which is proved in Section 5. If $r = 0$ one gets part (a) of Theorem 3.2. as a special case.

QUESTION [B] . Is it true that es $(X) \leq 2$ sd X if X is movable ?

The answer is positive if sd $(X) \leq 1$, as we have already commented on this in Theorem 3.2.

A general question that may be posed concerning the shape embedding index is to find some other sufficient conditions which imply se $(X) \leq 2$ sd (X) .

4. Continua in E^n of Stable Shape

Let $X \subset E^q$ be a compactum which has the stable shape, i.e. the shape of a finite polyhedron K . Here we wish, under some restrictions, give the positive answer to the following question :

Is there an ANR $L \subset E^q$ such that Sh X = Sh L .

The positive answer to this question was first given in $[H - I]_2$ under the assumption that dim $X \leq 2/3\ (q-1)$ and dim $K \geq 3$ (see also Corollary 2.3.).

First we state the following theorem from [V] which was obtained there as a corollary of the main approximation theorem (see also Venema's paper in these proceedings).

THEOREM 4.1. _If_ $X \subset E^q$ _has the shape of a finite polyhedron_ K _of dimension_ $k \leq q-3$, _then there exists a polyhedron_ $K' \subset E^q$ _such that_ dim $K' = k$ _and_ K' _has the simple homotopy type of_ K .

S.Ferry has an example which shows that Theorem 4.1. cannot be extended to the case when $k = q-1$.

In order to state the next theorem which was proved in [I - S] we need the notion of relative shape introduced in [C] . Since we need only the notion of relative shape equivalence when the ambient space is a PL manifold (which is an ANR) it can be defined in this case equivalently as:

Let X and Y compacta, M an ANR and $X,Y \subset M$. X and Y have the same *relative shape* in M if there exists a shape equivalence $\underline{f}:X \to Y$ such that the following diagram

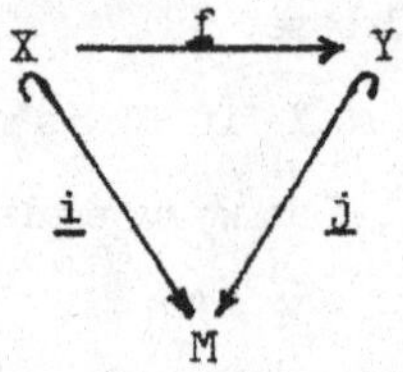

(1)

commutes (in the shape category), where $\underline{i}$ and $\underline{j}$ are shape morphisms induced by inclusions.

THEOREM 4.2. *Let* X *be a compactum in the interior of a* PL *q-manifold* M , K *be a k-dimensional polyhedron, and* $\underline{f}:K \to X$ *be a shape equivalence, where* $q \geq 5$ *and* $k \leq q-3$. *Then for each neighborhood* U *of* X *in* M *there exists a k-dimensional polyhedron* $K' \subset U$ *and a simple homotopy equivalence* $h:K \to K'$ *so that the following diagram*

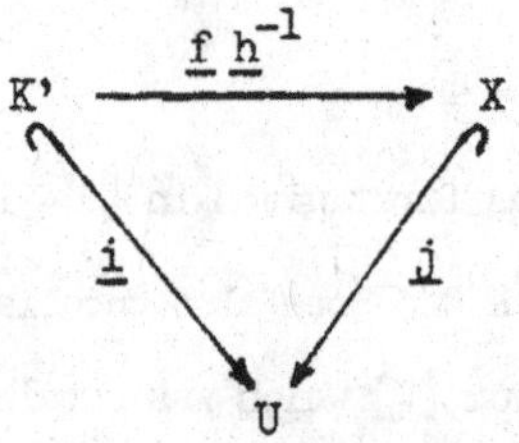

(2)

commutes, where $\underline{i}$ *and* $\underline{j}$ *are shape morphisms induced by inclusions, or in other words,* K' *and* X *have the same relative shape in* U *which we also say that is induced by* $\underline{f}\,\underline{h}^{-1}$ ($\underline{h}^{-1}$ is the shape equivalence induced by the inverse of h).

Outline of the proof. By the approximation theorem of [V] there exists $X' \subset \operatorname{int} M$ such that X' is ILC embedded (ILC stands for "inessential loop condition" see e.g. R.B. Sher these proceedings) and such that X and X' have the same relative shape in U , obtaining also the information on the shape equivalence in the diagram (1). Then a technical lemma, which we omit, applies to X' obtaining a smaller neighborhood V , a k-dimensional polyhedron K' and a simple homotopy equivalence $h:K \to K'$ such that diagram (2) , in which U is replaced by V , commutes. But a relative shape equivalence in V is clearly a relative shape equivalnece in U .

5. Generalization of Stallings Embedding Theorem in the Shape Category

The aim of this Section is to prove the generalization of Theorem 1.2. in the shape category. In this sense one would like to have a commutative diagram of the form (*) in the statement of Theorem 1.2. but in the shape category, where h should be replaced by a shape equivalence $\underline{h}$, f by a shape morphism $\underline{f}$ having a property which generalizes the r-connectedness of a map and the inclusion j is replaced by the shape morphism $\underline{j}$ induced by it . A shape morphism $\underline{f}:X \to Y$ is r-<u>shape connected</u> if it induces an isomorphism of homotopy pro-groups pro-$\pi_j(X)$ and pro-$\pi_j(Y)$ in dimensions $1 \leq j < r$ and an epimorphism in dimension r . Equivalently, $\underline{f}:X \to Y$ is r-shape connected if pro-$\pi_j(\underline{f})$ is trivial for $1 \leq i \leq r$. We shall now describe the homotopy pro-groups pro-$\pi_j(\underline{f})$ of a shape morphism $\underline{f}$. They are introduced in an analogous way as the homotopy groups $\pi_j(f)$ of a map f . Namely, if $\{X_i,p_i\}$ and $\{Y_i,q_i\}$ are pointed inverse sequences (or objects in pro-ANR_o) and $\{f_i\} : \{X_i\} \to \{Y_i\}$ is a morphism in pro-$HANR_o$ (we may assume that we have homotopy commutative diagrams of the form

$$\begin{array}{ccc} X_i & \xleftarrow{\ p_i\ } & X_{i+1} \\ \downarrow{\scriptstyle f_i} & \simeq & \downarrow{\scriptstyle f_{i+1}} \\ Y_i & \xleftarrow[\ q_i\ ]{} & Y_{i+1} \end{array}$$

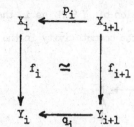

for every i and that homotopy preserves the base points), then one constructs the inverse sequence $\{M(f_i),\ \lambda_i\}$, where $M(f_i)$ is the reduced mapping cylinder of f_i and $\lambda_i:M(f_{i+1}) \to M(f_i)$ is constructed from the bonding maps $p_i:X_{i+1} \to X_i$ $q_i:Y_{i+1} \to Y_i$ and a homotopy between f_ip_i and q_if_{i+1} (see [M], Part I for a description). One obtains an inverse sequence of groups $\{\pi_j(M(f_i),X_i),\ \lambda_i\}$ which is called the jth homotopy pro-group of the shape morphism $\underline{f}$ and denoted by pro-$\pi_j(\underline{f})$.

The following generalization of Theorem 1.2. was proved in $[H - I]_2$:

THEOREM 5.1. Let X be a continuum of shape dimension $sd(X) = k$ which is pointed $(2k-q+1)$-movable. If there exists a $(2k-q+1)$-shape connected shape morphism $\underline{f}:X \to M$, of X into a q-dimensional PL manifold and if $q-k \geq 3$, then there exists a k-dimensional continuum $Y \subsetneq M$ and a shape equivalence $\underline{h}:X \to Y$ such that $\underline{j}\,\underline{h} = f$ where $\underline{j}$ is the shape map induced by the inclusion $Y \hookrightarrow M$.

The proof of Theorem 5.1. was based on the following theorem:

THEOREM 5.2. Let $\underline{f}:X \to K$ be an r-shape connected shape morphism of a pointed r-movable continuum X to a pointed CW-complex K . Then $X = \varprojlim \{X_i, p_i\}$ where all the bonding maps p_i are r-connected. Furthermore, if $sd\,(X) > r$, then X_i can be chosen such that $\dim X_i \leq sd\,X$ for all i .

This theorem has been proved in $[H - I]_2$. Its hypotheses imply that the homotopy pro-groups pro-$\pi_j(X)$ are stable (i.e. pro-$\pi_j(X)$ is isomorphic to a group) for $j < r$, so one can find a simpler proof of it in [F] (Theorem 4).

Proof of Theorem 5.1. Let $\{X_i, p_i\}$ be the inverse sequence. Then $\underline{f}$ is represented by a sequence of maps $\{f_i\}$, $f_i:X_i \to K$ such that $f_i p_i = f_{i+1}$ (a shape morphism into an ANR is always representated by a map), and we may assume that f_1 is $(2k-q+1)$-connected (since each p_i is $(2k-q+1)$-connected so is each f_i). Now, one can perform the inductive construction of $Y \subsetneq M$ as in the proof of Theorem 1.3. This gives us a shape equivalence and the commutativity in the shape categroy of the diagram

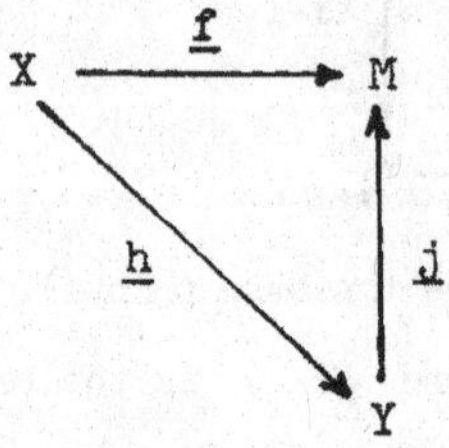

where $\underline{j}$ is the shape morphism induced by the inclusion $Y \hookrightarrow M$.

If $M = E^q$ then every shape morphism $\underline{f}:X \to E^q$ induces epimorphisms of all homotopy pro-groups, so part (b) of Theorem 3.3. follows from Theorem 5.1.

For each n even >2, there is an example described in $[H-I]_2$ which shows that the hypothesis that X be (2k-q+1)-movable cannot be removed. The above mentioned example from $[D-H]_1$ provides an alternate example displaying this property.

References

[B] K.Borsuk, Theory of shape, Mathematical Monographs Vol. 59, Polish Scientific Publishers, Warsaw 1975.

[C] T.A.Chapman, On some applications of finite-dimensional manifolds to the theory of shape, Fund. Math. 76 (1972), 181-193.

[Cu] M.L.Curtis, On 2-complexes in 4-spaces, Topology of 3-manifolds and related topics, Prentice Hall, 1962.

[D] M.J.Dunwoody, The homotopy type of two-dimensional complex, Bull. London Math. Soc. 8 (1976), 282-285.

$[D-H]_1$ P.F.Duvall and L.S.Husch, A continuum of dimension n which does not embedd up to shape in 2n-space, to appear in the Proceedings of 1978 Warsaw Topology Conference.

$[D-H]_2$ P.F.Duvall and L.S.Husch, Embedding finite covers into bundles with an application to embedding manifold-like continua up to shape, preprint.

[D-S] J.Dydak and J.Segal, Shape theory, An Introduction, Lecture Notes in Math. No 688, Springer-Verlag, Berlin-Heidelberg-New-York 1978.

[E-D] D.A.Edwards and R.Geoghegan, Shapes of complexes, ends of manifolds, homotopy limits and the Wall obstruction, Ann. Math. 101(1975), 521-535. Correction to "Shapes .... ", Ann. Math. 104(1976), 389.

[F] S.Ferry, A stable converse to the Vietoris-Smale theorem with applications to shape theory, Trans. Amer. Math. Soc. 261 (1980), 369-386.

[Fl] A.Flores, Über die Existenz n-dimensionaler Komplexe die nicht in den R_{2n} topologisch einbettbar sind, Ergebnisse eines Math. Kolloquium 5 (1932-33), 17-24.

[H] J.F.P.Hudson, Piecewise linear topology, W.A.Benjamin, Inc., New York, 1969.

$[H-I]_1$ L.S.Husch and I.Ivanšić, Shape domination and embedding up to shape, Compositio Math. 40(1980), 153-166

$[H-I]_2$ L.S.Husch and I.Ivanšić, Embeddings and concordances of embeddings up to shape, Preprint, University of Zagreb 1977.

[H-W] W.Hurewicz and H.Wallman, Dimension Theory, Princeton University Press,

Princeton 1948.

[I] I.Ivanšić, Embedding compacta up to shape, Bull. Acad.Polon.Sci.Sér.Sci.Math. Astronom. Phys. 25(1977), 471-475.

[I-S] I.Ivanšić and R.B.Sher, A complement theorem for continua in a manifold, preprint.

[K] A.Kadlof, An example resolving Borsuk's problem concerning the index e(X), Bull.Acad.Polon.Sci.Sér.Sci.Math.Astronom.Phys. 26(1978),9o5-9o7.

[Kr] J.Krasinkiewicz, Continuous images of continua and 1-movability, Fund.Math. 98(1978), 141-164.

[M] S.Mardešić, On the Whitehead theorem in shape theory I ; II , Fund. Math. 91 (1976), 51-64 ; 93-1o3.

[Mi] J.W.Milnor, Characteristic classes, Princeton Univ.Press, Princeton 1974.

[M-S] S.Mardešić and J.Segal, Shape Theory, ANR System Approach, In preparation.

[P] P.Peterson, Some non-embedding problems, Bol.Soc.Mat.Mexicana 2(1957), 9-15.

[St] J.Stallings, The embedding of homotopy types into manifolds, Mimeo notes, Princeton University 1965.

[T] A.Trybulec, On shape of movable curves, Bull.Acad.Polon.Sci.Sér.Sci.Math. Astronom.Phys. 21(1973), 727-733.

[V] G.A.Venema, An approximation theorem in the shape category, preprint.

University of Tennessee
Knoxville, TN 37916

University of Zagreb
Zagreb, Yugoslavia

ON SHAPE CONCORDANCES

L.S.Husch and I.Ivanšić

In [6], Stallings showed that if M is a compact PL q-dimensional manifold, if X and Y are compact connected n-dimensional subpolyhedra of M such that the inclusion $X \hookrightarrow M$ is $(2n-q+2)$-connected and if there exists a simple homotopy equivalence $f:X \to Y$ such that

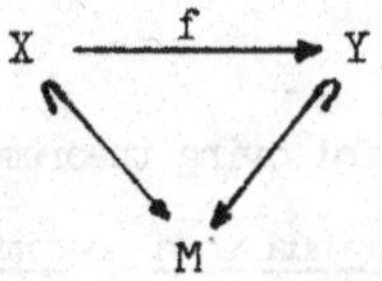

is homotopy commutative, then there exists a $(n+1)$-dimensional subpolyhedron L of $M \times [0,1]$ such that $L \cap (M \times \{0\}) = X \times \{0\}$, $L \times (M \times \{1\}) = Y \times \{1\}$, $X \times \{0\} \hookrightarrow L$ and $Y \times \{1\} \hookrightarrow L$ are simple homotopy equivalences and

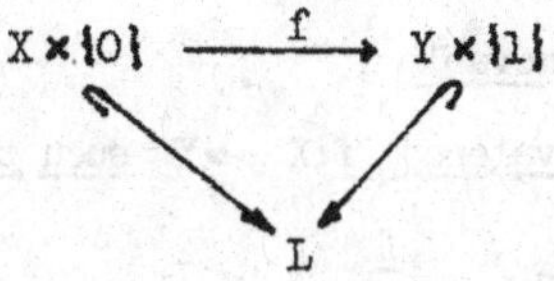

is homotopy commutative. L is said to be a <u>concordance</u> <u>between</u> X <u>and</u> Y <u>in</u> M <u>rel</u> f.

In this paper, we extend Stallings' results to the shape category.

DEFINITION. Suppose $X,Y \subseteq M$ are continua in a manifold M and suppose $\underline{f}:X \to Y$ is a shape equivalence. X is shape <u>concordance</u> <u>to</u> Y <u>relative</u> $\underline{f}$ if there exists a continuum $L \subseteq M \times [0,1]$ such that

(i) $L \cap (M \times \{0\}) = X \times \{0\}$;

(ii) $L \cap (M \times \{1\}) = Y \times \{1\}$;

(iii) $X \times \{0\} \hookrightarrow L$ and $Y \times \{1\} \hookrightarrow L$ are shape equivalences;

(iv) $$\begin{array}{ccc} X \times \{0\} & \xrightarrow{f} & Y \times \{1\} \\ & \searrow \quad \swarrow & \\ & L & \end{array}$$ is commutative in the shape category.

Recall [8] that $X \subseteq M$ satisfies ILC (inessential loops condition) if, given a neighborhood U of X in M, there exists a neighborhood V of X in $M, V \subseteq U$, such that any loop in $V - X$ which is inessential in V is inessential in $U - X$. If $X \subseteq E^q$ satisfies the more familiar 1-ULC condition for embeddings, then X satisfies SLC (small loops condition). Also, if $\dim X \leq q-2$ then ILC is equivalent to SLC [8].

The main result of this paper is the following theorem.

THEOREM 1. *Suppose* $X,Y \subseteq M$ *are continua in a* q-*dimensional* PL *manifold such that*

1.1. X *and* Y *satisfy* ILC *in* M;

1.2. *shape (fundamental) dimension* $\mathrm{sd}(X) = k \leq q-3$;

1.3. $X \hookrightarrow M$ *is* $(2k-q+2)$-*shape connected*;

1.4. X *is pointed* $(2k-q+2)$-*movable*;

1.5. *there exists a shape equivalence* $f: X \to Y$ *such that*

$$\begin{array}{ccc} X & \xrightarrow{f} & Y \\ & \searrow \quad \swarrow & \\ & M & \end{array}$$

is commutative in the shape category.

Then X *and* Y *are shape concordant relative to* f.

COROLLARY 2. *Suppose* $X,Y \subseteq E^{2k-r+1}$ *are continua such that* X *and* Y *satisfy* ILC *in* E^{2k-r+1}, X *and* Y *have the same shape, dimension* $\mathrm{sd}(X) = k \leq 2k-r-2$. *If* X *is* r-*shape connected and pointed* $(r+1)$-*movable, then* X *and* Y *are concordant relative to any shape equivalence.*

In particular we have a "trivial range" theorem.

COROLLARY 3. *Let* X *and* Y *be continua in* E^q *which have the same shape, satisfy* ILC, *and shape dimension* $\mathrm{sd}(X) = k$ *with* $q \geq 2k+2$. *Then* X *and* Y *are shape concordant relative to any shape equivalence.*

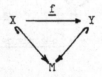

Since we use Stallings's results from [6] and since [6] has not appeared in print we shall give a short account of his work. The crucial result is the following.

THEOREM 4. *Let* $f:K \to M$ *be a PL-mapping of a compact connected n-dimensional polyhedron into a q-dimensional connected PL manifold* M *such that the dimension of the singular set of* f *is* $\leq s$ *and* f *is* $(s+1)$*-connected. If* $n \leq q-3$ *and* $s \leq q-5$ *, then there exists a compact connected n-dimensional subpolyhedron* $K_1 \subseteq M$ *such that* $f(K) \subseteq K_1$ *and the composition*

$$K \xrightarrow{f} f(K) \hookrightarrow K_1$$

is a simple homotopy equivalence.

In particular, if f is in general position, then the singular set has dimension $\leq 2n-q$.

4.1. *If* f *is in general position and if* f *is* $(2n-q+1)$*-connected, then the conclusion of Theorem* 4 *is true.*

Now suppose that X and Y are compact connected n-dimensional polyhedra in M such that $X \hookrightarrow M$ is $(2n-q+2)$-connected and suppose that there exists a simple homotopy equivalence $f:X \to Y$ such that

$$\begin{array}{ccc} X & \xrightarrow{f} & Y \\ & \searrow \quad \swarrow & \\ & M & \end{array}$$

is homotopy commutative. Let $F:X \times [0,1] \to M \times [0,1]$ be a homotopy such that $F(x,0) = (x,0)$ and $F(x,1) = f(x)$. By using relative simplicial approximation theorem and relative general position, we may assume that the dimension of the singular set of F is $2(n+1) - (q+1) = 2n-q+1$. If we let L be the subpolyhedron of $M \times [0,1]$ given by 4.1 which contains the image of F and slightly move $L - [(X \times \{0\}) \cup (Y \times \{1\})]$ off $M \times \{0,1\}$ we have the following.

4.2. L *is an* $(n+1)$*-dimensional subpolyhedron of* $M \times [0,1]$ *such that* $L \cap (M \times \{0\}) = X \times \{0\}$, $L \cap (M \times \{1\}) = Y \times \{1\}$, $X \times \{0\} \hookrightarrow L$ *and* $Y \times \{1\} \hookrightarrow L$ *are simple homotopy equivalences and*

$$\begin{array}{ccc} X\times\{0\} & \xrightarrow{\ f\ } & Y\times\{1\} \\ & \searrow \quad \swarrow & \\ & L & \end{array}$$

is homotopy commutative.

By using essentially the same proof, one can show the following.

4.3. *Let* W *be a connected* $(q+1)$*-dimensional PL manifold and let* X,Y *be disjoint compact connected* n*-dimensional subpolyhedra of* bdry W . *Suppose that there exists a simple homotopy equivalence* $f:X \to Y$ *such that*

$$\begin{array}{ccc} X & \xrightarrow{\ f\ } & Y \\ & \searrow \quad \swarrow & \\ & W & \end{array}$$

is homotopy commutative. If $X \hookrightarrow W$ *is* $(2n-q+2)$*-connected, then there exists a compact* $(n+1)$*-dimensional subpolyhedron* $L \subsetneq W$ *such that* $L \cap \text{bdry}\, W = X \cup Y$, $X \hookrightarrow L$ *and* $Y \hookrightarrow L$ *are simple homotopy equivalences and*

$$\begin{array}{ccc} X & \xrightarrow{\ f\ } & Y \\ & \searrow \quad \swarrow & \\ & L & \end{array}$$

is homotopy commutative.

We shall also use the extension of the Blakers-Massey Theorem to shape theory which was proved by Š.Ungar in his thesis (see [7]):

THEOREM 5. *Let* $\{X_i, A_i, B_i\}$ *object* pro-CW_0^{triad} *such that* $X_i = A_i \cup B_i$ *for each* i. *If the inclusion induced morphisms* (in pro-HCW_0) $\{A_i \cap B_i\} \hookrightarrow \{A_i\}$ *and* $\{A_i \cap B_i\} \hookrightarrow \{B_i\}$ *are* n*-shape connected and* m*-shape connected, respectively with* $n,m \geq 1$, *then the inclusion induced morphisms of relative pro-groups (pro-sets if* $k = 1$)

$$\pi_k(\underline{A}, \underline{A} \cap \underline{B}) \longrightarrow \pi_k(\underline{X}, \underline{B})$$

is an isomorphism for $1 \leq k \leq n+m-1$ *and is an epimorphism for* $k = n+m$.

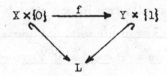

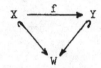

PROPOSITION 6. *Suppose that* X,Y *and* Z *are continua in a manifold* M *such*

that X *is shape concordant to* Y *relative to* f *and* Y *is shape concordant to* Z *relative to* g. *Then* X *is shape concordant to* Z *relative to* g f.

Proof. Let $L_1 \subseteq M\times[0,1/2]$ and $L_2 \subseteq M\times[1/2,1]$ be continua such that

(i) $L_1 \cap M\times\{0\} = X\times\{0\}$

$L_1 \cap M\times\{1/2\} = Y\times\{1/2\} = L_2 \cap M\times\{1/2\}$

$L_2 \cap M\times\{1\} = Z\times\{1\}$;

(ii) $X\times\{0\} \hookrightarrow L_1$, $Y\times\{1/2\} \hookrightarrow L_1$, $Y\times\{1/2\} \hookrightarrow L_2$ and $Z\times\{1\} \hookrightarrow L_2$ are shape equivalences ;

(iii) the diagrams

$$\begin{array}{ccc} X\times\{0\} & \xrightarrow{\ f\ } & Y\times\{1/2\} \\ & \searrow \quad \swarrow & \\ & L_1 & \end{array} \qquad \begin{array}{ccc} Y\times\{1/2\} & \xrightarrow{\ g\ } & Z\times\{1\} \\ & \searrow \quad \swarrow & \\ & L_2 & \end{array}$$

are commutative in the shape category.

Let $L = L_1 \cup L_2$; conditions (i), (ii) and (iv) of the definition of shape concordance are clearly satisfied.

Consider the exact homotopy sequence of a triple (in pro-groups and pro-sets)

$$\ldots \to \pi_k(\underline{L_1}, \underline{X\times\{0\}}) \to \pi_k(\underline{L}, \underline{X\times\{0\}}) \to \pi_k(\underline{L}, \underline{L_1}) \to \ldots$$

$\underline{L}$, $\underline{L_1}$, etc. denote appropriate objects in pro-CW_0 whose inverse limits are L, L_1, etc., respectively. Since $X\times\{0\} \hookrightarrow L_1$ is a shape equivalence $\pi_k(\underline{L_1}, \underline{X\times\{0\}})$ is equivalent to the trivial group in pro-groups for all $k \geq 2$ and to the trivial set in pro-sets for $k = 1$. Hence, $\pi_k(\underline{L}, \underline{X\times\{0\}})$ and $\pi_k(\underline{L}, \underline{L_1})$ are isomorphic in pro-groups for all $k \geq 2$ and in pro-sets for $k = 1$.

Since $Y\times\{1/2\} \hookrightarrow L_1$ and $Y\times\{1/2\} \hookrightarrow L_2$ are shape equivalences, by Theorem 5, $\pi_k(\underline{L_2}, \underline{Y\times\{1/2\}}) \to \pi_k(\underline{L}, \underline{L_1})$ is an isomorphism in pro-groups for $k \geq 2$ and in pro-sets for $k = 1$. Therefore, $\pi_k(\underline{L}, \underline{L_1})$, and thus $\pi_k(\underline{L}, \underline{X\times\{0\}})$, is equivalent to the trivial group or trivial set. By Mardešić [5], $X\times\{0\} \hookrightarrow L$ is a shape equivalence. Similarly, $Z\times\{1\} \hookrightarrow L$ is a shape equivalence.

The proof of the following is straightforward.

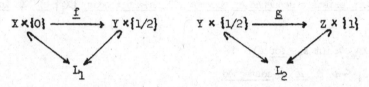

PROPOSITION 7. *If* X *and* Y *are continua in a manifold* M *such that* X *is*

shape concordant to Y relative to f , then Y is shape concordant to X relative to f^{-1} .

The following was proved by Ivanšić and Sher [4] :

THEOREM 8. Suppose X is a continuum in the interior of a q-dimensional PL manifold M such that

8.1. X satisfies ILC in M ;

8.2. $sd(X) = k \leq n-3$;

8.3. the pro-group $\pi_i(X)$ is stable for $0 \leq i \leq r-1$ and satisfies Mittag--Leffler condition for $i = r < n-3$.

Then there exists a sequence of compact PL neighborhoods $\{X_i\}$ of X in M such that

8.4. $X_{i+1} \subseteq \operatorname{int} X_i$ for all i ;

8.5. $X_i \hookrightarrow M$ is r-connected ;

8.6. X_i has a spine of dimension at most $\max(k, r+1)$;

8.7. $\bigcap X_i = X$.

PROPOSITION 9. Let M be a q-dimensional PL manifold and let $\{A_i\}$ and $\{B_i\}$ be sequences of q-dimensional PL submanifolds of M such that $A_{i+1} \subseteq \operatorname{int} A_i$ and $B_{i+1} \subseteq \operatorname{int} B_i$ for each i . Suppose that $\{K_i, p_i\}$ is an inverse sequence in pro-CW_0 with $\dim K_i \leq n \leq q-3$ for each i and each p_i is $(2n-q+2)$-connected. Suppose that there exist equivalences $\{h_i\} : \{K_i\} \to \{A_i\}$ and $\{h_i'\} : \{K_i\} \to \{B_i\}$ in pro-$HANR_0$ such that each h_i is $(2n-q+2)$-connected, each h_i' is a homotopy equivalence and

$$\begin{array}{ccc} K_i & \xrightarrow{\ h_i\ } & A_i \\ {\scriptstyle h_i'}\downarrow & & \downarrow \\ B_i & \hookrightarrow & M \end{array}$$

is homotopy commutative for each i .

Then $\bigcap A_i$ and $\bigcap B_i$ are shape concordant relative to $h' h^{-1}$, where h and h' are the shape maps induced by $\{h_i\}$ and $\{h_i'\}$ respectively.

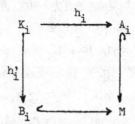

Proof. Let $\{g_i\}$ and $\{g_i'\}$ be inverses of $\{h_i\}$ and $\{h_i'\}$, respectively. By choosing subsequences, if necessary, we may assume that domain $h_i = K_i =$ domain h_i', domain $g_i = A_{i+1}$, domain $g_i' = B_{i+1}$ and the diagram

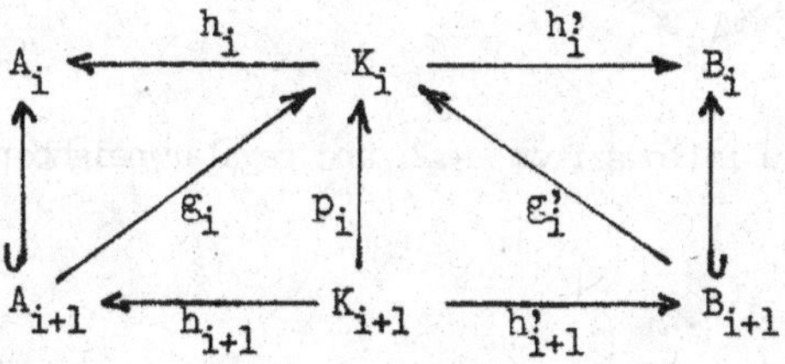

is homotopy commutative for each i.

Since h_i and h_i' are $(2n-q+1)$-connected, and without loss of generality we may assume that h_i and h_i' are PL maps in general position, by Theorem 4 there exist compact polyhedra $R_i \subseteq A_i$ and $S_i \subseteq B_i$ such that $h_i : K_i \to R_i$ and $h_i' : K_i \to S_i$ are simple homotopy equivalences. Let $k_i : R_i \to K_i$ be a homotopy inverse for h_i.

Consider the following statement for positive integer m.

$\mathcal{T}(m)$: *There exist compact polyhedra* $T_1, T_2, \ldots, T_m \subseteq M \times [0,1]$ *and PL manifolds* $M_m \subseteq M_{m-1} \subseteq \ldots \subseteq M_1 \subseteq M \times [0,1]$ *such that*

9.1. $T_i \cap (M \times \{0\}) = R_i \times \{0\}$;

9.2. $T_i \cap (M \times \{1\}) = S_i \times \{1\}$;

9.3. $R_i \times \{0\} \hookrightarrow T_i$ *and* $S_i \times \{1\} \hookrightarrow T_i$ *are simple homotopy equivalences;*

9.4.

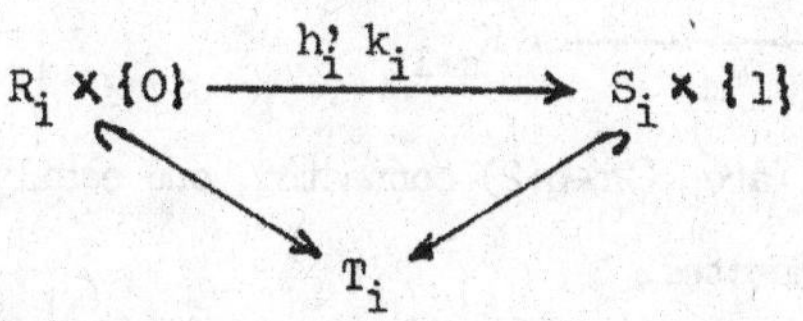

is homotopy commutative;

9.5. M_i *is a regular neighborhood of* $(A_i \times \{0\}) \cup T_i \cup (B_i \times \{1\})$ *in* M_{i-1} *mod* $[(\text{bdry } A_i \times \{0\}) \cup (\text{bdry } B_i \times \{1\})]$ *such that* $M_i \cap \text{bdry } M_{i-1} = (A_i \times \{0\}) \cup (B_i \times \{1\})$.

LEMMA 10. $\mathcal{T}(1)$ *is true.*

Proof. Since $h_1 : K_1 \to A_1$ is $(2n-q+2)$-connected and $h_1 : K_1 \to R_1$ is homotopy equivalence, $R_1 \hookrightarrow A_1$ is $(2n-q+2)$-connected. It is easily checked that

$$\begin{array}{ccc} R_1 & \xrightarrow{h_1'k_1} & S_1 \\ & \searrow \quad \swarrow & \\ & M & \end{array}$$

is homotopy commutative. $\mathcal{T}(1)$ now follows from 4.2. and regular neighborhood theory [1].

LEMMA 11. $\mathcal{T}(m)$ <u>implies</u> $\mathcal{T}(m+1)$.

<u>Proof</u>. Adopt the notation of $\mathcal{T}(m)$.

SUBLEMMA 12. $R_{m+1} \times \{0\} \hookrightarrow M_m$ <u>is</u> $(2n-q+2)$-<u>connected</u>.

<u>Proof</u>. Since M_m is a regular neighborhood of $(A_m \times \{0\}) \cup T_m \cup (B_m \times \{1\})$, the latter is a strong deformation retract of M_m. Since $S_m \hookrightarrow B_m$ and $R_m \times \{0\} \hookrightarrow T_m$ are homotopy equivalences, S_m and $R_m \times \{0\}$ are strong deformation retracts of B_m and T_m, respectively. Hence $A_m \times \{0\} \hookrightarrow M_m$ is a homotopy equivalence and to prove the sublemma it suffices to show that $R_{m+1} \hookrightarrow A_m$ is $(2n-q+2)$-connected. From the commutativity of the diagram

$$\begin{array}{ccc} A_m & \hookleftarrow & A_{m+1} \\ \uparrow {\scriptstyle h_m} & & \uparrow {\scriptstyle h_{m+1}} \\ K_m & \xleftarrow[p_m]{} & K_{m+1} \end{array}$$

and the fact that $h_m\ h_{m+1}$ and p_m are $(2n-q+2)$-connected, one easily checks that $R_{m+1} \hookrightarrow A_m$ is $(2n-q+2)$-connected.

SUBLEMMA 13.

$$\begin{array}{ccc} R_{m+1} \times \{0\} & \hookrightarrow & M_m \\ & {\scriptstyle h_{m+1}'k_{m+1}} \searrow & \cup\!\!\uparrow \\ & & S_{m+1} \times \{1\} \end{array}$$

<u>is homotopy commutative</u>.

<u>Proof</u>. The inclusion $R_{m+1} \times \{0\} \hookrightarrow M_m$ is equal to the composition

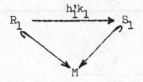

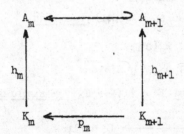

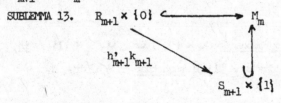

$R_{m+1}\times\{0\}\hookrightarrow A_{m+1}\times\{0\}\hookrightarrow A_m\times\{0\}\hookrightarrow M_m$. But this composition is homotopic to the composition

$$R_{m+1}\times\{0\}\hookrightarrow A_{m+1}\times\{0\}\xrightarrow{g_m} K_m\times\{0\}\xrightarrow{h_m} R_m\times\{0\}\hookrightarrow A_m\times\{0\}\hookrightarrow M_m$$

which, in turn is homotopic to

$$R_{m+1}\times\{0\}\xrightarrow{k_{m+1}} K_{m+1}\xrightarrow{p_m} K_m\xrightarrow{h_m} R_m\times\{0\}\xrightarrow{h'_m k_m} S_m\times\{1\}\hookrightarrow T_m\hookrightarrow M_m .$$

Since k_m is the homotopy inverse of h_m , this is homotopic to

$$R_{m+1}\times\{0\}\xrightarrow{k_{m+1}} K_{m+1}\xrightarrow{p_m} K_m\xrightarrow{h'_m} S_m\times\{1\}\hookrightarrow M_m$$

which, in turn, is homotopic to $h'_{m+1}k_{m+1}$.

Now we can apply 4.3. to obtain $T_{m+1}\subseteq M_m$ such that 9.1.-9.4. are satisfied. By [1] we can choose a regular neighborhood M_{m+1} of $(A_{m+1}\times\{0\})\cup T_{m+1}\cup(B_{m+1}\times\{1\})$ in M_m so that 9.5. is satisfied. This completes the proof of $\mathcal{T}(m)$.

By induction and the maximality principle, we can find a sequence $\{T_i\}$ of compact subpolyhedra and a sequence $\{M_i\}$ of PL submanifolds of $M\times[0,1]$ such that 9.1. - 9.4. hold for all i .

Now we claim that $L=\bigcap M_i$ is the desired shape concordance. Clearly $L\cap(M\times\{0\})=\bigcap A_i\times\{0\}$ and $L\cap(M\times\{1\})=\bigcap B_i\times\{1\}$. Recall that in the proof of Sublemma 12 it was shown that $A_i\times\{0\}\hookrightarrow M_i$ is a homotopy equivalence for each i . Hence the inclusion induced map $\bigcap A_i\times\{0\}\hookrightarrow\bigcap M_i$ is a shape equivalence.

We now show that

$$(*)\qquad \begin{array}{ccc} \bigcap A_i\times\{0\} & \xrightarrow{\underline{h'}\,\underline{h}^{-1}} & \bigcap B_i\times\{0\} \\ & \searrow \quad \swarrow & \\ & \bigcap M_i & \end{array}$$

is homotopy commutative in the shape category. Given i , consider

$$A_{i+1}\times\{0\}\hookrightarrow A_i\times\{0\}\hookrightarrow M_i$$

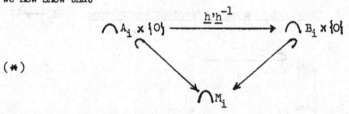

which is homotopic to

$$A_{i+1} \times \{0\} \xrightarrow{g_i} K_i \times \{0\} \xrightarrow{h_i} R_i \times \{0\} \hookrightarrow A_i \times \{0\} \hookrightarrow M_i$$

which is equel to

$$A_{i+1} \times \{0\} \xrightarrow{g_i} K_i \times \{0\} \xrightarrow{h_i} R_i \times \{0\} \hookrightarrow T_i \hookrightarrow M_i \ .$$

By 9.4. this is homotopic to

$$A_{i+1} \times \{0\} \xrightarrow{g_i} K_i \times \{0\} \xrightarrow{h_i} R_i \times \{0\} \xrightarrow{h_i' k_i} S_i \times \{1\} \hookrightarrow T_i \hookrightarrow M_i \ .$$

Since k_i is the homotopy inverse of h_i , this is homotopic to

$$A_{i+1} \times \{0\} \xrightarrow{h_i' g_i} B_i \times \{1\} \hookrightarrow M_i \ .$$

From the commutativity of $(*)$ and the fact that $\cap A_i \times \{0\} \hookrightarrow \cap M_i$ and $\underline{h}'\underline{h}^{-1}$ are shape equivalences, we get that $\cap B_i \times \{1\} \hookrightarrow \cap M_i$ is also a shape equivalence. This completes the proof of Proposition 9 .

From Ivanšić [3] , we have the following.

PROPOSITION 14. Suppose M is q-dimensional PL manifold and $\{R_i, r_i\}$ is an inverse sequence in pro-CW_0 such that $\dim R_i \leq n$, $q-n \geq 3$, and r_i is $(2n-q+1)$-connected for all i . Suppose $s_1 : R_1 \to R_1'$ is a $(2n-q+1)$-connected map of R_1 into an n-dimensional subpolyhedron of M and N_1 is a regular neighborhood of R_1' in M. Then

14.1. there is a sequence of compact PL submanifolds $\{N_i\}$ of M such that $N_{i+1} \subsetneq \operatorname{int} N_i$ for all i ;

14.2. $N_i \searrow R_i'$, where $\dim R_i' \leq n$;

14.3. for each i there exists a simple homotopy equivalence $s_i : R_i \to R_i'$ such that

$$\begin{array}{ccc} N_{i+1} & \hookrightarrow & N_i \\ \uparrow & & \uparrow \\ R_{i+1}' & & R_i' \\ s_{i+1} \uparrow & & \uparrow s_i \\ R_{i+1} & \xrightarrow{r_i} & R_i \end{array}$$

<u>is homotopy commutative.</u>

Let $X \subseteq M$ satisfy the hypotheses of Theorem 1 . Since $X \hookrightarrow M$ is shape (2k-q+2)-connected and X is pointed (2k-q+2)-movable, it is easily seen that the pro-group $\pi_i(\underline{X})$ is stable for $0 \leq i \leq 2k-q+1$ and satisfies Mittag-Leffler condition for $i = 2k-q+2$. By Theorem 8, there exists a sequence of compact PL neighborhoods $\{X_i\}$ of X in M satisfying 8.4. - 8.7. From the proof of Theorem 4 in [7] , there exists an inverse sequence $\{K_i, p_i\}$ in pro-CW_0 such that shape $(\varprojlim K_i)$ = shape (X) , the bonding maps p_i are (2k-q+2)-connected and $\dim K_i \leq k$. Since shape $(\varprojlim K_i)$ = shape (X), there exists an isomorphism $\{h_i\} : \{K_i\} \rightarrow \{X_i\}$ and its inverse $\{g_i\} : \{X_i\} \rightarrow \{K_i\}$ (in pro-$HANR_0$). By cho - -osing subsequences, if necessary, we may assume that domain $g_i = X_{i+1}$, domain $h_i = K_i$ and the diagrams

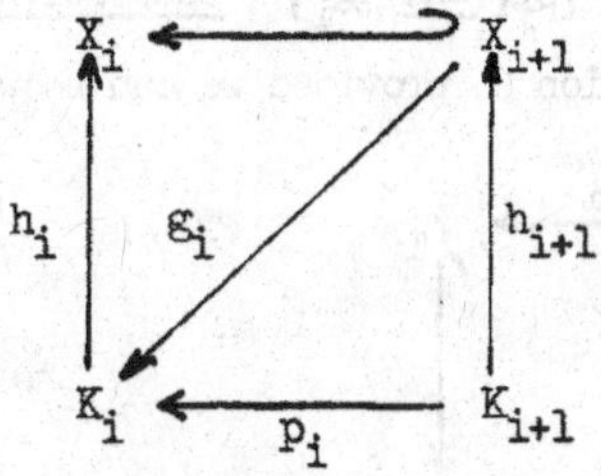

are homotopy commutative for each i .

LEMMA 15. <u>For each</u> i , h_i <u>is</u> (2k-q+2)-<u>connected.</u>

<u>Proof of Lemma</u> 14. Suppose $\lambda \in \pi_j(X_i)$, $j \leq 2k-q+2$. Since $X_i \hookrightarrow M$ and $X_{i+1} \hookrightarrow M$ are (2k-q+2)-connected, there exists $\lambda' \in \pi_j(X_{i+1})$ which is sent to λ by the inclusion induced homomorphism. Then $h_{i*}(g_{i*}(\lambda')) = \lambda$; hence, $h_{i*} : \pi_j(K_i) \rightarrow \pi_j(X_i)$ is an epimorphism for $j \leq 2k-q+2$.

Suppose $\lambda \in \pi_j(K_i)$ such that $h_{i*}(\lambda) = 0$ for $j < 2k-q+2$. If $i > 1$, then $p_{i-1*}(\lambda) = g_{i-1*}\, h_{i*}(\lambda) = 0$. Hence $\lambda = 0$ by connectivity of p_{i-1} . The proof that $\lambda = 0$ for $i = 1$ is slightly more complicated and is left to the reader.

By 4.1., there exists compact polyhedron $L_1 \subseteq X_1$ such that $h_1 : K_1 \rightarrow L_1$ is a simple homotopy equivalence. By Proposition 14, if N_1 is a regular neighborhood of L_1 in X_1, then there exists a sequence of compact PL submanifolds

$\{N_i\}$ of X_1 such that $N_{i+1} \subsetneq \operatorname{int} N_i$ for each i, $N_i \searrow R_i$ where $\dim R_i \leq k$ and for each i there exists a simple homotopy equivalence $s_i : K_i \longrightarrow R_i$ such that

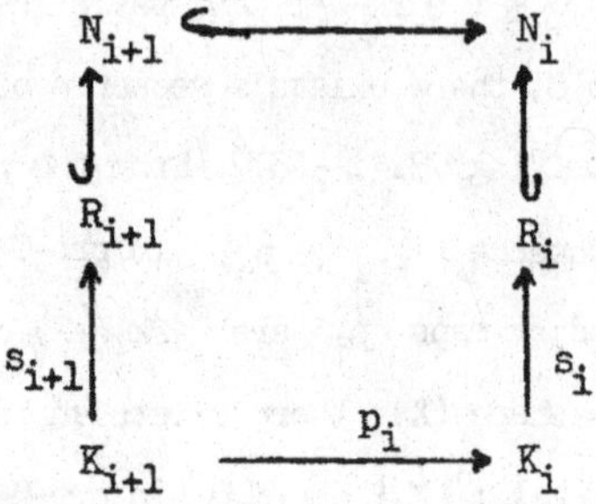

is homotopy commutative.

PROPOSITION 16. X <u>and</u> $\cap N_i$ <u>are shape concordant relative to</u> sg <u>where</u> s <u>and</u> g <u>are the shape maps induced by</u> $\{s_i\}$ <u>and</u> $\{g_i\}$, <u>respectively</u>.

<u>Proof</u>. This follows from Proposition 9. provided we can show that

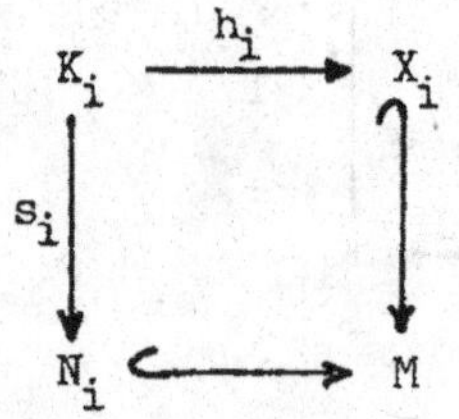

is homotopy commutative for each i. But this follows from the commutativity of

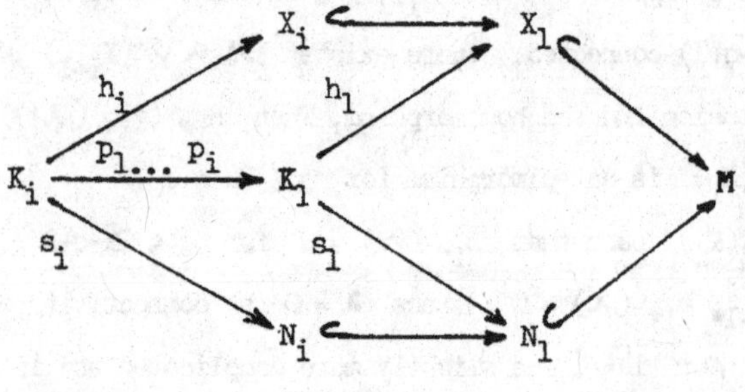

and the fact that $h_1 = s_1$.

Now let Y be a continuum in M and let $\underline{f}: X \rightarrow Y$ be a shape equivalence which satisfies the hypotheses of Theorem 1. Let $\{Y_i\}$ be an inverse sequence of compact PL neighborhoods of Y in M satisfying 8.4. - 7. Let $\{f_i\} : \{X_i\} \rightarrow \{Y_i\}$

represent $\underline{f}\ \underline{h} = \underline{h}'$; without loss of generality, we may assume that domain $f_i = X_i$. Note that $\{f_i\}\{h_i\} : \{K_i\} \to \{Y_i\}$ is an equivalence in pro-$HANR_0$. Let $\{g_i'\} : \{Y_i\} \to \{K_i\}$ represent the inverse of the latter ; by choosing subsequences, if necessary, we may assume that

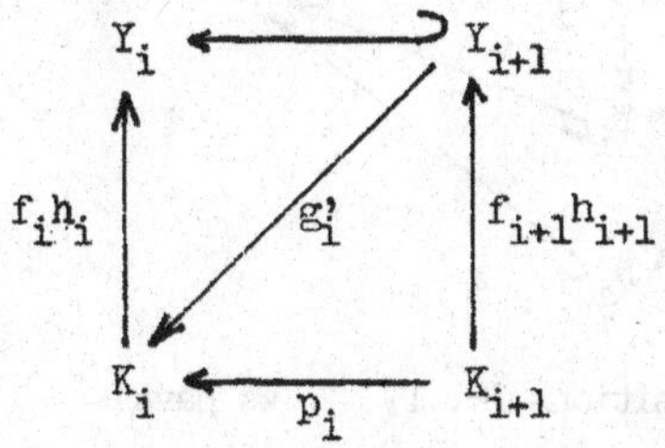

is homotopy commutative for each i.

Analogous to Lemma 5, we have

LEMMA 17. <u>For each</u> i, $h_j' = f_i h_i$ <u>is</u> $(2k-q+2)$-<u>connected</u>.

By 4.1., there exists a compact polyhedron $L_1' \subseteq Y_1$ such that $h_1' : K_1 \to L_1'$ is a simple homotopy equivalence. Let N_1' be a regular neighborhood of L_1' in Y_1; let $\{N_i'\}$ be a sequence of compact PL submanifolds, $\{R_i'\}$ be a sequence of compact polyhedra and let $s_i' : K_i \to N_i'$ be maps which satisfy 14.1. - 3. Analogous to Proposition 16. we have the following.

PROPOSITION 18. Y <u>and</u> $\cap N_i'$ <u>are shape concordant relative to</u> $s'g'$ <u>where</u> s' <u>and</u> g' <u>are the shape maps induced by</u> $\{s_i'\}$ <u>and</u> $\{g_i'\}$, <u>respectively</u>.

PROPOSITION 19. $\cap N_i$ <u>and</u> $\cap N_i'$ <u>are shape concordant relative to</u> $s'\,\underline{s}^{-1}$.

<u>Proof</u>. Recall that $s_i : K_i \to N_i$ and $s_i' : K_i \to N_i'$ are homotopy equivalences. Hence, in order to apply Proposition 9., all we need to show is that

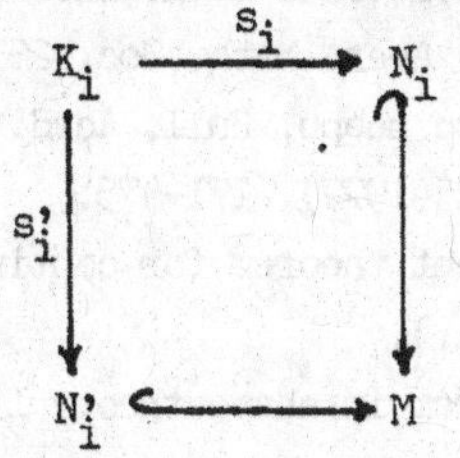

is homotopy commutative for each i. This follows from

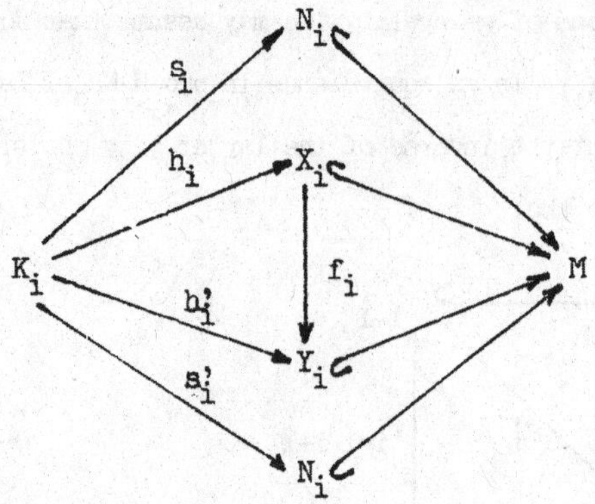

Proof of Theorem 1. By Propositions 16, 19, 7, we have :

X_i and $\bigcap N_i$ are shape concordant relative to $\underline{s}\,\underline{g}$;

$\bigcap N_i$ and $\bigcap N'_i$ are shape concordant relative to $\underline{s}'\,\underline{s}^{-1}$;

$\bigcap N'_i$ and Y are shape concordant relative to $\underline{g}'^{-1}\underline{s}'^{-1}$.

By Proposition 6 , X and Y are shape concordant relative to

$$\underline{g}'^{-1}\,\underline{s}'^{-1}\underline{s}'\underline{s}^{-1}\,\underline{s}\,\underline{g} = \underline{g}'^{-1}\underline{g} \quad .$$

But $\underline{g}'^{-1} = \underline{h}' = \underline{f}\,\underline{h}$ and , hence, $\underline{g}'^{-1}\,\underline{g} = \underline{f}\,\underline{h}\,\underline{g} = \underline{f}$. This completes the proof of Theorem 1 .

References

[1] M.M.Cohen, A general theory of relative regular neighborhoods, Trans. Amer. Math. Soc. 136(1969), 189-230.

[2] S.Ferry, A stable converse to the Vietoris-Smale theorem with applications to shape theory, Trans. Amer. Math. Soc. 261(1980), 369-386.

[3] I.Ivanšić, Embedding compacta up to shape, Bull. Acad. Polon. Sci. Sér. Sci. Math. Astronom. Phys. 25(1977), 171-475.

[4] I.Ivanšić and R.B.Sher, A complement theorem for continua in a manifold, (to appear in Topology Proc.).

[5] S.Mardešić, On the Whitehead theorem in shape theory I ; II, Fund. Math. 91 (1976), 51-64 ; 93-103 .

[6] J.Stallings, The embedding of homotopy types into manifolds, Mimeographed notes

Princeton University 1965.

[7] Š.Ungar, n-connectdness of inverse systems and applications to shape theory, Glasnik Mat. Ser III 13(33)(1978), 371-396.

[8] G.A.Venema, Embeddings of compacta with shape dimension in the trivial range, Proc. Amer. Math. Soc. 55(1976), 443-448.

University of Tennessee
Knoxville, TN 37916

University of Zagreb
Zagreb, Yugoslavia

COMPLEMENT THEOREMS IN SHAPE THEORY

by

R.B. Sher

1. INTRODUCTION. In his seminal paper of 1968 [Bo_1], Karol Borsuk laid the foundations for what has become known as <u>shape theory</u>. Since then shape theory has become a rather well developed area of geometric topology and has proved to be useful for the understanding of the global geometric structure of spaces, particularly metric compacta. In 1972, T.A. Chapman published a result linking shape theory to another exceptionally active and fruitful area of geometric topology, that of Q-manifold theory, the study of manifolds modeled on the Hilbert cube Q. The link is Chapman's celebrated Complement Theorem, which we now state in a form only slightly different from, and equivalent to, that of [Ch_1].

CHAPMAN'S COMPLEMENT THEOREM. *If X and Y are Z-sets in Q, then X and Y have the same shape if and only if* $Q-X \cong Q-Y$.

In a companion paper [Ch_2], Chapman also proved a version of the Complement Theorem in which the role of Q is played by E^n, n-dimensional euclidean space. It is the purpose of the present paper to survey some of the ideas and results which grew from these two theorems.

For our purposes, we shall understand "complement theorem" to mean a result which establishes conditions under which the shape equivalence of two compacta X and Y, each embedded in the space M, is equivalent to the topological equivalence of M-X and M-Y. (For an account of a complement theorem for pairs, see [Sa].)

In retrospect, one can find a number of examples of complement theorems in the literature prior to 1972. Two of particular noteworthiness are:

(1) Borsuk's result, essentially established in [Bo_1], that for continua X and Y in the euclidean plane E^2, X and Y have the same shape if and only if $E^2-X \cong E^2-Y$. The proof of this result, which gives insight into the intuitive meaning of "having the same shape" for plane continua, is quite elementary. By attempting a proof of this fact, one may find many ideas which might be useful in the proofs of later, more sophisticated, complement theorems. The result does not hold for plane compacta.

(2) McMillan's Cellularity Criterion [Mc_1]. This result predates shape theory, of course, but easily implies that for a continuum X satisfying the cellularity criterion (see § 2 for the definition) in E^n, $n \geq 5$, X has the shape of a point if and only if $E^n-X \cong E^n$-pt. This result is related to Morton Brown's notion of <u>cellular set</u> which plays so important a role in the proof of the Generalized Schoenflies Theorem

[Br]. The result also holds, by [Mc_2], when $n=3$.

We also note that the concept of "weak flatness" is related to our notion of complement theorem. Recall that a topological k-sphere Σ in E^n is said to be weakly flat ([Ro], [Du]) if $E^n-\Sigma \cong E^n-S^k$, where S^k is the "standard" k-sphere in E^n. While results involving weak flatness obviously bear on our topic, we shall not attempt to treat them here.

We will restrict our attention to complement theorems for which the ambient space M is E^n or Q, although we shall take note in Section 5 of a partial complement theorem in which the role of M is played by a piecewise linear n-manifold. Of course, when stating a complement theorem, we shall have to be quite explicit about any special properties of X and Y, as well as the way in which X and Y are embedded in M. This is indicated by well known examples involving knotting or "wild" embeddings.

For background material, see [Bo_2] or [DS_1] for shape theory, [Hud] or [Ze] concerning piecewise linear topology, and [Ch_3] concerning Q-manifold theory. Our definitions, notations, and conventions are as found in these and other standard references, and should cause the reader no difficulty.

In Section 2 we shall lay down some groundwork involving loop conditions, fundamental dimension, and shape connectivity. In Section 3 we sketch the proof of a basic finite-dimensional complement theorem, while in Section 4 we treat a generalization of Chapman's Complement Theorem in Q. Section 5 deals with a partial complement theorem in a PL manifold, while Section 6 briefly touches on Chapman's description of the shape category by way of weak proper homotopy theory on Z-set complements in Q. Our bibliography, while not all encompassing, should serve well the reader who is interested in further pursuing these matters.

We have made no systematic attempt to discuss applications of, or work influenced by, complement theorems. Instead, the paper concentrates on those results and techniques which seem likely to prove useful in further investigations. While any attempt to reasonably examine a large body of work must necessarily involve some omissions, the author apologizes for any serious oversights.

2. PRELIMINARIES. LOOP CONDITIONS, FUNDAMENTAL DIMENSION, AND SHAPE CONNECTIVITY. Even well-behaved spaces (such as the Cantor set, an arc, etc.) may be badly embedded in a manifold, say with nonsimply connected complement; see, e.g., [FA], [Bl], or [Wo_1]. It follows that a complement theorem must include some sort of hypothesis to take this fact into account, in much the same sort of way that a taming theorem must.

In the case of Chapman's Complement Theorem, the "niceness of embedding" hypothesis is that X and Y be Z-sets in Q, a hypothesis which actually implies, via the Z-set

Unknotting Theorem [Ch_3; Theorem 19.4], tameness and unknottedness. Such a hypothesis on the compactum $X \subset M$ essentially allows the construction of certain mappings into M which are close to, but do not hit, the compactum X. A hypothesis of this sort is frequently stated as a loop condition.

Loop conditions come in two varieties, global and local, the difference in essence being how far one is willing to move to adjust the mappings. Applying local loop conditions, one exercises control by making very small movements; in the case of global loop conditions, one is able to make adjustments within a neighborhood of X, but within such a neighborhood quite large (e.g. on the order of Diam(X)) movements may occur. Generally speaking, local loop conditions are appropriate for taming theorems, while global loop conditions are more suitable for complement theorems.

Four global loop conditions which have been of particular interest in the study of complement theorems are as follows, where in each X is a compact subset of the manifold M.

DEFINITION 2.1. [Mc_1] X satisfies the cellularity criterion (CC) if for each neighborhood U of X there exists a neighborhood V of X such that each loop in V-X is nullhomotopic in U-X.

DEFINITION 2.2. [CDD] X satisfies the small loops condition (SLC) if for each neighborhood U of X there exist a neighborhood V of X and $\varepsilon > 0$ such that each ε-loop in V-X is nullhomotopic in U-X.

DEFINITION 2.3. [Da] X is globally 1-alg (embedded) if for each neighborhood U of X there exists a neighborhood V of X such that each loop in V-X which is nullhomologous in V-X is nullhomotopic in U-X.

DEFINITION 2.4. [Ve_1] X satisfies the inessential loops condition (ILC) if for each neighborhood U of X there exists a neighborhood V of X such that each loop in V-X which is nullhomotopic in V is nullhomotopic in U-X.

In dealing with complement theorems in euclidean spaces, we shall focus our attention on the inessential loops condition. In a large number of cases, complement theorems which have used a different loop condition have carried an additional hypothesis which implies that the compacta in question satisfy the inessential loops condition anyway. As we shall see momentarily, this is particularly so in codimension (or co-fundamental-dimension) three; otherwise the global 1-alg condition seems more appropriate.

The explicit relationship between the inessential loops condition and each of the other three global loop conditions stated above is provided by the following three easily proved propositions. The setting for each stated proposition consists of the compactum X embedded in a piecewise linear n-manifold M. Definitions of "approximatively 1-connected" and "fundamental dimension" appear later.

PROPOSITION 2.5. If X satisfies CC, then X satisfies ILC. The converse holds provided X is approximatively 1-connected.

PROPOSITION 2.6. If X satisfies ILC, then X satisfies SLC. The converse holds provided dim(X) does not exceed n-2.

PROPOSITION 2.7. If X is globally 1-alg and the fundamental dimension of X does not exceed n-3, then X satisfies ILC. If X satisfies ILC and is shape dominated by the inverse limit of a sequence of polyhedra each having abelian fundamental group, then X is globally 1-alg.

If X is a compactum, the fundamental dimension of X, denoted Fd(X), may be defined to be min {dim Y: Sh(X) = Sh(Y), Y a compactum}.

If $k \geq 1$ and the homotopy pro-group pro-$\pi_k(X,x)$ of the compactum X is trivial for all $x \in X$, then X is said to be approximatively k-connected. If X is approximatively k-connected for $1 \leq k \leq r$, then X is said to be r-shape connected. When $X \in ANR$ and $r \geq 0$, X is said to be r-connected if X is connected and $\pi_k(X) = 0$ for $1 \leq k \leq r$.

An important result linking the notions of fundamental dimension and the inessential loops condition is the following, established by Venema in [Ve_3].

THEOREM 2.8 (Venema). Suppose X is a compactum in the interior of the PL n-manifold M^n, $n \geq 5$, and $Fd(X) \leq k \leq n-3$. Then X satisfies ILC if and only if X has arbitrarily close compact PL manifold neighborhoods with k-dimensional spines.

Combining this with Theorem 2 of [ISV], we obtain the following generalization which proves useful for co-fundamental-dimension three complement theorems.

THEOREM 2.9 (Ivanšić-Sher-Venema). Suppose X is an r-shape connected compactum in the interior of the PL n-manifold M^n, $n \geq 5$, and $Fd(X) \leq k \leq n-3$. Then X satisfies ILC if and only if X has arbitrarily close compact PL manifold neighborhoods having r-connected components with k-dimensional spines.

3. FINITE DIMENSIONAL COMPLEMENT THEOREMS. In this section we sketch the proof of a basic co-fundamental-dimension three complement theorem in E^n, $n \geq 5$. The theorem we choose to concentrate on, which we now state, appears as Theorem A in [ISV].

THEOREM 3.1 (Ivanšić-Sher-Venema). Let X and Y be r-shape connected continua in E^n of fundamental dimension at most k and satisfying ILC, where $n \geq \max(2k+2-r, k+3, 5)$. Then Sh(X)=Sh(Y) implies $E^n - X \cong E^n - Y$. The converse holds if $n \geq k+4$.

We remark that, because of linking phenomena, the hypothesis that X and Y be continua is necessary in Theorem 3.1. If, however, when $r=0$ we replace "0-shape connected continua" in its statement by "compacta," we recover Venema's trivial range complement theorem [Ve_1]. (Since $n \geq 2k+2$ and $n \geq 5$, the condition $n \geq k+4$ is automatic.) Theorem 3.1, with this additional interpretation, may thus be regarded as a generalization and natural extention of Venema's fundamental result.

The statement of Theorem 3.1 subsumes many of its antecedents. It is, for example, more general than finite-dimensional complement theorems established in [Ch_2], [CD], [GS], [HR_1], [Li_1], [Li_2], [Si_2], and [Ve_1]. Its proof, however, really goes back to the proof of Chapman's original complement theorem in Q. In this sense it is the natural descendent of a whole family of results, each of which may be proved by a refinement of the basic ancestral technique. Our proof is sketched in I and II below.

I. <u>Building a homeomorphism between two complements</u>. Suppose X and Y are shape equivalent r-shape connected continua in E^n of fundamental dimension at most k and satisfying ILC, where $n \geq \max(2k+2-r, k+3, 5)$. Our task is to construct a homeomorphism $h:E^n-X \to E^n-Y$. We shall accomplish this by constructing a sequence $\{U_i\}_{i=1}^{\infty}$ of closed neighborhoods of X and a sequence $\{h_i\}_{i=1}^{\infty}$ of PL homeomorphisms of E^n onto itself such that

(i) $U_1 \supset U_2 \supset \cdots \supset \cap_{i=1}^{\infty} U_i = X$,

(ii) $h_{i+1}|E^n-U_i = h_i|E^n-U_i$, $i=1,2,\cdots$, and

(iii) $Y = \cap_{i=1}^{\infty} h_i(U_i)$.

It is then easy to see that the function $h:E^n-X \to E^n-Y$, defined by $h(x)=\lim_{i\to\infty} h_i(x)$ for all $x \in E^n-X$, is a homeomorphism.

Of course, great care must be exercised while constructing the U_i's and h_i's. It is necessary to proceed inductively, keeping careful track of how the fact that $Sh(X)=Sh(Y)$ is related to that portion of the construction already carried out. The following lemma will facilitate the construction. Its statement looks rather unwieldy, but its utility lies in the fact that it is designed to deal precisely with our inductive construction. The proof, which is given in [ISV], involves a relatively straightforward application of general position, piping, and Stalling's Engulfing Theorem.

LEMMA 3.2. <u>Let</u> A <u>and</u> B <u>be continua in</u> E^n <u>such that</u> A <u>has arbitrarily close compact</u> PL <u>manifold neighborhoods with k-dimensional spines and</u> B <u>has arbitrarily close open</u> (2k+2-n)-<u>connected neighborhoods, where</u> $k \leq n-3$. <u>Let</u> $\underline{f} = \{f_i, A, B\}$ <u>and</u> $\underline{f}' = \{f'_i, B, A\}$ <u>be fundamental sequences in</u> E^n <u>such that</u> $\underline{f}'\,\underline{f} \simeq id_A$. <u>Let</u> U <u>be an open</u> (2k+2-n)-<u>connected neighborhood of</u> A, <u>and</u> $f:E^n \to E^n$ <u>be a</u> PL <u>homeomorphism such that there exists a neighborhood</u> W <u>of</u> B <u>with</u> $f^{-1}|W \simeq f'_i|W$ <u>in</u> U <u>for almost all</u>

i. Then for every neighborhood N of B, there exist an open $(2k+2-n)$-connected neighborhood V of B lying in $N \cap f(U)$, a PL homeomorphism $q:E^n \to E^n$, and a neighborhood M of A such that

(1) $q|E^n - U = f|E^n - U$,

(2) $q(A) \subset V$, and

(3) $q|M \simeq f_i|M$ in V for almost all i.

With the above lemma in hand, we are now ready to carry out our construction. Let $\underline{g} = \{g_i, X, Y\}$ and $\underline{g}' = \{g_i', Y, X\}$ be fundamental sequences in E^n such that $\underline{g}'\,\underline{g} \simeq \underline{id}_X$ and $\underline{g}\,\underline{g}' \simeq \underline{id}_Y$. Recall that, by Theorem 2.9, X and Y each have arbitrarily close compact PL manifold neighborhoods with k-dimensional spines whose interiors are $(2k+2-n)$-connected.

We may begin by letting $U_1 = E^n$ and h_1 be the identity. Inductively, suppose we have constructed closed neighborhoods $U_1, U_2, \cdots, U_m$ of X and PL homeomorphisms $h_1, h_2, \cdots, h_m$ of E^n onto itself such that if $m > 1$,

(1) $U_1 \supset U_2 \supset \cdots \supset U_m$ and $U_m \subset N_{1/m}(X)$,

(2) $h_{i+1}|E^n - U_i = h_i|E^n - U_i$, $i = 1, 2, \cdots, m-1$, and

(3) $Y \subset h_m(U_m) \subset N_{1/m}(Y)$.

We also assume that

(4) $h_m(U_m)$ contains a $(2k+2-n)$-connected open neighborhood R_m of Y, and

(5) there exists a neighborhood S_m of X such that $h_m|S_m \simeq g_i|S_m$ in R_m for almost all i.

We now will construct U_{m+1} and h_{m+1} so that the appropriate analogues of (1)-(5) hold. Continuing inductively, we obtain $\{U_i\}_{i=1}^{\infty}$ and $\{h_i\}_{i=1}^{\infty}$ so that (i)-(iii) hold. This will complete the proof.

The construction of U_{m+1} and h_{m+1} is carried out by a two-fold application of Lemma 3.2.

Step 1. Let $A = Y$, $B = X$, $f_i = g_i'$, $f_i' = g_i$, $U = R_m$, $f = h_m^{-1}$, and $W = S_m$. Then the hypothesis of Lemma 3.2 is satisfied. Letting N be a sufficiently small neighborhood of X, the conclusion of Lemma 3.2 yields an open $(2k+2-n)$-connected neighborhood V of X lying in $h_m^{-1}(R_m) \cap N_{1/m+1}(X)$, a PL homeomorphism $q:E^n \to E^n$, and a neighborhood M of Y such that $q|E^n - R_m = h_m^{-1}|E^n - R_m$, $q(Y) \subset V$, and $q|M \simeq g_i'|M$ in V for almost all i.

Step 2. We now assign new meaning to the variables A, B, etc. and apply Lemma 3.2 once again. This time, let $A = X$, $B = Y$, $f_i = g_i$, $f_i' = g_i'$, $U = V$, $f = q^{-1}$, and $W = M$. Once more, the hypothesis of Lemma 3.2 is satisfied. Letting $\tilde{N}$ be a sufficiently small neighborhood of Y, the conslusion of Lemma 3.2 yields an open (2k+2-n)-connected neighborhood $\tilde{V}$ of Y whose closure lies in $q^{-1}(V) \cap N_{1/m+1}(Y)$ a PL homeomorphism $\tilde{q}:E^n \to E^n$, and a neighborhood $\tilde{M}$ of X such that $\tilde{q}|E^n - V = q^{-1}|E^n - V$, $\tilde{q}(X) \subset \tilde{V}$, and $\tilde{q}|\tilde{M} \simeq \tilde{g}_i|\tilde{M}$ in $\tilde{V}$ for almost all i. Now, let U_{m+1} be the closure of $\tilde{q}^{-1}(\tilde{V})$, h_{m+1} be $\tilde{q}$, R_{m+1} be $\tilde{V}$, and S_{m+1} be $\tilde{M}$. It is easily verified that the analogues of (1)-(5) hold with m replaced by m+1. This completes the proof.

II. <u>Building a shape equivalence from homeomorphic complements</u>. Suppose X and Y are r-shape connected continua in E^n of fundamental dimension at most k and satisfying ILC, where $n \geq \max(2k+2-r, k+4, 5)$. Given $E^n - X \cong E^n - Y$, we wish to show $Sh(X) = Sh(Y)$. First, assume $r = 0$. In this case, let $\tilde{X}$ and $\tilde{Y}$ be k-dimensional continua such that $Sh(X) = Sh(\tilde{X})$ and $Sh(Y) = Sh(\tilde{Y})$ Since $n \geq 2k+2$, we may assume by [Št] that $\tilde{X}$ and $\tilde{Y}$ are embedded in E^n as 1-ULC subsets. By I above, $E^n - \tilde{X} \cong E^n - X \cong E^n - Y \cong E^n - \tilde{Y}$; it follows from Theorem 1.2 of [GS] that $Sh(\tilde{X}) = Sh(\tilde{Y})$, and so $Sh(X) = Sh(Y)$.

The above does not reveal the need for the hypothesis $n \geq k+4$. Of course when $r = 0$, k is in the trivial range with respect to n, and co-fundamental-dimension four is automatic since $n \geq 5$. The need for the condition $n \geq k+4$ when $r \geq 1$ is due to our dependence on the following hyperplane result, which appears as Theorem 4 in [ISV].

THEOREM 3.3 (Ivanšić-Sher-Venema). <u>Let</u> X <u>and</u> Y <u>be 1-shape connected continua in</u> E^n <u>of fundamental dimension at most</u> k <u>and satisfying</u> ILC, <u>where</u> $k \leq n-3$. <u>Then</u> $E^{n+1} - X \cong E^{n+1} - Y$ implies $Sh(X) = Sh(Y)$.

Theorem 3.3 generalizes Theorem 1' of [Li_2], in which it is assumed that each of X and Y has the shape of a finite simplicial complex. Briefly, the idea of the proof of Theorem 3.3 is as follows: Write $X = \cap_{i=1}^{\infty} M_i$, where M_i is a compact PL n-manifold in E^n, $M_{i+1} \subset \text{int } M_i$, $\pi_1(M_i) = 0$, and M_i has a k-dimensional spine, $i = 1,2,\cdots$. Let $N_i = M_i \times [-1/i, 1/i] \subset E^n \times E^1 = E^{n+1}$. Let $h: E^{n+1} - X \to E^{n+1} - Y$ be a homeomorphism, where it may be assumed that h induces a homeomorphism of the quotient space E^{n+1}/X onto E^{n+1}/Y. If $i = 1,2,\cdots$, let $N_i' = E^{n+1} - h(E^{n+1} - N_i)$. Then $\underline{X} = \{N_i\}_{i=1}^{\infty}$ and $\underline{Y} = \{N_i'\}_{i=1}^{\infty}$ form, along with the inclusion maps, inverse sequences of ANR's associated with X and Y, respectively. Let $p_i: N_i \to M_i \times \{1/i\}$ denote the natural projection and define $f_i: N_i \to N_i'$ by $f_i(x) = h(p_i(x))$, $x \in N_i$, $i = 1,2,\cdots$. Then (1) $\underline{f} = (f_i, \text{id}): \underline{X} \to \underline{Y}$ is a level preserving system map and (2) $f_i: N_i \to N_i'$ is a homotopy equivalence. From this it easily follows that $Sh(X) = Sh(Y)$.

To use Theorem 3.3 to complete our proof when $r \geq 1$, we need an embedding (up to shape) theorem (in E^{n-1}) to use as Štan'ko's result was used in the case $r = 0$. Such a result is provided in [Iv]. Since $k \leq n-4$, $k-(2k+2-n) \geq 2$, and so Theorem 5 of [Iv] shows that X and Y may be embedded up to shape in E^{n-1}, say as $\tilde{X}$ and $\tilde{Y}$. In fact, the proof of Theorem 5 of [Iv] shows that we may obtain such $\tilde{X}$ and $\tilde{Y}$ satisfying ILC in E^{n-1}. Then, by I above, $E^n - \tilde{X} \cong E^n - X \cong E^n - Y \cong E^n - \tilde{Y}$, and it follows from Theorem 3.3 that $Sh(X) = Sh(\tilde{X}) = Sh(\tilde{Y}) = Sh(Y)$.

Having completed our sketch of the proof of Theorem 3.1, we remark that it is possible to weaken the hypothesis a bit by replacing connectivity in dimension r by pointed r-movability. We note that it follows from Theorem 4 of [Fe] that an (r-1)-shape connected continuum A is pointed r-movable if and only if A has the shape of some locally (r-1)-connected continuum.

THEOREM 3.4 (Ivanšić-Sher-Venema). *Let X and Y be (r-1)-shape connected, pointed r-movable continua in E^n of fundamental dimension at most k and satisfying ILC, where $n \geq \max(2k+2-r, k+3, 5)$. Then $Sh(X) = Sh(Y)$ implies $E^n - X \cong E^n - Y$. The converse holds if $n \geq k+4$ and X and Y are 1-shape connected.*

In case X and Y have polyhedral shape the co-fundamental dimension four condition required by Theorem 3.4 may be eliminated. The result then takes the following form, which generalizes some of the results of [Li_1] and [Li_2].

THEOREM 3.5 (Ivanšić-Sher-Venema). *Let X and Y be continua in E^n satisfying ILC and each having the shape of an r-connected finite complex of dimension at most k, where $n \geq \max(2k+1-r, k+3, 5)$. Then $Sh(X) = Sh(Y)$ implies $E^n - X \cong E^n - Y$. The converse holds if $r \geq 1$.*

As previously mentioned, Theorem 3.1 encompasses most of the finite-dimensional complement theorems which appear in the literature. There are, however, exceptions. These have typically hypothesized that the compacta in question have very special shape representatives. For example, Rushing [Ru] has established the following result about compacta having the shape of a sphere. (We give here Rushing's precise statement, which is for S^n rather than E^n; this is, of course, an unimportant technicality.)

THEOREM 3.6 (Rushing). *Let $X \subset S^n$, $n \geq 5$, be compact. Then, for $k \neq 1$, $Sh(X) = Sh(S^k)$ is equivalent to $S^n - X \cong S^n - S^k$ if X is globally 1-alg (and if $S^n - X$ has the homotopy type of S^1 when $k = n-2$).*

The added condition when $k = n-2$ is necessary in order to take into account knotting. If $2 \leq k \leq n-3$, then Theorem 3.6 follows from Theorem 3.5. Rushing's proof is somewhat different though. When $Sh(X) = Sh(S^k)$, he uses Stalling's Engulfing Theorem, in a manner adapted from some work on weak flatness of spheres [HR_2], to display $S^n - X$ as a topological copy of $S^{n-k-1} \times R^{k+1}$. The proofs in the cases not covered by Theorem 3.5 have a similar flavor. For the converse, a "Černavskiĭ meshing technique" is used to construct the required fundamental sequences. Of course, the advantage one has in a theorem of this type is that the topological type of one of the complements, $S^n - S^k$ in this case, is "standard."

A generalization of the above has been established by Venema [Ve_2]. Let $P = \{p_i\}_{i=1}^{\infty}$ denote a sequence of primes, or the constant sequence having value 0, or the constant sequence having value 1. Let S_P^1 denote the solenoid determined by the sequence P; that is, S_P^1 is the inverse limit of the sequence $S^1 \xleftarrow{p_1^*} S^1 \xleftarrow{p_2^*} S^1 \xleftarrow{p_3^*} \cdots$, where $p_i^*: S^1 \to S^1$ is the map defined by $p_i^*(z) = z^{p_i}$ for $z \in S^1$, $i=1,2,\cdots$. If $k \geq 2$, denote by S_P^k the (k-1)-fold suspension of S_P^1. Mardešić and Segal [MS] have shown that every S^k-like continuum has the shape of S_P^k for some sequence P as described above. Venema's result thus generalizes Rushing's from spheres to sphere-like continua. In its statement, S_P^k is "standardly embedded" in S^n.

THEOREM 3.7 (Venema). Let $X \subset S^n$, $n \geq 5$, be a globally 1-alg compactum. Then $S^n - X \cong S^n - S_P^k$, $k \neq 1$, if and only if $Sh(X) = Sh(S_P^k)$ and in case $k = n-2$, $\pi_1(S^n - X)$ is abelian and $\pi_i(S^n - X) = 0$, $i \geq 2$.

Rushing has shown that the condition $k \neq 1$ is required in Theorems 3.6 and 3.7. Specifically, he has given an example of a compact ANR X globally 1-alg embedded in S^n, $n \geq 5$, for which $S^n - X \cong S^n - S^1$, but $Sh(X) \neq Sh(S^1)$. Of course, it is the case that if X is globally 1-alg embedded in S^n, $n \geq 5$, and $Sh(X) = Sh(S_P^1)$, then $S^n - X \cong S^n - S_P^k$.

We conclude this section by mentioning that Rushing and Venema [RV] have established a version of the trivial range complement theorem in E^4. In place of the inessential loops condition, there appears the "disk pushing property." The compactum $X \subset E^4$ is said to have the disk pushing property if for each PL disk $D \subset E^4$ with $BdD \cap X = \emptyset$ and for every $\varepsilon > 0$, there exists a PL homeomorphism $h: E^4 \to E^4$ such that h is fixed outside the ε-neighborhood of X and $h(D) \cap X = \emptyset$.

THEOREM 3.8 (Rushing-Venema). Let X and Y be compacta in E^4 of fundamental dimension at most 1 and satisfying the disk pushing property. Then $Sh(X) = Sh(Y)$ if and only if $E^4 - X \cong E^4 - Y$.

It is not known if the disk pushing property may be replaced in Theorem 3.8 by the inessential loops condition.

4. A GENERALIZATION OF CHAPMAN'S COMPLEMENT THEOREM IN Q. The proof of Theorem 3.1 sketched in the last section can easily be modified to yield a proof of Chapman's Complement Theorem. In fact, the proof is somewhat easier in Q because of the strong

hypothesis that X and Y are Z-sets. Dimension restrictions vanish, since the Z-sets have infinite deficiency; and the Z-set approximation and unknotting theorems [Ch_3] make it possible to avoid the general position, piping, and engulfing arguments of the finite-dimensional case. Of ocurse, it can be argued that a Z-set hypothesis is not the most appropriate type for a complement theorem anyway, because the Z-set condition is local and essentially implies tameness.

In this section we present a complement theorem in Q which applies to a larger class of sets than the Z-sets. We call these sets the _weak Z-sets_, and the theorem is as follows.

THEOREM 4.1. _Suppose_ X _and_ Y _are weak_ Z-_sets in_ Q. _Then_ $Sh(X) = Sh(Y)$ _if and only if_ $Q-X \cong Q-Y$.

In essence, this result is established in Appendix 3 of [CS]. The proof we suggest here is less sophisticated and is strongly related to, and motivated by, the proof of Theorem 3.1. In fact, the definition of weak Z-set is so manufactured that the same proof applies. The idea is to find the embedding condition in Q-manifolds that is related to the Z-set condition in the same way that the inessential loops condition is related to the 1-ULC condition. The decisive idea is to be able to obtain, for a weak Z-set X in a Q-manifold, arbitrarily small closed neighborhoods of X each of which has a Z-set "spine." Actually, the existence of such neighborhoods characterizes the weak Z-sets, a fact which is the exact analogue of Theorem 2.8.

Suppose X is a compactum in the Q-manifold M. Then X is a _weak_ Z-_set_ if for each closed neighborhood U of X and closed set $A \subset U$ there exists a homotopy $H: A \times I \to U$ such that $H(A \times \{1\}) \cap X = \phi$ and $H(x,t) = x$ if $t = 0$ and $X \in A$ or if $t \in I$ and $x \in FrU$. Note that the definition may be altered to give the definition of (compact) Z-_set_ if it is additionally stipulated that the homotopy H may be chosen to be arbitrarily small. Thus a compact Z-set is a weak Z-set. We also remark that the notion of weak Z-set is related to that of globally unstable set introduced in [$\check{C}e_1$].

By standard results involving Z-sets, it may be shown that the set A in the above definition may be assumed to be a Z-set.

THEOREM 4.2. _Suppose_ X _is a compactum in the_ Q-_manifold_ M. _Then_ X _is a weak_ Z-_set if and only if for each closed neighborhood_ V _of_ X _and_ Z-_set_ $B \subset V$ _there exists a homotopy_ $G: B \times I \to V$ _such that_ $G(B \times \{1\}) \cap X = \phi$ _and_ $G(x,t) = x$ _if_ $t = 0$ _and_ $x \in B$ _or if_ $t \in I$ _and_ $x \in FrV$.

Now, Theorem 4.2 may be used to obtain yet another characterization of the weak Z-sets, here stated as Theorem 4.3. Theorem 4.3 may, in turn, be used to establish the crucial characterization of Theorem 4.4.

By an <u>ambient isotopy</u> of the space Y we mean a homotopy $h:Y\times I\to Y$ such that $h_0=id_Y$ and such that the map $h^*:Y\times I\to Y\times I$ defined by $h^*(y,t)=(h(y,t),t)$ for all $(y,t)\in Y\times I$ is a homeomorphism. Generally we speak of the level maps of h and write, e.g., "the ambient isotopy $h_t:Y\to Y$." If $C\subset Y$ and $h_t(y)=y$ for all $t\in I$ and $y\in C$ we say that the ambient isotopy $h_t:Y\to Y$ is <u>fixed</u> on C.

THEOREM 4.3. <u>Suppose X is a compactum in the Q-manifold M. Then X is a weak Z-set if and only if for each neighborhood U of X and Z-sets $A, B\subset M$ with $A\cap B\cap X=\phi$, there exists an ambient isotopy $h_t:M\to M$ fixed on $B\cup cl(M-U)$ such that $h_1(X)\cap A=\phi$.</u>

Suppose now that X is a compactum lying in the Q-manifold M, U is a closed neighborhood of X, and $K\subset int\ U$ is a compact Z-set. Then K is a <u>spine of U relative to X</u> if for each neighborhood V of K there exists an ambient isotopy $h_t:M\to M$ fixed on $K\cup cl(M-U)$ such that $h_1(X)\subset V$. If for each neighborhood W of X, there exist U and K as above for which $U\subset W$, we say that "X has arbitrarily small Z-spine neighborhoods."

THEOREM 4.4. <u>Suppose X is a compactum in the Q-manifold M. Then X is a weak Z-set if and only if X has arbitrarily small Z-spine neighborhoods.</u>

Proof. First suppose that X is a weak Z-set. Since X is compact we may assume M is compact. We may then identify M with $N^m\times Q=N^m\times I_1\times I_2\times\cdots$, where N^m is a compact piecewise linear manifold with boundary and $I_1=I_2=\cdots=[-1,1]$. If $k=1, 2,\cdots$, let $I^k=\Pi_{i=1}^k I_i$, let $Q_{k+1}=\Pi_{i=k+1}^{\infty} I_i$, and let $\pi_k:M\to I_k$ denote the projection. Note that X has arbitrarily small closed neighborhoods of the form $U=N_0^{m+n}\times Q_{n+1}\subset(N^m\times I^n)\times Q_{n+1}=M$, where N_0^{m+n} is a regular neighborhood in $N^m\times I^n$ of the polyhedron $K_0\subset N^m\times I^n$ meeting $\partial(N^m\times I^n)$ regularly. Denote by K the Z-set $K_0\times\{(-1, -1, -1,\cdots)\}\subset int\ U$. We claim that K is a spine of U relative to X. To establish this claim, let V be a neighborhood of K. Interior to V we may find a closed neighborhood of K of the form $(N_1^{m+n}\times Q_{n+1})\cap(\cap_{i=n+1}^p\pi_i^{-1}([-1,\varepsilon]))$, where $-1<\varepsilon<1$, $p>n$, and $N_1^{m+n}\subset int\ N_0^{m+n}$ is a regular neighborhood in $N^m\times I^n$ of K_0. Denote by A the Z-set $\cup_{i=n+1}^p\pi_i^{-1}(1)$ in M. Then by Theorem 4.3. there exists, since X is a weak Z-set, an ambient isotopy $h_t^1:M\to M$ such that h_t^1 is fixed on $K\cup cl(M-U)$ and $h_1^1(X)\cap A=\phi$. Now choose δ, $\varepsilon<\delta<1$, so that $h_1^1(X)\cap(\cup_{i=n+1}^p\pi^{-1}([\delta,1]))=\phi$, and choose a regular neighborhood N_2^{m+n} of K_0 lying in $int\ N_0^{m+n}$ so that $h_1^1(X)\subset N_2^{m+n}\times Q_{n+1}$. In $L=(N^m\times I^n)\times I_{n+1}\times\cdots\times I_p$, let L_1 denote $N_1^{m+n}\times[-1,\varepsilon]\times\cdots\times[-1,\varepsilon]$ and L_2 denote $N_2^{m+n}\times[-1,\delta]\times\cdots\times[-1,\delta]$. Then L_1 and L_2 are regular neighborhoods in L of $K_0\times\{(-1,\cdots,-1)\}\subset L$ meeting ∂L regularly. By Add. 2.16.3 of [Hud], there exists an ambient isotopy $k_t:L\to L$ such that k_t is fixed on $(K_0\times\{(-1,\cdots,-1)\})\cup$ $\cup cl(L-(N_0^{m+n}\times I_{n+1}\times\cdots\times I_p))$ and $k_1(L_2)=L_1$. Noting that $M=L\times Q_{p+1}$, if $t\in I$ let $h_t^2=k_t\times id_{Q_{p+1}}:M\to M$. Finally, if $t\in I$, let

$$h_t = \begin{cases} h^1_{2t} & \text{if } 0\le t\le 1/2 \\ h^2_{2t-1}h^1_1 & \text{if } 1/2\le t\le 1. \end{cases}$$

It follows that h_t is an ambient isotopy of M fixed on $K\cup \mathrm{cl}(M-U)$ such that $h_1(X)\subset V$.

Now suppose that X has arbitrarily small Z-spine neighborhoods. Let U be a closed neighborhood of X in M and let $A\subset U$ be a Z-set. Interior to U we may find a closed neighborhood W of X and a Z-set $K\subset \mathrm{int}\ W$ such that K is a spine of W relative to X. There exists a homotopy $H:A\times I\to U$ such that $K\ H(A\times\{1\})=\phi$ and $H(x,t)=x$ if $t=0$ and $x\in A$ or if $t\in I$ and $x\in \mathrm{Fr}U$. Let V be a neighborhood of K such that $V\cap H(A\times\{1\})=\phi$, and let $h_t:M\to M$ be an ambient isotopy fixed on $K\cup \mathrm{cl}(M-W)$ such that $h_1(X)\subset V$. Then, if $F:A\times I\to U$ is defined by $F(x,t)=h_t^{-1}(H(x,t))$ for $x\in A$ and $t\in I$, we note that $F(A\times\{1\})\cap X=\phi$ and that $F(x,t)=x$ if $t=0$ and $x\in A$ or if $t\in I$ and $x\in \mathrm{Fr}U$. It follows that X is a weak Z-set.

The neighborhood U constructed in the first portion of the above proof is a compact Q-manifold with Z-set frontier whose spine relative to X is a polyhedron. Thus we have the following corollary of the proof.

COROLLARY 4.5. *Suppose X is a compactum in the Q-manifold M having arbitrarily small Z-spine neighborhoods. Then X has arbitrarily small Z-spine neighborhoods each of which is a compact Q-manifold with Z-set frontier and polyhedral Z-spine relative to X.*

By Corollary 17.3 of [Ch_3], each Z-set in a Q-manifold has a neighborhood which embeds as an open subset of Q. This along with Theorem 4.4 yields the following.

COROLLARY 4.6. *Suppose X is a weak Z-set in the Q-manifold M. Then some neighborhood of X is homeomorphic to an open subset of Q.*

We are now prepared to sketch the proof of Theorem 4.1.

Proof of Theorem 4.1. Suppose first that X and Y are weak Z-sets in Q and that $Sh(X)=Sh(Y)$. We note that a version of Lemma 3.2 holds for X and Y, where the use of regular neighborhoods and piecewise linear embeddings in the proof of that lemma is replaced by the use of neighborhoods with Z-spines and Z-embeddings. Such a lemma allows us, as in the proof of Theorem 3.1, to prove that $Q-X\cong Q-Y$.

If, on the other hand, X and Y are weak Z-sets in Q and $Q-X\cong Q-Y$, let $\tilde{X}$ and $\tilde{Y}$ be Z-embedded copies in Q of X and Y respectively. By the above, $Q-X\cong Q-\tilde{X}$ and $Q-Y\cong Q-\tilde{Y}$, so $Q-\tilde{X}\cong Q-\tilde{Y}$. By Chapman's Complement Theorem, $Sh(\tilde{X})=Sh(\tilde{Y})$, so $Sh(X)=Sh(Y)$.

According to $[\check{C}e_2]$ or [Ge], a compact set X in the Q-manifold M is <u>cellular</u> if $X=\cap_{i=1}^{\infty}K_i$, where if $i=1,2,\cdots$, $K_i\cong Q$, $FrK_i\cong Q$, FrK_i is a Z-set in K_i, and $X\subset \operatorname{int} K_i$. Evidently a cellular set must have trivial shape; that is, the shape of a singleton set. The following states that for compacta of trivial shape, the cellular sets are precisely the weak Z-sets.

COROLLARY 4.7. *Suppose X is a compactum of trivial shape in the Q-manifold M. Then X is cellular if and only if X is a weak Z-set.*

Proof. Suppose first that X is cellular. Then inside any neighborhood of X we may find a neighborhood N of X such that $(vQ,Q)\cong(N,FrN)$, where vQ is the cone over Q. The image of the cone point under this homeomorphism is a spine of N relative to X, and so X is a weak Z-set by Theorem 4.4.

Now suppose that X is a weak Z-set. Then $Q-X\cong Q-\{pt\}$ by Theorem 4.1, and hence for any neighborhood U of X, we may find a bicollared copy Q' of Q lying in $U-X$. By the Schoenflies Theorem for Q (see $[Wo_2]$; also note §22 on pg. 38 of $[Ch_3]$, there is a neighborhood V of X such that $Q'=FrV$, $V\subset U$, and $(V,FrV)\cong(Q\times I,Q\times\{0\})$. It follows that X is cellular.

5. A PARTIAL COMPLEMENT THEOREM FOR CONTINUA IN A MANIFOLD. The complement theorems we have stated up until now have dealt with compacta embedded in E^n or Q. For compacta lying in other manifolds, the situation is a bit more complicated. The technique used in the first part (I) of the proof of Theorem 3.1 has, however, been modified in [IS] to give a partial result, stated as Theorem 4.1 below.

Throughout this section our results are concerned only with pointed 1-movable continua. It follows from Theorem 7.1.3 of $[DS_1]$ that shape morphisms between such continua may be regarded as <u>pointed</u> morphisms. For the remainder of this section we assume, then, that all shape morphisms are pointed; we shall, however, surpress base points from our notation.

Recall that a map $f:X\to Y$ between ANR's is r-<u>connected</u> if $f_{\#}:\pi_i(X)\to\pi_i(Y)$ is an isomorphism when $0\le i\le r-1$ and an epimorphism when $i=r$. With this in mind, we say that the shape morphism $\underline{f}:X\to Y$ between pointed 1-movable continua is <u>shape</u> r-<u>connected</u> if $\underline{f}_{\#}$: pro-$\pi_i(X)\to$ pro-$\pi_i(Y)$ is an isomorphism of pro-groups for $0\le i\le r-1$ and an epimorphism for $i=r$. We also recall that a pro-group $\underline{G}=\{G_\alpha,g_{\alpha\beta},A\}$ is <u>stable</u> if $\underline{G}$ is isomorphic in the category pro-groups to a group, and that $\underline{G}$ satisfies the <u>Mittag-Leffler condition</u> if for each $\alpha\in A$ there exists $\beta\ge\alpha$ such that for all $\gamma\ge\beta$ $g_{\alpha\gamma}(G_\gamma) = g_{\alpha\beta}(G_\beta)$.

We now state our partial complement theorem. It uses the notion of <u>relative shape equivalence</u>, rather than shape equivalence. Relative shape equivalence was introduced in $[Ch_1]$, where the basic definitions may be found.

THEOREM 5.1 (Ivanšić-Sher). Let X_1 and X_2 be continua in the interior of the piecewise linear n-manifold M such that for j=1 or 2, X_j has fundamental dimension at most k, X_j satisfies ILC, and pro-$\pi_i(X_j)$ is stable for $0\le i\le r-1$ and satisfies the Mittag-Leffler condition for $i=r<n-3$, where $n\ge\max(2k+2-r, k+3, 5)$. Suppose the inclusion of X_1 into M is shape r-connected and that X_1 and X_2 have the same shape relative to M. Then $M-X_1\cong M-X_2$.

We note that the hypothesis that the inclusion of X_1 into M be shape r-connected is necessary in Theorem 5.1. In Counterexample 6, Chapter 8 of [Ze] is shown the existence, for $m\ge 2$, of an m-sphere S_1^m inessentially (piecewise linearly) embedded in $\text{int}(B^{2m}\times S^1)$ so that S_1^m does not bound an (m+1)-cell in $B^{2m}\times S^1=M$. If S_2^m is a piecewise linear m-sphere lying in the interior of a (2m+1)-cell in M, then it is easily seen that $M-S_1^m\not\cong M-S_2^m$.

As we have come to expect, the proof of such a theorem reduces to verifying a suitable analogue of Lemma 3.2. This is done in [IS]. To accomplish this, an analogue of Theorem 2.9 is required. This is provided by Theorem 5.2, which we state below after giving the appropriate definitions.

If X is a compactum lying in the interior of the piecewise linear n-manifold M, a defining sequence for X is a sequence $\{U_i\}_{i=1}^{\infty}$ of compact piecewise linear n-manifolds in M such that $X=\cap_{i=1}^{\infty}U_i$ and, if $j=1,2,\cdots$, $U_{j+1}\subset\text{int }U_j$. We call a defining sequence $\{U_i\}_{i=1}^{\infty}$ r-connected if for $j=1,2,\cdots$, the inclusion of U_{j+1} into U_j is an r-connected mapping.

THEOREM 5.2. Suppose X is a continuum of fundamental dimension at most k lying in the interior of the piecewise linear n-manifold M and satisfying ILC, where $n\ge 5$ and $k\le n-3$. Suppose pro-$\pi_i(X)$ is stable for $0\le i\le r-1$ and satisfies the Mittag-Leffler condition for $i=r<n-3$. Then there exists an r-connected defining sequence $\{U_i\}_{i=1}^{\infty}$ for X such that if $j=1,2,\cdots$, then U_j has a spine of dimension at most $k'=\max(k,r+1)$.

We note in the following that Theorem 5.2 may be improved in the case $k=1=r$ by obtaining $k'=1$.

THEOREM 5.3. Suppose X is a pointed 1-movable continuum of fundamental dimension $k\le 1$ lying in the interior of the piecewise linear orientable n-manifold M and satisfying ILC, where $n\ge 5$. Then there exists a 1-connected defining sequence $\{U_i\}_{i=1}^{\infty}$ for X such that if $j=1,2,\ldots$, then U_j has a spine of dimension k.

It would seem as though Theorems 2.8, 2.9, 4.4, 5.2, and 5.3 should be of interest in many problems dealing with embedded compacta. In this connection, we remark that

Venema [Ve_4] has established the following approximation theorem up to shape. In its conclusion, "dem" denotes demension in the sense of Štan'ko; see [Ed]. When $Fd(X) \leq n-3$, $Fd(X) = dem(X)$ is equivalent to ILC.

THEOREM 5.4 (Venema). Suppose X is a compactum in the interior of the PL n-manifold M^n. Then for every neighborhood U of X in M there exists a compactum $X' \subset U$ such that

a). $dem(X') = Fd(X)$, and

b). X and X' have the same relative shape in U.

A similar result, which we now state, was established in [IS].

THEOREM 5.5 (Ivanšić-Sher). Let X be a continuum in the interior of the piecewise linear n-manifold M, K be a k-dimensional polyhedron, and $f:K \to X$ be a shape equivalence, where $n \geq 5$ and $k \leq n-3$. Then for each neighborhood U of X in M there exists a k-dimensional polyhedron $K' \subset U$ and a simple homotopy equivalence $H:K \to K'$ so that the homotopy inverse of h and f induce a relative shape equivalence of K' and X in U.

As an application of Theorems 5.1 and 5.5, one can almost immediately obtain Theorem 5.6 below, which appears in [IS]. It generalizes Theorem 2.4 of [CDD] and Theorem 3 of [Li_1]. Since deleted product neighborhoods are I-regular neighborhoods [Si_1], Theorem 5.6 also yields the fact, established in [SGH], that I-regular neighborhoods exist for ILC embedded continua in the interior of a piecewise linear n-manifold, $n \geq 5$, which have the shape of a codimension three polyhedron. Recall that if X is a compactum in the interior of the PL n-manifold M, a deleted product neighborhood of X is a compact PL manifold neighborhood N of X in M such that $N - X \cong \partial N \times [0,1)$.

THEOREM 5.6. Let X be a continuum in the interior of the piecewise linear n-manifold M such that X satisfies ILC and has the shape of a k-dimensional polyhedron K, where $n \geq 5$ and $k \leq n-3$. Then X has a deleted product neighborhood in M.

6. CHAPMAN'S CATEGORY ISOMORPHISM THEOREM. In Chapman's fundamental paper [Ch_1] there appears another sort of "complement theorem" upon which we have not yet remarked. This result, which appears as Theorem 1 in [Ch_1], actually gives an alternate description of shape theory via a homotopy category whose objects are certain contractible open subsets of Q.

Following Chapman's notation, let S denote the category whose objects are Z-sets in Q and whose morphisms are fundamental equivalence classes of fundamental sequences in Q between Z-sets. This constitutes a subcategory of Borsuk's fundamental category, and it is well known that S is sufficient to completely describe shape theory for metric compacta. If M and N are spaces, then maps $f,g:M\to N$ are said to be <u>weakly properly homotopic</u> if for each compactum $B\subset N$ there exist a compactum $A\subset M$ and a homotopy $F:M\times I\to N$ joining f and g so that $F((M-A)\times I)\cap B=\phi$; the maps f and g are <u>properly homotopic</u> if there exists a proper map $F:M\times I\to N$ which is a homotopy joining f and g. We let P denote the category whose objects are complements of Z-sets in Q and whose morphisms are weak proper homotopy classes of proper maps.

THEOREM 6.1 (Chapman). <u>There exists a category isomorphism</u> T <u>from</u> P <u>onto</u> S <u>such that</u> $T(M)=Q-M$ <u>for each object</u> M <u>of</u> P.

Theorem 6.1 undoubtedly led Edwards and Hastings [EH] to formulate the notion of "strong shape theory" by examining the category $\tilde{P}$ whose objects are the objects of P and whose morphisms are <u>proper</u> homotopy classes of proper maps. Strong shape theory has been described in different terms by Dydak and Segal [DS_2] and Kodama and Ono [KO] (and was anticipated earlier by others, such as Christie [Chr]). It is not yet known whether every shape equivalence is a strong shape equivalence, and there is much current work that aims at settling this question. We also remark that explicit finite-dimensional versions of Theorem 6.1 do not seem to have been considered.

The University of North Carolina at Greensboro
Greensboro, NC 27412

BIBLIOGRAPHY

[Bl] W. A. Blankinship, *Generalization of a construction of Antoine*, Ann. of Math. 53 (1951), 276-291.

[Bo_1] K. Borsuk, *Concerning homotopy properties of compacta*, Fund. Math. 62 (1968), 223-254.

[Bo_2] K. Borsuk, *Theory of Shape*, Monografie Matematyczne, Tom 59, Polish Scientific Publishers, Warsaw, 1975.

[Br] Morton Brown, *A proof of the generalized Schoenflies Theorem*, Bull. Amer. Math. Soc. 66 (1960), 74-76.

[$Če_1$] Z. Čerin, *Homotopy properties of locally compact spaces at infinity-triviality and movability*, Glas. Mat. Ser. III 13 (1978), 347-370.

[$Če_2$] Z. Čerin, *On cellular decompositions of Hilbert cube manifolds*, preprint.

[Ch_1] T. A. Chapman, *On some applications of infinite-dimensional manifolds to the theory of shape*, Fund. Math. 76 (1972), 181-193.

[Ch_2] T. A. Chapman, *Shapes of finite-dimensional compacta*, Fund. Math. 76 (1972), 261-276.

[Ch_3] T. A. Chapman, *Lectures on Hilbert Cube Manifolds*, CBMS Regional Conference Series in Mathematics No. 28, American Mathematical Society, Providence, 1976.

[CS] T. A. Chapman and L. C. Siebenmann, *Finding a boundary for a Hilbert cube manifold*, Acta Math. 137 (1976), 171-208.

[Chr] D. Christie, *Net homotopy for compacta*, Trans. Amer. Math. Soc. 56 (1944), 275-308.

[CDD] D. Coram, R. J. Daverman, and P.F. Duvall, Jr., *A loop condition for embedded compacta*, Proc. Amer. Math. Soc. 53 (1975), 205-212.

[CD] D. Coram and P. Duvall, Jr., *Neighborhoods of sphere-like continua*, General Topology and Appl. 6 (1976), 191-198.

[Da] Robert J. Daverman, *On weakly flat 1-spheres*, Proc. Amer. Math. Soc. 38 (1973), 207-210.

[Du] Paul F. Duvall, Jr., *Weakly flat spheres*, Michigan Math. J. 16(1969), 117-124.

[DS_1] J. Dydak and J. Segal, *Shape Theory, An Introduction*, Lecture Notes in Mathematics No. 688, Springer-Verlag, New York, 1978.

[DS_2] J. Dydak and J. Segal, *Strong shape theory*, Dissertationes Mathematicae, to appear.

[EH] D. A. Edwards and H.M. Hastings, *Čech and Steenrod Homotopy Theory with Applications to Geometric Topology*, Lecture Notes in Mathematics No. 542, Springer-Verlag, New York, 1976.

[Ed] Robert D. Edwards, *Demension theory, I*, in Geometric Topology (L. C. Glaser and T. B. Rushing, editors), Lecture Notes in Mathematics No. 438, Springer-Verlag, New York, 1975, pp. 195-211.

[Fe] S. Ferry, *A stable converse to the Vietoris-Smale theorem with applications to shape theory*, Trans. Amer. Math. Soc. 261 (1980), 369-386.

[FA] R. H. Fox and Emil Artin, *Some wild cells and spheres in three-dimensional space*, Ann. of Math. 49 (1948), 979-990.

[Ge] Ross Geoghegan, ed., *Open problems in infinite-dimensional topology*, Topology Proceedings 4 (1979), 287-338.

[GS] Ross Geoghegan and R. Richard Summerhill, *Concerning the shapes of finite-dimensional compacta*, Trans. Amer. Math. Soc. 179 (1973), 281-292.

[HR_1] J. G. Hollingsworth and T. B. Rushing, *Embeddings of shape classes of compacta in the trivial range*, Pacific J. Math. 60 (1975), 103-110.

[HR_2] J. G. Hollingsworth and T. B. Rushing, *Homotopy characterizations of weakly flat codimension 2 spheres*, Amer. J. Math. 98 (1976), 385-394.

[Hud] J. F. P. Hudson, *Piecewise Linear Topology*, W. A. Benjamin, New York, 1969.

[Iv] I. Ivanšić, *Embedding compacta up to shape*, Bull. Acad. Polon. Sci. Sér. Sci. Math. Astronom. Phys. 25 (1977), 471-475.

[IS] I. Ivanšić and R. B. Sher, *A complement theorem for continua in a manifold*, Topology Proceedings, to appear.

[ISV] I. Ivanšić, R. B. Sher, and G. A. Venema, *Complement theorems beyond the trivial range*, Illinois J. Math., to appear.

[KO] Y. Kodama and J. Ono, *On fine shape theory*, Fund. Math. 105 (1979), 29-39.

[Li_1] Vo-Thanh-Liem, *Certain continua in* S^n *of the same shape have homeomorphic complements*, Trans. Amer. Math. Soc. 218 (1976), 207-217.

[Li_2] Vo-Thanh-Liem, *Certain continua in* S^n *with homeomorphic complements have the same shape*, Fund. Math. 97 (1977), 221-228.

[MS] S. Mardesic and J. Segal, *Shapes of compacta and ANR systems*, Fund. Math. 72 (1971), 41-59.

[Mc_1] D. R. McMillan, Jr., *A criterion for cellularity in a manifold*, Ann. of Math. 79 (1964), 327-337.

[Mc_2] D. R. McMillan, Jr., *Strong homotopy equivalence of 3-manifolds*, Bull. Amer. Math. Soc. 73 (1967), 718-722.

[Ro] Ronald H. Rosen, *The five dimensional polyhedral Schoenflies Theorem*, Bull. Amer. Math. Soc. 70 (1964), 511-516.

[Ru] T. B. Rushing, *The compacta* X *in* S^n *for which* $Sh(X)=Sh(S^k)$ *is equivalent to* $S^n-X\approx S^n-S^k$, Fund. Math. 97 (1977), 1-8.

[RV] T. B. Rushing and Gerard A. Venema, *A weak flattening criterion for compacta in 4-space*, in Geometric Topology (James C. Cantrell, editor), Academic Press, New York, 1979, pp. 649-654.

[Sa] Katsuro Sakai, *Replacing maps by embeddings of* $[0,1)$*-stable Q-manifold pairs*, Math. Japon. 22 (1977), 93-98.

[Si_1] L. C. Siebenmann, *Regular (or canonical) open neighborhoods*, General Topology and Appl. 3 (1973), 51-61.

[Si_2] L. Siebenmann, *Chapman's classification of shapes: A proof using collapsing*, Manuscripta Math. 16 (1975), 373–384.

[SGH] L. C. Siebenmann, L. Guillou, and H. Hähl, *Les voisinages ouvert réguliers: critères homotopiques d'existence*, Ann. Sci École Norm. Sup, (4) 7 (1974), 431-462.

[Št] M. A. Štan'ko, *Approximation of imbeddings of compacta in codimensions greater than two*, Soviet Math. Dokl. 12 (1971), 906-909.

[Ve_1] Gerard A. Venema, *Embeddings of compacta with shape dimension in the trivial range*, Proc. Amer. Math. Soc. 55 (1976), 443-448.

[Ve_2] Gerard A. Venema, *Weak flatness for shape classes of sphere-like continua*, General Topology and Appl. 7 (1977), 309-319.

[Ve_3] Gerard A. Venema, *Neighborhoods of compacta in Euclidean space*, preprint.

[Ve_4] Gerard A. Venema, *An approximation theorem in the shape category*, preprint.

[Wo_1] Raymond Y. T. Wong, *A wild Cantor set in the Hilbert cube*, Pacific J. Math. 24 (1968), 189-193.

[Wo_2] Raymond Y. T. Wong, *Extending homeomorphisms by means of collarings*, Proc. Amer. Math. Soc. 19 (1968), 1443-1447.

[Ze] E. C. Zeeman, *Seminar on Combinatorial Topology*, Mimeographed Notes, Institut Hautes Études Sci., Paris, 1963.

EMBEDDINGS IN SHAPE THEORY

Gerard A. Venema*
Department of Mathematics
Calvin College
Grand Rapids, Michigan 49506/USA

In this paper we will study several results from the theory of topological embeddings and their relationships with shape theory. Most of the material was motivated by proofs of the finite dimensional complement theorems and is not new, but we intend to examine it from a somewhat different point of view.

Since Chapman proved his famous complement theorem for Z-sets in the Hilbert cube, a number of authors have formulated and proved versions of the theorem for finite dimensional spaces (see the article by R. B. Sher in these proceedings). It has become evident that the finite dimensional theorems are really shape theory versions of unknotting theorems. Unknotting theorems in turns are an important part of the theory of topological embeddings (the study of the way in which one space is situated in another). We will review the major kinds of theorems on topological embeddings and discuss the analogous results in shape theory.

The goal is to understand the strong parallel which exists between the results on topological embeddings and the results in shape theory. This should lead to a deeper understanding of the shape theory results as well as those on topological embeddings. The basic observation is that if all definitions in a topological theorem are globalized by replacing local conditions with conditions about homotopies in neighborhoods, then a shape theorem results.

Theorems and definitions will be stated in three different settings: the piecewise linear (PL) setting consisting of finite dimensional polyhedra and PL maps, the topological setting consisting of compacta and continuous maps, and the shape setting consisting of compacta and shape morphisms. In the past the parallel between the PL and shape settings has been emphasized as a motivation for the theorems in shape theory. The point of view taken here is that the relationship

---

*Research partially supported by National Science Foundation grant number MCS 7902661.

between the topological and shape settings is just as strong and that is the relationship we will emphasize. We hope that the shape theorems can now serve as motivation for some new theorems in the topological setting and we raise some such questions in the last two sections of the paper. References will be given for all theorems, but the proofs will be given of the shape theorems only. A good reference for the others is T. B. Rushing's book [16], although the final versions of some of the theorems were proved after the book was written.

There are four basic kinds of problems to be described. The first kind is taming in which the problem is to distinguish a class of tame embeddings about which interesting theorems can be proved and to give a property which characterizes embeddings in that class. The second kind of problem is approximation. There we wish to show that every embedding can be approximated by a tame one. The third problem is unknotting. We show that, for certain spaces, any two tame embeddings are equivalent. The fourth problem, embedding, is to determine which compacta can be embedded in a given manifold.

All spaces in this paper are assumed to be finite dimensional and metric. The *dimension* of a space X (dim X) is the *covering dimension*. The *fundamental dimension* of X (Fd(X)) is the minimum of $\{\dim Y \mid Sh(X) = Sh(Y)\}$. The basic properties of fundamental dimension are described in [15]. A *map* is a continuous function. An *embedding* is a map which is a homeomorphism onto its image. An *isotopy* is a homotopy h_t, $0 \le t \le 1$, such that each h_t is a homeomorphism. A *pseudoisotopy* is a homotopy h_t, $0 \le t \le 1$, such that h_t is a homeomorphism as long as $t<1$. A *polyhedron* is the underlying set of a finite dimensional simplicial complex. The symbol M^n will always denote an n-manifold (without boundary) which will be assumed to be piecewise linear and to be equipped with a metric d. The notation $f \simeq g$ means that f is homotopic to g.

§1. TAMING.

The first goal of a theory of topological embeddings is to distinguish the tame embeddings from the wild ones.

DEFINITION 1.1. An embedding $g:K \to M^n$ of a polyhedron K into the PL manifold M^n is *tame* if there exists an isotopy h_t of M such that h_0 = id and $h_1g(K)$ is a subpolyhedron of M.

DEFINITION 1.2. A compactum $X \subset M^n$ is *locally simply co-connected* (1-LCC) if for every $\varepsilon > 0$ and for every $x \in X$ there exists a $\delta > 0$ such that the inclusion induced homomorphism $\pi_1(N_\delta(x)-X) \to \pi_1(N_\varepsilon(x)-X)$ is the zero map.

THEOREM 1.3 (Bryant and Seebeck [4]). *If* $g:K^k \to M^n$ *is an embedding of a compact k-dimensional polyhedron into a* PL *n-manifold*, $k \leq n-3$, $n \geq 5$, *then* g *is tame if and only if* g(X) *is* 1-LCC.

Thus the 1-LCC property characterizes the tame embeddings of polyhedra in codimension three. We shall now see that the same property distinguishes the tame embeddings of compacta. The theory for compacta was worked out by M. A. Šťaňko [19] and is known as "demension theory".

DEFINITION 1.4 [19]. If $X \subset M^n$ is a compactum, the *demension* of X (which stands for "dimension of embedding" of X) is less than or equal to k (abbreviated dem $X \leq k$) if for every $\varepsilon > 0$ there exists a compact PL manifold neighborhood N of X such that N has a k-dimensional spine and ε-collapses to that spine. (Recall that a polyhedron $K \subset N$ is a *spine* of N if N simplicially collapses to K and that the collapse is an *ε-collapse* if the deformation of N to K induced by the collapse moves no point more than ε .

The demension of X depends on how X is embedded in M. Šťaňko goes on to define a compactum to be tame if its embedding dimension equals its intrinsic dimension.

DEFINITION 1.5 [19]. A compactum $X \subset M$ is *tame* if dem X = dim X.

THEOREM 1.6 (Šťaňko [19]) *Suppose* $X \subset M^n$ *is compact.*

Part 1. *If* $n \neq 3$ *and* dim $X \geq n-2$, *then* dem X = dim X.

Part 2. *If* $n \neq 4$ *and* dim $X \leq n-3$, *then either*

a) dem X = dim X *and* X *is* 1-LCC, *or*

b) dem X = n-2 *and* X *is not* 1-LCC.

Again the tame embeddings are characterized by the 1-LCC property. The justification for making the above definition of tame comes from the proofs of the unknotting theorems to be stated in §3 below.

We next turn to the shape category. The theorems there are virtually the same as those in the topological category, providing that all definitions are suitably globalized. Each condition about pointwise closeness is dropped and replaced by a condition which requires only closeness in the global sense of homotopy in neighborhoods of X.

DEFINITION 1.7 [21]. A compactum $X \subset M^n$ satisfies the *inessential loops condition* (abbreviated ILC) if for every neighborhood U of X there exists a neighborhood V of X in U such that each loop in V-X which is homotopically inessential in V is also inessential in U-X.

DEFINITION 1.8 [22]. Suppose $X \subset M^n$ is compact. The *fundamental dimension of embedding* of X is less than or equal to k (abbreviated $FDE(X) \leq k$) If for every neighborhood U of X there exists a PL manifold neighborhood N of X in U such that N has a k-dimensional spine. (The retraction of N to the spine is not assumed to be small.)

THEOREM 1.9 [22]. *Suppose* $X \subset M^n$ *is compact.*

Part 1. *If* $n \neq 3$ *and* $Fd(X) \geq n-2$, *then* $FDE(X) = Fd(X)$.

Part 2. *If* $n \neq 4$ *and* $Fd(X) \leq n-3$, *then either*

a) $FDE(X) = Fd(X)$ *and* X *satisfies* ILC, *or*

b) $FDE(X) = n-2$ *and* X *does not satisfy* ILC.

Thus the tame embeddings in the shape category are characterized by the ILC. The case n=3 must be excluded from Part 1 of both Theorems 1.6 and 1.9 since the 1-dimensional compact subsets of E^3 constructed in [1] and [13] have fundamental dimension (dimension) one and fundamental dimension of embedding (demension) two. It is not known whether Part 2 of either theorem is true when n=4.

We now prove Theorem 1.9. The same method could also be used to prove Theorem 1.6; it would just be necessary to introduce epsilons. Such a proof of Theorem 1.6 would be simpler than the one in [19] and perhaps even than that in [7].

LEMMA 1.10 (Characterization of fundamental dimension [22]). *Suppose* X *is a compact subset of* M^n. *Then* $Fd(X) \leq k$ *if and only if for every neighborhood* U *of* X *there exists a neighborhood* V *of* X *in* U *and a* k*- dimensional polyhedron* $K \subset U$ *such that the inclusion of* V *into* U *is homotopic in* U *to a map of* V *into* K.

REMARK. It follows from Lemma 1.10 that $FDE(X) \geq Fd(X)$ for every $X \subset M^n$.

PROOF. If X satisfies the UV condition, then X has the shape of an inverse limit of k-dimensional polyhedra and thus has fundamental dimension at most k.

Suppose $Fd(X) \leq k$. Then there exists a compactum Y such that $Sh(X)=Sh(Y)$ and $\dim Y \leq k$. We consider first the special case in which $M^n = E^n$. Embed Y in E^{2k+1}. There exist fundamental sequences $\underline{f} = \{f_i,X,Y\}_{E^n,E^{2k+1}}$ and $\underline{g} = \{g_i,Y,X\}_{E^{2k+1},E^n}$ such that $\underline{g}\,\underline{f} = \underline{id_X}$.

Let U be a neighborhood of X. There exists a neighborhood W of Y and an integer i_0 such that $g_i(W) \subset U$ for all $i \geq i_0$. Choose a neighborhood V of X in U and an integer $i_1 \geq i_0$ such that $f_i(V) \subset W$ and $g_i f_i|V \simeq j$ for all $i \geq i_1$, where $j: V \to U$ is the inclusion map. Since $\dim Y \leq k$, there exists a polyhedron $L \subset W$ of dimension at most k and a map $p:Y \to L$ such that $p \simeq$ inclusion in W [14, Theorem IV.9]. It may be assumed that $g_{i_1}|K$ is PL. Then $j \simeq g_{i_1} f_{i_1}|V \simeq g_{i_1} p f_{i_1}|V$, and $K = g_{i_1}(L)$ is the polyhedron needed to finish the proof of the special case.

Now, if $M^n \neq E^n$, we first embed M^n as a PL subset of E^{2n+1}. We can apply the proof of the special case to X (considered as a subset of E^{2n+1}) and then use a PL retraction of a neighborhood of M^n in E^n onto M^n to push the homotopies into M^n.

We use Δ^n to denote an n-dimensional simplex; $S^n = \partial\Delta^{n+1}$.

LEMMA 1.11 [22]. *Suppose* $X \subset M^n$ *is compact and* $Fd(X) \leq n-2$. *Then* $\pi_1(U,U-X)=0$ *for every neighborhood* U *of* X *in* M.

PROOF. Let $U \supset X$ and $f\colon (\Delta^1, \partial\Delta^1) \to (U, U-X)$ be given. Lemma 1.10 implies that there exists a compact PL manifold neighborhood V of X and a polyhedron $K \subset U$ such that the inclusion $V \hookrightarrow U-f(\partial\Delta^1)$ is homotopic in $U-f(\partial\Delta^1)$ to a map of V into K. We assume that U is connected by concentrating only on the component of U containing $f(\Delta^1)$ if necessary.

Let $\tilde{U}$ denote the universal covering space of U and $p\colon \tilde{U} \to U$ the projection. Let $\overline{V} = p^{-1}(V)$, $\overline{X} = p^{-1}(X)$, and let $\tilde{f}\colon (\Delta^1, \partial\Delta^1) \to (\tilde{U}, \tilde{U}-\overline{X})$ denote a lift of f. The homotopy lifting property can be used to lift the homotopy of V downstairs to a homotopy $H\colon \overline{V}\times I \to \tilde{U}$ such that H_0=inclusion and $H_1(\overline{V}) \subset p^{-1}(K)$. Throw H and $\tilde{f}$ into general position.

We claim that $\partial\overline{V}$ does not separate $\tilde{f}(0)$ from $\tilde{f}(1)$ in $\tilde{U}$. If it does, there must be a component B of $\partial\overline{V}$ and a map $g\colon \Delta^1 \to \tilde{U}$ such that $g(0) = \tilde{f}(0)$, $g(1) = \tilde{f}(1)$ and $g(\Delta^1) \cap B$ consists of exactly one point. Then $(H|B\times I)^{-1}(g(\Delta^1))$ is a compact 1-manifold with one end point. Since no such 1-manifold exists, $\partial\overline{V}$ does not separate $\tilde{f}(0)$ from $\tilde{f}(1)$. Join $\tilde{f}(0)$ and $\tilde{f}(1)$ with an arc A which lies in $\tilde{U}-\partial\overline{V}$.

Notice that, since $A \cap \partial\overline{V} = \phi$, we must have $A \cap \overline{V} = \phi$ and so $A \cap X = \phi$. Because $\tilde{U}$ is simply connected, $\tilde{f}(\Delta^1) \cup A$ is null-homotopic in $\tilde{U}$. Projecting that homotopy down we see that f is homotopic (rel $\partial\Delta^1$) to $f'\colon \Delta^1 \to U-X$. Thus $\pi_1(U, U-X)=0$.

PROOF OF THEOREM 1.9, Part 1. The cases in which $n \leq 2$ or $Fd(X) \geq n-1$ are easy, so we concentrate on $Fd(X)=n-2$ and $n \geq 4$. By the remark following Lemma 1.10, it is enough to show that $FDE(X) \leq n-2$. Given a neighborhood U of X, choose a PL manifold neighborhood N of X such that $N \subset U$. Let N^1 denote the 1-skeleton of N and let N_*^{n-2} denote the dual (n-2)-skeleton. (N_*^{n-2} is the union of all simplices in the second barycentric subdivision of N which do not intersect N^1.) It follows from Lemma 1.11 and Stallings' engulfing theorem [17] that there is a PL homeomorphism $h\colon M \to M$ such that $h|M^n-N = id$ and $h(N^1) \cap X = \phi$. But $X \subset R$ for some regular neighborhood R of $h(N_*^{n-2})$ in U and so the proof is complete.

LEMMA 1.12 [21]. *Suppose* $X \subset M^n$ *is compact and satisfies* ILC. *Let* $k = Fd(X)$. *Then* $\pi_i(U, U-X)=0$, $0 \leq i \leq n-k-1$, *for every open neighborhood* U *of* X *in* M^n.

PROOF. The case $k=n-2$ is taken care of by Lemma 1.11, so we may assume that $k\leq n-3$. We also assume that U is connected since if each component of U satisfies the conclusion of the Lemma, then U does. Consider the universal cover $p:\tilde{U} \to U$. Denote $p^{-1}(X)$ by $\bar{X}$. The proof of Lemma 1.11 shows that $\pi_1 (\tilde{U},\tilde{U}-\bar{X}) = 0$ and of course $\pi_1(\tilde{U}) = 0$. We wish to show that $\pi_1(\tilde{U}-\bar{X})=0$.

Let $f:S^1 \to \tilde{U}-\bar{X}$ be a loop. Extend f to $f: \Delta^2 \to \tilde{U}$. Choose a neighborhood V of X using the ILC hypothesis. There are disks $D_1,\ldots,D_\ell$ in Δ^2 such that $pf(D_i)\subset V$ for each i and $(pf)^{-1}(X)\subset \text{int } D_1 \cup\ldots\cup \text{int } D_\ell$. By the choice of V, each $pf|D_i$ can be replaced by a map which agrees with pf on D_i and maps D_i into U-X. Lifting that map, we see that $f\sim *$ in $\tilde{U}-\bar{X}$. Thus $\pi_1(\tilde{U}-\bar{X})=0$.

We claim that $H_i(\tilde{U},\tilde{U}-\bar{X})=0$ for $2\leq i \leq n-k-1$. If so, the proof will be complete since the relative Hurewicz theorem then gives $\pi_i(\tilde{U},\tilde{U}-\bar{X})=0$ for $2\leq i\leq n-k-1$ and thus we see that $\pi_i (U,U-X)=0$ for the sames values of i.

Lemma 1.10 implies that the inclusion of X into any neighborhood factors, up to homotopy, through a map onto a k-dimensional polyhedron. The homotopy lifting property shows that $\bar{X}$ has the same property. The continuity axiom for Čech cohomology then gives $H^q_c(\bar{X})=0$ for $q\geq k+1$. Finally Alexander duality proves $H_i(\tilde{U},\tilde{U}-\bar{X})=H^{n-i}_c(\bar{X}) = 0$ for $n-i\geq k+1$ as claimed.

LEMMA 1.13 [21]. *Let* $X\subset M^n$ *be a compactum such that* $Fd(X)\leq n-3$. *For every neighborhood* U *of* X *there exists a neighborhood* V *of* X *in* U *such that for any compact polyhedron* $K\subset V$ *with* $\dim K\leq n-3$ *there is a polyhedron* P *and a regular neighborhood* N *of* P *such that* $\dim P\leq Fd(X)$ *and* $K\subset N\subset U$.

PROOF. The proof is by induction on $k=\dim K$. If $k\leq Fd(X)$, then P=K and V=U will work. So we assume that $k>Fd(X)$ and that the Lemma is true for polyhedra of dimension less than k.

Choose $V'\subset U$ using the inductive hypothesis. By Lemma 1.10 there is a neighborhood V of X in V' and a polyhedron $P'\subset V'$ such that $\dim P' \leq Fd(X)$ and the inclusion $V \hookrightarrow V'$ is homotopic in V' to a map of V to P'. Let $f:K\times[0,1] \to V'$ be a homotopy such that $f_0=id$ and $f_1(K)\subset P'$. By Zeeman's Piping Lemma [24, Lemma 48], there is a subpolyhedron L of $K\times I$ such that dim $L\leq k-1$, $S(f)\subset L$, and $K\times I \searrow K\times\{1\}\cup L$. By induction it may be assumed that $P'\cup f(L)\subset N\subset U$, where N is a regular neighborhood of some polyhedron of dimension $\leq Fd(X)$. Since $K\times I\searrow L\cup K\times\{1\}$, N can be pushed out to cover all of $f(K\times I)$.

PROOF OF THEOREM 1.9, Part 2. The case $n\leq 3$ follows from standard techniques (see §4 of [22]). Suppose $n\geq 5$. Let $U\supset X$ be given. Choose $V\subset U$ using Lemma 1.13. Let W be a compact PL manifold neighborhood of X in V. Denote the (n-3)-skeleton of W by W^{n-3} and the dual 2-skeleton by W^2_*. By Lemma 1.13 there is a polyhedron P with dim $P\leq Fd(X)$ and a regular neighborhood N of P such that $W^{n-3}\subset N\subset U$. Exactly as in the proof of Part 1, we can use Stallings' engulfing theorem to find a PL homeomorphism h of M^n such that $h|M-W=id$ and $h(W^2_*)\cap X=\phi$. But then $X\subset R$ for some regular neighborhood R of h(P) in U.

§2. APPROXIMATION.

There are codimension three approximation theorems in all three settings.

THEOREM 2.1 (Bryant [3]). *Suppose* K^k *is a compact* k- *dimensional polyhedron,* $k\leq n-3$, *and* g: $K\to M^n$ *is a topological embedding. For every* $\varepsilon>0$ *there exists a PL embedding* $h:K\to M$ *such that* $d(g(x),h(x))<\varepsilon$ *for every* $x\in K$.

THEOREM 2.2 (Štaňko [20]). *Suppose* $X\subset M^n$ *is compact and* dim $X\leq n-3$. *For every* $\varepsilon>0$ *there exists an embedding* g: $X\to M$ *such that* $d(x,g(x))<\varepsilon$ *for each* $x\in X$ *and* g(X) *is tame.*

ADDENDUM. *There exists a pseudoisotopy* h_t *of* M^n *such that* $h_0=id$, $h_1(g(x))=x$ *for each* $x\in X$ *and* h_1 *is a homeomorphism over* M-X.

To say that h_1 is a *homeomorphism over* M-X means that $h_1|h_1^{-1}(M-X)$ is a homeomorphism of $h_1^{-1}(M-X)$ onto M-X.

DEFINITION 2.3 [5]. Suppose X, $Y\subset M^n$ are compact. A sequence of maps $\{f_k\}_{k=1}^{\infty}$ is a *relative fundamental sequence* from X to Y in M if there exists a neighborhood U of X in M such that, for each k, $f_k:U\to M^n$ and f_k is homotopic to the inclusion of U into M and such that for every neighborhood V of Y there exists a neighborhood U' of X in U such that $f_i|U'\simeq f_j|U'$ in V for almost all i and j. We say that X and Y have the same *relative shape* in M if there exist relative fundamental sequences from X to Y and from Y to X such that each composite sequence is equivalent to the appropriate identity sequence.

Relative shape equivalence in a neighborhood is used to measure "closeness" for embeddings in the shape category.

DEFINITION 2.4. Suppose again that $X,Y \subset M^n$ are compact. An *approaching isotopy* from X to Y is a map $H:M\times[0,\infty) \to M$ such that each H_t (defined by $H_t(x) = H(x,t)$) is a homeomorphism, $\{H_i\}_{i=1}^{\infty}$ is a relative shape equivalence from X to Y in M and for every neighborhood U of Y there exists a neighborhood V of X and a number t_0 such that $H(V\times[t_0,\infty))\subset U$.

THEOREM 2.5 [23]. *Suppose* $X\subset M^n$ *is compact and* $Fd(X)\leq n-3$. *For every neighborhood* U *of* X *in* M *there exists a compactum* $X'\subset U$ *such that* X *and* X' *have the same relative shape in* U *and* dem X' = Fd(X).

ADDENDUM. *The relative shape equivalence from* X' *to* X *is realized by an approaching isotopy* H_t *such that* H_0 = id *and* H_t *converges to a homeomorphism over* M−X *as* $t\to\infty$.

To begin the proof of Theorem 2.5, let k=Fd(X). By Lemma 1.10 there exists a sequence $\{U_i\}_{i=0}^{\infty}$ of neighborhoods of X and finite k-dimensional polyhedra $K_i\subset U_i$ such that $U_0 = U$, $U_{i+1}\subset U_i$ and for each $i\geq 0$ there exists a homotopy $f_i:U_{i+1}\times[0,1] \to U_i$ such that $f_i(x,0)=x$ and $f_i(x,1)\in K_i$ for every $x\in U_{i+1}$. For each $i\geq 0$, we amalgamate $K_{i\cdot k}, K_{i\cdot k+1},\ldots,K_{i\cdot k+(k-1)}$ to form a k-dimensional polyhedron L_i as follows:

(1) Let $L_{i,1} = K_{(i+1)\cdot k-1}$.

(2) Inductively let $L_{i,j}=K_{(i+1)\cdot k-j} \cup f_{(i+1)\cdot k-j}(L_{i,j-1}^{(k-1)} \times [0,1]) \cup L_{i,j-1}$ for $2\leq j\leq k$.

(3) Define L_i to be $L_{i,k}$.

Here the superscript (k-1) denotes the (k-1)-dimensional skeleton. The maps f_ℓ can be adjusted slightly so that each $L_{i,j}$ is a polyhedron. Let $U'_i = U_{i\cdot k}$ and define $\beta_i:U'_{i+1} \to L_i$ by $\beta_i(x)=f_{i\cdot k+(k-1)}(x,1)$ for each $i\geq 0$.

LEMMA 2.6 [23]. *For every polyedron* $P \subset U'_{i+1}$ *with* $\dim P \leq k$ *and for every regular neighborhood* N *of* L_i, *there exists a* PL *isotopy* h_t *of* U'_i *with compact support such that* $h_0 = \mathrm{id}$ *and* $h_1(N) \supset P$. *Furthermore,* $h_1^{-1}|P$ *is homotopic to* $\beta_i|P$ *in* N.

PROOF. Let $L_{i,0} = \phi$. We actually prove the following inductive statement.

If $p \leq k$, *a finite* p- *dimensional polyhedron* $P \subset U_{i \cdot k+p}$ *can be engulfed with* N *keeping* $L_{i,k-p}$ *fixed.*

If p=0 or if p=k=1, the result is easy. Suppose p=1<k. We must construct a homotopy of P into N which keeps $P \cap L_{i,k-1}$ fixed. There is a homotopy of P into N which keeps $P \cap L_{i,k-1}$ in N: first push $P \cap L_{i,k-1}$ along $L_{i,k-1}$ until it is near $L^{(k-1)}_{i,k-1}$ and then use the homotopy $f_{i \cdot k}$ to pull P into a neighborhood of $K_{i \cdot k}$. By squeezing out the fibers of that homotopy which lies over $P \cap L_{i,k-1}$, we get a homotopy $g: P \times [0,1] \to U'_i$ such that $g_0 = \mathrm{id}$, $g_1(P) \subset N$ and $g(x,t) = x$ for every $x \in P \cap L_{i,k-1}$. Put g in general position on $(P - L_{i,k-1}) \times [0,1]$, keeping $g((P \cap L_{i,k-1}) \times [0,1]$ fixed. Then g will embed $(P - L_{i,k-1}) \times [0,1]$ in $U'_i - L_{i,k-1}$. Push N out along that embedded homotopy to engulf P.

Now suppose that p<k and that the inductive statement above is true for polyhedra of dimension p'<p. First construct a homotopy $g: P \times [0,1] \to U_{i \cdot k+(p-1)}$ such that $g_0 = \mathrm{id}$, $g|(P \cap L_{i,k-p+1}) \times [0,1] = \mathrm{id}$, and $g(P \times \{1\}) \subset L_{i,k-p+1}$ (just as above). Put $g|(P - L_{i,k-p}) \times [0,1]$ into general position, keeping $g((P \cap L_{i,k-p}) \times [0,1] \cup g(P \times \{1\}))$ fixed and let $S = S(g|(P - L_{i,k-p}) \times [0,1]) \cup g^{-1}(g((P - L_{i,k-p}) \times [0,1)) \cap L_{i,k-p})$. Then $\dim S < (p+1)+k-n \leq (p+1) + (n-3) - n = p-2$. Let Σ denote the shadow of S. We have $\dim \Sigma \leq p-1$. By induction, there exists an isotopy h'_t of U'_i such that $h'_1(N) \supset g(\Sigma)$ and $h'_t|L_{i,k-p+1} = \mathrm{id}$. Thus $g(P \times \{1\} \cup \Sigma) \subset h'_1(N)$. But $g(P \times [0,1]) \searrow g(P \times \{1\} \cup \Sigma \cup (P \cap L_{i,k-p}) \times [0,1])$ and $g(P \times \{1\} \cup \Sigma \cup (P \cap L_{k,k-p}) \times [0,1]) \supset g(P \times [0,1]) \cap L_{i,k-p}$, and so we can find the isotopy needed to finish the proof of this case by simply following the inverse of the collapse.

Finally, suppose that $p=k\leq n-3$. Let $g:P\times[0,1]\to U_{(i+1)\cdot k-1}$ be a homotopy such that $g_0=id$ and $g_1(P)=\beta_i(P)\subset K_{(i+1)\cdot k-1}$. Put g in general position keeping $g(P\times\{1\})$ fixed. (Recall that $L_{i,k-k}=\emptyset$.) By Zeeman's Piping Lemma [24, Lemma 48], g can be adjusted so that there is a polyhedron $J\subset P\times[0,1]$ such that

(1) $S(g)\subset J$,

(2) $\dim J\leq k-1$,

(3) $\dim(J\cap(P^{(k-1)}\times[0,1]))\leq k-2$, and

(4) $P\times[0,1]\searrow J\cup P^{(k-1)}\times[0,1]\cup P\times\{1\}$.

Let Σ denote the shadow of $J\cap(P^{(k-1)})\times[0,1])$. Note that $P\times[0,1]\searrow P^{(k-1)}\times[0,1]\cup P\times\{1\}\cup J\searrow P\times\{1\}\cup\Sigma\cup J$ by property (4) and the definition of Σ. Properties (2) and (3) imply that $\dim(\Sigma\cup J)\leq k-1$. By induction, there exists an isotopy h'_t of U'_i such that $h'_0=id$, $h'_1(N)\supset\Sigma\cup J$ and $h'_t|L_{i,1}=id$. Thus $g(P\times\{1\}\cup\Sigma\cup J)\subset h'_1(N)$ and we can engulf the rest of $g(P\times[0,1])$ by following the inverse of the collapse of $P\times[0,1]$ to $P\times\{1\}\cup\Sigma\cup J$.

The proof of the inductive statement is now complete. To finish the proof of the lemma, we merely observe that in the last case (in which $P\subset U'_{i+1}$) the isotopy h_t constructed has the properties that $h_1(N)\supset g(P\times[0,1])$ and $h_t|g_1(P)=id$. Since $g_1|P=_i|P$, we have that $h_1^{-1}|P$ is homotopic to $h_1\beta_i^{-1}|P=\beta_i|P$ in N.

PROOF OF THEOREM 2.5. Let N_i be a regular neighborhood of L_i for each i. By the lemma (with $P=L_{i+1}$), there exists a homeomorphism $g_i:U'_i\to U'_i$ such that $g_i(L_{i+1})\subset N_i$ and $g_i|L_{i+1}$ is homotopic to $\beta_i|L_{i+1}$ in N_i. By making N_{i+1} smaller if necessary, we can arrange that $g_i(N_{i+1})\subset N_i$ and that $g_i|N_{i+1}$ is homotopic to $\beta_i|N_{i+1}$ in N_i. Define $G_i:N_{i+1}\to N_0$ by $G_i=g_0\circ g_1\circ\ldots\circ g_i$ and let $X'=\bigcap_{i=1}^{\infty}G_i(N_{i+1})$. The regular neighborhoods N_i can be chosen inductively, as the lemma is applied, so that dem $X'=k$. We must check that X and X' have the same relative shape in $U=U_0$.

Now $X=\bigcap_{i=0}^{\infty}U'_i$ and $X'=\bigcap_{i=0}^{\infty}G_i(N_{i+1})$. We define maps $f_i:G_i(N_{i+1})\to U'_{i+1}$ by $f_i=G_{i-1}\circ\beta_i$ and $f'_i=G_i^{-1}$. Since all the maps g_i, g_i^{-1} and β_i are homotopic to the inclusions in U, we have that f_i and f'_i are homotopic to the inclusions in U.

Furthermore, $f_i \circ f'_1 : G_i(N_{i+1}) \to G_{i-1}(N_i)$ is equal to $g_0 \circ g_1 \circ \ldots \circ g_{i-1} \circ \beta_i \circ g_i^{-1} \circ g_{i-1}^{-1} \circ \ldots \circ g_1^{-1} \circ g_0^{-1}$ and thus is homotopic to the inclusion in $G_{i-1}(N_i)$ by construction of g_i. The map $f'_{i-1} \circ f_i : U'_{i+1} \to U'_i$ is just β_i and consequently is homotopic to the inclusion in U'_i. Since $\beta_{i+1} : U'_{i+2} \to N_{i+1} \subset U'_{i+1}$ is homotopic to the inclusion and $\beta_i | N_{i+1}$ is homotopic to $g_i | N_{i+1}$, we have that f_{i+1} is homotopic to $f'_1 | G_{i+1}(N_{i+2})$ in U'_{i+1} because g_1^{-1} is homotopic to the identity.

The relative fundamental sequences needed to complete the proof are now constructed by carefully extending f_i and f'_i in the usual way (using the homotopy extension property).

PROOF OF THE ADDENDUM. In the proof above, $f'_i = G_i^{-1} = g_i^{-1} \circ g_{i-1}^{-1} \circ \ldots \circ g_0^{-1}$ and each g_j^{-1} is isotopic to the identity in U'_j via an isotopy with compact support. Hence we can define a map $H : M \times [0,\infty) \to M$ by letting $H | M \times [i,i+1]$ be g_i^{-1} composed with the isotopy between the identity and g_{i+1}^{-1}. The map H has the property that each $H_t : M \to M$ defined by $H_t(x) = H(x,t)$ is a PL homeomorphism.

§3. UNKNOTTING.

In this section we consider the problem of showing that two tame embeddings are equivalent in various settings. The shape theorems are stated elsewhere as complement theorems but are stated here in terms of approaching isotopies to emphasize the fact that they are unknotting theorems. In fact, a careful reading of the proofs of Theorems 3.2 and 3.3 will show that the basic construction is the same for both proofs. All the results stated except Theorem 3.5 have appeared previously in different form. We begin with a consideration of unknotting in the trivial range where the problem is much easier.

THEOREM 3.1 (Gugenheim [9]). *If* $f,g : K^k \to M^n$ *are two* PL *embeddings of a compact polyhedron of dimension* k, $2k+2 \leq n$, *such that* $f \simeq g$ *in* M, *then there exists a* PL *isotopy* h_t *of* M *such that* $h_0 = id$ *and* $h_1 f = g$.

THEOREM 3.2 (Bryant [2]). *Suppose* X *is a* k- *dimensional compactum with* $2k+2 \leq n$, $n \geq 5$, *and* $f,g : X \to M^n$ *are two embeddings such that* $f \simeq g$ *and both* $f(X)$ *and* $g(X)$ *are*

1-LCC. *Then there exists an isotopy* h_t *of* M^n *such that* $h_0=id$ *and* $h_1 f = g$.

THEOREM 3.3 [21]. *Suppose* $X,Y \subset M^n$ *are compacta of fundamental dimension* k, $2k+2 \leq n$, $n \leq 5$, *and both* X *and* Y *satisfy* ILC. *Then every relative shape equivalence from* X *to* Y *can be realized by an approaching isotopy which converges on* M-X *to a homeomorphism of* M-X *onto* M-Y.

The situation in codimension three is not so clear. We have a theorem in each of the PL and shape settings, but only a question in the topological setting.

THEOREM 3.4 (Edwards [6]). *Suppose* $g: K^k \to M^n$ *is a topological embedding of the compact,* k- *dimensional polyhedron* K *into* M^N, $k \leq n-3$. *For every* $\varepsilon>0$ *there exists a* $\delta>0$ *such that any two* PL *embeddings within* δ *of* g are ε- *equivalent by ambient isotopy.*

THEOREM 3.5. *Suppose* $X \subset M^n$ *is a compactum with* $Fd(X)=k \leq n-3$ *and* $n \geq 5$. *Let* $r=2k+2-n$. *Suppose* pro-$\pi_i(X)$ *is stable for* $0 \leq i \leq r-1$ *and Mittag-Leffler for* $i=r$. *For every neighborhood* U *of* X *there exists a neighborhood* V *of* X *in* U *such that if* X_1 and X_2 *are two compacta in* V *which satisfy* ILC *and have the same relative shape as* X *in* V, *then there is an approaching isotopy from* X_1 *to* X_2 *with support in* U.

We shall see in §4 that the conditions on pro-$\pi_i(X)$ of Theorem 3.5 are the very same ones which are needed in the embedding theorems there. Ferry [8] has proved that a compactum satisfies those conditions if and only if it has the shape of some LC^{r-1} compactum. There is no codimension three unknotting theorem in the topological setting. In view of Ferry's theorem and Theorem 3.5 it seems reasonable to ask whether there is an unknotting theorem for locally connected compacta.

Let us now turn to the proof of Theorem 3.5. We will not give a complete proof, but will merely show how to reduce the theorem to the closely related result of Ivanšić and Sher [11, Theorem 3].

The following hypotheses will be assumed for the remainder of §3.

$X \subset M^n$ is compact.

$Fd(X) = k \leq n-3$.

$r = 2k+2-n$

pro-$\pi_i(X)$ is stable for $i \leq r-1$.

pro-$\pi_r(X)$ satisfies the Mittag-Leffler condition.

LEMMA 3.6. *For every neighborhood* U *of* X *there exists a neighborhood* U_1 *of* X *in* U *such that if* $X_1 \subset U_1$ *has the same relative shape as* X *in* U_1 *and* W *is any neighborhood of* X_1, *then there is a neighborhood* W_1 *of* X_1 *in* W *such that the inclusion induced homomorphism* $\pi_i(U_1,W_1) \to \pi_i(U,W)$ *is the zero map,* $i \leq r$.

PROOF. Fix i. Use the Mittag-Leffler condition to choose a neighborhood U_1 of X in U such that if $g:S^i \to U_1$ is a map and U' is a neighborhood of X, then g_1 is homotopic in U (rel base point) to a map of S^i into U'. Use the stability of pro-$\pi_{i-1}(X)$ to choose U_1 to have the further property that for every neighborhood V of X in U_1 there is a neighborhood V_1 of X such that if $g:S^{i-1} \to V_1$ and $g \simeq *$ in U_1 then we also have that $g \simeq *$ in V. Let X and W be as in the statement of the Lemma. Let $\{f_j\}$ and $\{f'_j\}$ be relative fundamental sequences from X_1 to X and from X to X_1 respectively. Choose a neighborhood V of X such that $f'_j(V) \subset W \cap U_1$ for almost all j. There is a neighborhood V_1 corresponding to this choice of V and U_1 as explained above. We finally choose W_1, a neighborhood of X_1 in W, such that $f_j(W_1) \subset V_1$ and $f'_j f_j|W_1 \simeq id$ in $W \cap U_1$ for almost all j.

Let $g:(\Delta^i,\partial\Delta^i) \to (U_1,W_1)$ be a map. Then $g|\partial\Delta^i \simeq f_j g|\partial\Delta^i$ in U_1. Now $f_j g(\partial\Delta^i) \subset V_1$ so $f_j g|\partial\Delta^i$ extends to a map of Δ^i into V by the choice of V_1. Thus $f'_j f_j g|\partial\Delta^i$ extends to a map of Δ^i into $W \cap U_1$. But $g|\partial\Delta^i \simeq f'_j f_j g|\partial\Delta^i$ in $W \cap U_1$, so $g|\partial\Delta^i$ extends to $g_1:\Delta^i \to W \cap U_1$. Let $g:S^i \to U_1$ be the map which agrees with g on the northern hemisphere of S^i and agrees with g_1 on the southern hemisphere.

Adjust $g(\Delta^i)$ so that $g(\Delta^i)$ contains the path traced by the base point of X during the homotopy from the inclusion $V \subset U_1$ to $f'_j|V$. By the choice of U_1, g is homotopic in U (rel base point) to a map $g_1:S^i \to V$. Now the adjustment made in $g(\Delta^i)$ just above makes it possible to conclude that g is homotopic (rel some base point in W) to the map $f'_j \circ g_1$. We then see that g is homotopic in U (rel $\partial\Delta^i$) to a map consisting of g_1 plus $f'_j \circ g_1$.

LEMMA 3.7. *For every neighborhood* U *of* X *there exists a neighborhood* V *of* X *in* U *such that if* $X_1 \subset V$ *has the same relative shape as* X *in* V, W *is a neighborhood of* X_1, $P \subset V$ *is a compact polyhedron of dimension* $\leq r$ *and* K *is a polyhedron in* W *of dimension* $\leq k$, *then* P *can be engulfed with* W *keeping* K *fixed.*

PROOF. The Lemma is proved by standard engulfing techniques (as in [17] for example), using Lemma 3.6 to construct the necessary homotopies.

PROOF OF THEOREM 3.5. Given U, choose V to be the neighborhood given by Lemma 3.7. Suppose X_1 and X_2 are as in the statement of the Theorem. Let N_1 be a PL manifold neighborhood of X_1 such that the inclusion of X_1 into N_1 is r-shape connected. Such an N_1 exists by [11, Theorem 1]. We need a PL isotopy h_t of M with support in U such that h_0=id, $h_1(X_2) \subset N_1$ and $h_1(X_2) \hookrightarrow N_1$ is r-shape connected. The proof of [11, Theorem 3] then finishes the proof of the present theorem.

Let N_2 be a PL manifold neighborhood of X_2 with k-dimensional spine K. Any isotopy pushing K into N_1 can easily be adjust to push X_2 into N_1 as well. There exists a homotopy of K into N_1 given by the relative shape equivalence from X_2 to X_1. By Zeeman's Piping Lemma again, there exists an r-dimensional subset P of $K \times I$ such that $K \times I \searrow P \cup K \times \{1\}$ and P contains the singular set of the homotopy. Engulf the image of P keeping the image of $K \times \{1\}$ fixed using Lemma 3.7. Since $K \times I - P$ is embedded, we can push out along the track of the homotopy to complete the construction of h_t. The fact that $h_1(X_2) \hookrightarrow N_1$ is r-shape connected follows from the same argument as was used in the proof of Lemma 3.6.

§4. EMBEDDING.

In this last section we mention some theorems which fit naturally into the scheme outlined here and raise a question about embedding compacta. We will not give any proofs. Interested readers are referred to the paper by Husch and Ivanšić on "Embedding up to shape" elsewhere in these proceedings.

THEOREM 4.1 (Stallings [18]). *Let* $f: K^k \to M^n$ *be a map of a compact* k- *dimensional polyhedron into a* PL n- manifold, $k \leq n-3$. *If* f *is a* (2k-n+1)- *connected map, then* f *is homotopic to a map which is a simple homotopy equivalence of* K *into some* k- *dimensional polyhedron* $K' \subset M$.

THEOREM 4.2 (Husch and Ivanšić [10]). *Suppose* X *is a continuum with* $Fd(X)=k\leq n-3$ *and let* $r = 2k+1-n$. *Suppose* $\text{pro-}\pi_i(X)$ *is stable for* $i<r$ *and Mittag-Leffler for* $i=r$. If $\underline{f}:X \to M$ *is an* r- *shape connected shape map, then there exists a* k- *dimensional continuum* $Y \subset M$ *such that* $\underline{f}$ *is shape equivalent to a shape equivalence* $\underline{g}:X \to Y$.

The main reason for mentioning Theorem 4.2 is to use it as a model for a possible theorem in the topological setting. The following seems to be the most reasonable.

QUESTION 4.3. *Suppose* $f:X \to E^n$ *is a map of a* k- *dimensional compactum into Euclidean* n- *space*, $k\leq n-3$. *If* X *is locally connected through dimension* $2k-n+1$ (i.e. X *is* LC^{2k-n+1}), *then is* f *homotopic to a cell-like map?*

REFERENCES

1. H. G. Bothe, *Ein eindimensionales Kompaktum im* E^3, *das sich nicht Lagetreu in die Mengersche Universalkurve einbetten lässt*, Fund. Math. 54(1964), pp. 251-258.

2. J. Bryant, *On embeddings of compacta in Euclidean space.* Proc. Amer. Math. Soc. 23(1969), pp. 46-51.

3. J. Bryant, *Approximating embeddings of polyhedra in codimension three*, Trans. Amer. Math. Soc. 170 (1972), pp. 85-95.

4. J. Bryant and C. L. Seebeck III, *Locally nice embeddings in codimension three*, Quart. J. Math. Oxford (2) 21 (1970), pp. 265-272.

5. T. A. Chapman, *Shapes of finite dimensional compacta*, Fund. Math. 76 (1972), pp. 261-276.

6. R. D. Edwards, *The equivalence of close piecewise linear embeddings*, Gen. Topology and its Appl. 5 (1975), pp. 147-180.

7. R. D. Edwards, *Demension theory*, I, in Geometric Topology, Lecture Notes in Mathematics, vol. 438, Springer-Verlag, New York (1975), pp. 195-211.

8. S. C. Ferry, *A stable converse to the Vietoris-Smale theorem with applications to shape theory*, Trans. Amer. Math. Soc. 261 (1980), pp. 369-386.

9. V.K.A.M. Gugenheim, *Piecewise linear isotopy and embeddings of elements and spheres* (I), Proc. London Math. Soc. 3(3)(1953), pp. 29-53.

10. L. Husch and I. Ivanšić, *Embeddings and concordances of embeddings up to shape*, preprint.

11. I. Ivanšić and R. B. Sher, *A complement theorem for continua in a manifold*, to appear in Topology Proceedings.

12. I. Ivanšić, R. B. Sher, and G. A. Venema, *Complement theorems beyond the trivial range*, to appear in Illinois J. Math.

13. D. R. McMillan, Jr. and H. Row, *Tangled embeddings of 1-dimensional continua*, Proc. Amer. Math. Soc. 22 (1969), pp. 378-385.

14. J. Nagata, *Modern Dimension Theory*, John Wiley and Sons, New York, 1965.

15. S. Nowak, *Some properties of fundamental dimension*, Fund. Math. 85 (1974), pp. 211-117.

16. T. B. Rushing, *Topological Embeddings*, Academic Press, New York, 1973.

17. J. R. Stallings, *The piecewise-linear structure of Fucliden space*, Proc. Cambridge Philos. Soc. 58 (1962), pp. 481-488.

18. J. R. Stallings, *The embedding of homotopy types into manifolds*, Mimeographed Notes, Princeton University, 1965.

19. M. A. Štańko, *The embedding of compacta in Euclidean space*, Math USSR Sbornik 12 (1970), pp. 234-254.

20. M. A. Štańko, *Approximation of compacta in* E^n *in codimensions greater than two*, Math. USSR Sbornik 19 (1973), pp. 625-636.

21. G. A. Venema, *Embeddings of compacta with shape dimension in the trivial range*, Proc. Amer. Math. Soc. 55 (1976), pp. 443-448.

22. G. A. Venema, *Neighborhoods of compacta in Euclidean space*, Fund. Math. CIX (1980) pp. 71-78.

23. G. A. Venema, *An approximation theorem in shape theory*, preprint.

24. E. C. Zeeman, *Seminar on combinatorial topology*, Mimeographed Notes, Institut des Hautes Etudes Scientifiques, Paris, 1963.

UNDER WHAT CONDITIONS ARE SHAPE HOMOLOGY $\overline{\underline{E}}_*$ AND STEENROD HOMOLOGY ${}^s\underline{E}_*$ ISOMORPHIC ?

Friedrich W. Bauer

0. Intorduction:

Homology $\underline{E}_*$ with coefficients in a spectrum $\underline{E}$ turned out to be of great importance, not only in topology, but in many different branches of applications (e.g. in algebra and analysis). In dealing with these applications it very soon became necessary to find an appropriate definition of such a homology for spaces which are not anymore CW-spaces.

The homology theory $\mathcal{E}_*$ defined on the category Com of based compacta by L.G. Brown, R.G. Douglas and P.A.Fillmore [6], [8], [9] serves as an impressive example because this $\mathcal{E}_*$ firstly appeared, far from all algebraic topology, in connection with problems of perturbations of linear operators in complex Hilbert space.

On compact polyhedra, $\mathcal{E}_*$ turned out to be isomorphic to $\underline{BU}_*$ (complex K-homology) but for arbitrary compacta, $\mathcal{E}_*$ seemed to be something essentially new.

Later on it was proved by the present author [6] that $\mathcal{E}_*$ is nothing else than (strong) shape K-homology $\overline{\underline{BU}}_*$ (at least for finite dimensional compacta). At this time the theory of shape homologies $\overline{\underline{E}}_*$ was already quite extensively developed (cf. [3], [5]).

Simultaneously and independently D.S.Kahn, J. Kaminker and C. Schochet invented so-called Steenrod homology theories ${}^s\underline{E}_*$ with coefficients in a spectrum $\underline{E}$ (cf.§2) and they also succeeded in proving that for finite dimensional compacta ${}^s\underline{E}_*$ and $\mathcal{E}_*$ became isomorphic.

So, quite naturally, the question comes up, under what conditions $\overline{\underline{E}}_*$ and ${}^s\underline{E}_*$ turn out to be isomorphic.

This problem is solved (at least for finite dimensional compacta) by theorem 3.3.. One of the main tools for proving this assertion becomes theorem 3.1. which is an extension of proposition 4.4. [6] and which appears here for the first time.

At this point some historical remarks are useful:

1) The first indication that Steenrod-Sitnikov homology ${}^{s}H_{*}(X;G)$ is related to strong shape theory was given in [3]. There we provided an isomorphism between ${}^{s}H_{*}(X;Z)$ and $H_{*}(|\bar{S}(X)|;Z)$, where $\bar{S}(X)$ denotes the shape singular complex of a continuum X. Then in [5], [6] the relations between $\bar{\underline{E}}_{*}(X) = \bar{\pi}_{*}(\underline{E}\bar{\wedge} X)$ and $\underline{E}_{*}(|\bar{S}(X)|)$ were investigated. As a biproduct we achieved a new proof of the fact that $H_{*}(|\bar{S}(X)|;G)$ becomes isomorphic to ${}^{s}H_{*}(X;G)$ for finitely generated G (here we denote by ${}^{s}H_{*}$ the homology in the way it was defined by K. Sitnikov, cf. [5], [10] for further references). Moreover it became clear that ${}^{s}H_{*}(X;G)$ became isomorphic to $\overline{\underline{K(G)}}_{*}(X)$ for any abelian group G.

2) Let $\underline{E}$ be an Ω-spectrum, then the first proof of

$$ {}^{s}\underline{E}_{*}(X) \approx \bar{\underline{E}}_{*}(X), $$

for finite dimensional compacta was given in [5] §8. However in order to accomplish this, we had to use both Alexander duality theorems (which are available for $\bar{\underline{E}}_{*}$ as well as for ${}^{s}\underline{E}_{*}$ but in each case with different cohomology functors as counterparts, cf. §4).

Therefore a direct proof seemed to be very desirable.

3) Theorem 3.2. is nothing else than a mild generalization of corollary 5.6. [6] which itself extends a result of J. Milnor [10], valid for ordinary homology. It exhibits again the powerful rôle of the clusteraxiom (cf. §2). The fact that J.Milnor's characterization of ${}^{s}H_{*}(X;G)$ appears as a corollary of theorem 3.2. was already notified in [6] §6).

4) The relation between $\bar{\underline{E}}_{*}$ and ${}^{s}\underline{E}_{*}$ can be briefly illustrated like this: To each CW-spectrum $\underline{E}$ there exists a Ω-spectrum $_{\Omega}\underline{E}$ as well as a function of spectra $\gamma : \underline{E} \longrightarrow {}_{\Omega}\underline{E}$ which induces an equivalence in the Boardman category (cf. §1) and therefore (cf. 2.4.) a natural isomorphism ${}^{s}\underline{E}_{*} \approx {}^{s}_{\Omega}\underline{E}_{*}$. On the other hand theorem 3.3. b) provides us with an isomorphism ${}^{s}_{\Omega}\underline{E}_{*} \approx \overline{{}_{\Omega}\underline{E}}_{*}$ (for finite dimensional compacta). Hence in this case we obtain ${}^{s}\underline{E}_{*}$ by the following recipe: Construct $_{\Omega}\underline{E}$ and then form $\overline{{}_{\Omega}\underline{E}}_{*}$.

As a consequence we obtain more non-isomorphic homology functors $\bar{\underline{E}}_{*}$ than we have homology functors ${}^{s}\underline{E}_{*}$.

In §6, we present an example which shows, that there exist two spectra $\underline{E}$, $\underline{E}'$ together with an isomorphism ${}^{s}\underline{E}_{*} \approx {}^{s}\underline{E}'_{*}$ but at the same time there exists a finite dimensional compactum X such that $\bar{\underline{E}}_{*}(X) \not\approx \bar{\underline{E}}'_{*}(X)$.

This phenomenon is of course reflected in the behaviour of the related cohomology theories ${}^{s}\underline{E}^{*}$ resp. $\check{\underline{E}}^{*}$ (the singular, resp. the Čech cohomology with coefficients in $\underline{E}$). It is well-known that for fixed X, $\underline{E}^{*}(X)$ (§1 (4)) does not depend functorially on $\underline{E}$ as an object of the Boardman category $\underline{B}_{h}$ (rather than on $\underline{E}$ being considered as an object of <u>Spec</u>). So it can be expected that the dual homologies $\bar{\underline{E}}_{*}$, ${}^{s}\underline{E}_{*}$ behave in the same way.

The present paper is organized as follows: The first section is devoted to the Boardman category and to the cohomology functors $\underline{E}^*$, ${}^s\underline{E}^*$. In §2 we introduce the two related homologies $\tilde{\underline{E}}_*$ and ${}^s\underline{E}_*$. The main theorem 3.3. is formulated in §3 and derived from theorem 3.1. and theorem 3.2. whose proofs are deferred to §5. Alexander duality for ${}^s\underline{E}_*$ and ${}^s\underline{E}^*$ (§4 (1)) reveals itself in §4 as a corollary of the Alexander duality, relating $\tilde{\underline{E}}_*$ and $\underline{E}^*$. Finally we give in §6 the already mentioned examples.

It should be mentioned that one is able to define $\tilde{\underline{E}}_*$ as well as ${}^s\underline{E}_*$ for spectra which are not anymore CW-spectra. This has applications (for example in S-duality [5] §8). However we are not dealing with this aspect of the theory within this paper.

1. The Boardman category and cohomology functors:

Although we are primarily concerned with two homology functors $\bar{\underline{E}}_*$ and ${}^s\underline{E}_*$, we must deal with the related cohomology functors $\underline{E}^*$ and ${}^s\underline{E}^*$, which we call Čech resp. singular cohomology.

To this end we introduce the category Spec of CW-spectra. The objects are spectra $\underline{E} = \{ E_n, \ \sigma_n : \Sigma E_n \longrightarrow E_{n+1}, \ n \in \mathbb{Z} \}$ where E_n is a based CW-complex and σ_n a cellular map. As usual, Σ denotes reduced suspension. One could equally well live with a more general definition (e.g. by using spaces E_n having only the homotopy type of a based CW-complex and arbitrary based continuous maps σ_n) as well as with a more restrictive definition (e.g. by assuming σ_n to be a cellular inclusion). The transition from one definition to the other would in any case cause merely some technical problems (see [1], [7]; in [7] our spectra are called "prespectra").

A morphism $\underline{f} = \{f_n\} : \underline{E} \longrightarrow \underline{E}'$ is what is usually called a function of spectra: i.e. $f_n : E_n \longrightarrow E'_n$ is a based continuous map which strictly commutes with the bonding maps σ_n. It turns out to be convenient to assume that f_n is also cellular.

In Spec we have special inclusions $i: \underline{A} \subset \underline{B}$ which are called dense by J. Boardman [7] (by other authors "full embeddings" or "cofinal" [1]):

We say that $\underline{A} \subset \underline{B}$ is a dense embedding, $\underline{A}, \underline{B} \in$ Spec whenever to any finite subcomplex $K \subset B_n$ there exists an index k and a $\tau : \ \Sigma^k K \longrightarrow A_{n+k}$, rendering the diagram

$$\begin{array}{ccccccc} \Sigma^k K \subset \Sigma^k B_n & \longrightarrow & \Sigma^{k-1} B_{n+1} & \longrightarrow & \cdots & \longrightarrow & B_{n+k} \\ & \searrow^{\tau} & & & & & \cup \\ & & & & & & A_{n+k} \end{array}$$

commutative.

If we would assume that $\sigma_n: \Sigma E_n \longrightarrow E_{n+1}$ is always a cellular inclusion, it would be sufficient to require that $\Sigma^k K$ is contained in A_{n+k} for sufficiently large k. The Boardman category $\underline{B}$ is defined to be the quotient category [2] $\underline{\text{Spec}}/\mathcal{S}$ where $\mathcal{S}$ denotes the class of all dense embeddings. In other words, we take the universal category $\underline{B}$ which originates from $\underline{\text{Spec}}$ by inverting all arrows in $\mathcal{S}$.

An Ω-spectrum $\underline{E} = \{E_n\} \in \underline{\text{Spec}}$ is one, having the property that the maps $E_n \longrightarrow \Omega E_{n+1}$, adjoint to σ_n are homotopy equivalences for all n.

A basic result in stable homotopy theory (cf. [7] §10) provides to each $\underline{E} \in \underline{\text{Spec}}$ an Ω-spectrum $_\Omega\underline{E}$ and a morphism $\gamma : \underline{E} \longrightarrow {_\Omega\underline{E}}$ in $\underline{\text{Spec}}$ such that γ becomes an equivalence in the homotopy category $\underline{B}_h$ (a category which is sometimes called the Boardman category instead of $\underline{B}$ and which we are not going to define in all details, referring to [1] ,[7]).
Sometimes we will, by an abuse of notation, speak of an Ω-spectrum $\underline{E}$ if $\Omega E_{n+1} \simeq E_n$ is only true for $n > 0$.
The explicit construction of $_\Omega\underline{E}$ is needed in §6; therefore we record it briefly (cf. [7] § 10):
Let $\delta_n: E_n \longrightarrow \Omega E_{n+1}$ be the adjoint of $\sigma_n: \Sigma E_n \longrightarrow E_{n+1}$. We consider the infinite mapping cylinder (i.e. the telescope) of the sequence of mappings:

$$(1) \quad E_n \xrightarrow{\delta_n} \Omega E_{n+1} \xrightarrow{\Omega\delta_{n+1}} \Omega^2 E_{n+2} \rightarrow \dots$$

which we call $_\Omega E_n$. We have an inclusion $i_n: E_n \subset {_\Omega E_n}$ and observe that the natural mapping $\Sigma \Omega^k X \rightarrow \Omega^{k-1} X$ (adjoint to the identity) induces a mapping $_\Omega\sigma_n$: $\Sigma\, {_\Omega E_n} \longrightarrow {_\Omega E_{n+1}}$ whose adjoint becomes a homotopy equivalence. In order to complete the construction of a spectrum $\underline{E} \in \underline{\text{Spec}}$ we have to realize that each telescope (1) is of the homotopy type of a CW-complex and therefore can be assumed to be a CW-complex itself (for example by applying the geometric realization of the singular complex functor $|S(\ )|$).

More generally there exists to each $\underline{E} = \{E_n, \sigma_n : \Sigma E_n \to E_{n+1}, n \in \mathbb{Z}\}$, where E_n is not necessarily a CW-complex, always a <u>CW-substitute</u>; see [8] §2 for details.

Let X be any based space, then we have the suspension-spectrum $\underline{X} = \{X_n\}$

$$X_n = \begin{cases} \Sigma^n X & n \geq 0 \\ * & n < 0. \end{cases}$$

We have $\underline{X} \in \underline{\text{Spec}}$ whenever X is a CW-complex. For arbitrary $X = (X, x_0)$, we consider again $|S(X)| = X'$.

We can define for any spectrum $\underline{A} \in \underline{\text{Spec}}$ and given coefficient spectrum $\underline{E} \in \underline{\text{Spec}}$ the cohomology group

$$(2) \qquad \underline{E}^n(\underline{A}) = \underline{B}_h(\Sigma^{-n} \underline{A}, \underline{E}) \qquad n \in Z$$

in particular

$$(3) \qquad {}^s\underline{E}^n(X) = \underline{E}^n(\underline{X})$$

for a based topological space X. This establishes the <u>singular cohomology</u> functor with coefficients in $\underline{E}$.

${}^s\underline{E}^*$ becomes a covariant functor with respect to morphisms of the coefficient spectra in $\underline{B}_h$. In particular an equivalence $\underline{E} \approx \underline{E}'$ in $\underline{B}_h$ induces a natural isomorphism

$${}^s\underline{E}^n(X) \approx {}^s\underline{E}'^n(X) .$$

<u>Čech cohomology</u> with coefficients in $\underline{E}$ is defined by:

$$(4) \qquad \underline{E}^n(X) = \varinjlim_k \left[\Sigma^k X, E_{n+k}\right]$$

with obvious bonding maps

$$\left[\Sigma^k X, E_{n+k}\right] \to \left[\Sigma^{k+1} X, \Sigma E_{n+k}\right] \to \left[\Sigma^{k+1} X, E_{n+k+1}\right].$$

This makes sense for all spaces $(X, x_o) \in \underline{Top}_o$.

In general $\underline{E}^n(X)$ and ${}^s\underline{E}^n(X)$ are not isomorphic, even not for X being a CW-space. However one has:

<u>1.1. Proposition:</u> (cf. [7] theorem 10.10.): Let $\underline{E}$ be a Ω-spectrum in <u>Spec</u> and X a CW-space, then there exists a natural isomorphism

$$(5) \qquad {}^s\underline{E}^n(X) \approx \underline{E}^n(X), \quad n \in \mathbb{Z}.$$

Moreover (5) remains valid for any $\underline{E} \in \underline{Spec}$ and compact CW-space X.

The reasons for calling $\underline{E}^*(\ )$ Čech cohomology are the following:

1) For $\underline{E} = \underline{K(G)}$ = Eilenberg-MacLane spectrum (G= abelian group) $\underline{K(G)}^*(X)$ turns out to be, for a very large class of spaces, Čech cohomology $\check{H}^*(X;G)$ (cf. [11] theorem 6.1.).

2) There is a continuity property holding for $\underline{E}^*(\ )$:

Let $\{X_\alpha\}$ be an inverse system of compacta with a limit $X = \varprojlim X_\alpha$, being a compactum, be given, then one has an isomorphism

$$\underline{E}^*(X) \approx \varinjlim \underline{E}^*(X_\alpha).$$

This can be easily proved in a standard way by using the fact that all E_i are ANEs.

The mapping $\gamma : \underline{E} \longrightarrow {}_{\Omega}\underline{E}$ induces a natural isomorphism

(6) $\qquad {}^{s}\underline{E}^{n}(\ ) \approx {}^{s}_{\Omega}\underline{E}^{n}(\ )$

but only a natural transformation

$$\gamma^{*} : \quad \underline{E}^{n}(\ ) \longrightarrow {}_{\Omega}\underline{E}^{n}(\ )$$

defined for all spaces X. For CW-spaces one has according to 1.1. and (6):

$$\begin{array}{ccc} \underline{E}^{n}(\ ) & \xrightarrow{\gamma^{*}} & {}_{\Omega}\underline{E}^{n}(\ ) \\ & \searrow & \wr\wr \\ & & {}^{s}\underline{E}^{n}(\ ) \end{array}$$

establishing a natural transformation between $\underline{E}^{*}$ and ${}^{s}\underline{E}^{*}$.

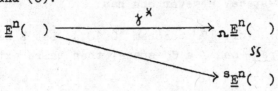

2. The homology functors:

Concerning the details of the strong shape category $\overline{\underline{K}}$ resp. its homotopy category $\overline{\underline{K}}_h$ we refer to [5]. Let X,Y be based topological spaces, then a strong shape mapping $\overline{f}: X \longrightarrow Y$ is a suitable 2-functor $\underline{P}_Y \longrightarrow \underline{P}_X$ where $\underline{P}_Y$ denotes the 2-category whose objects are continuous based mappings $g: Y \longrightarrow P \in \underline{P}$ = subcategory of all ANEs. The 1-morphisms $(r,\omega): g_1 \rightarrow g_2$, $g_i: Y \longrightarrow P_i$ are pairs, where $r: P_1 \longrightarrow P_2$ is a continuous map and $\omega: r g_1 \simeq g_2$ a fixed homotopy. The description of 2-morphisms in $\underline{P}_Y$ involves homotopies between homotopies. We are exclusively concerned with spaces X,Y which are either based compacta or based CW-complexes (namely spaces E_n, appearing in spectra $\underline{E} = \{E_n\} \in \underline{\text{Spec}}$). We denote by $\underline{\text{Com}}$ the category of based compacta, with based continuous maps.

Let $\overline{f} \in \overline{\underline{K}}(X,Y)$ and $Z \in \underline{K}$ be given, then the description of $\overline{f} \wedge 1_Z: X \wedge Y \longrightarrow Y \wedge Z$ causes some trouble because a continuous $g: Y \wedge Z \longrightarrow P \in \underline{P}$ is in general not related to any continuous $g': Y \longrightarrow P' \in \underline{P}$ unless Z is of a very special kind. In order to settle this problem one is obliged to enrich $\overline{\underline{K}}$ by new objects $\underline{P}_Y \bar{\wedge} \underline{P}_Z = Y \bar{\wedge} Z$ which are no longer topological spaces but categories, whose objects are given decompositions

$$g = (Y \wedge Z \xrightarrow{\quad g_1 \wedge g_2 \quad} P_1 \wedge P_2 \xrightarrow{\quad r \quad} P), \quad P, P_i \in \underline{P},$$

of those $g \in \underline{P}_{Y \wedge Z}$ which allow such a decomposition. The 1- resp. 2-morphisms are defined similarly. Again, for details see [5], where in particular the problem is solved under what circumstances the natural transformation $\alpha: Y \wedge Z \longrightarrow Y \bar{\wedge} Z$ (determined by the obvious forgetful functor $\underline{P}_Y \bar{\wedge} \underline{P}_Z \rightarrow \underline{P}_{Y \wedge Z}$) becomes an equivalence.

This turns out to be true in the following cases:
1) Y,Z both being compacta, 2) Y (or Z) is a finite CW-space while the other factor is arbitrary, 3) Both Y,Z are CW-spaces ([5] proposition 1.4.).

The technical details in developing $\overline{\underline{K}}$, $\overline{\underline{K}}_h$ and the $\bar{\wedge}$ - product are somewhat involved but they do not cause more than technical difficulties.

Let $\underline{E} \in$ Spec be a CW-spectrum, then the <u>shape homology functor</u>

$$\overline{\underline{E}}_*: \ \overline{\underline{Com}}_h \longrightarrow \underline{Ab}^Z$$

is defined in complete analogy to classical homology with coefficients in $\underline{E}$ by:

$$(1) \qquad \overline{\underline{E}}_n(X) = \varinjlim \overline{\pi}_{n+1}(E_1 \bar{\wedge} X),$$

where $\overline{\pi}_m(X,x_o) = \overline{\underline{K}}_h((S^m,*), (X,x_o))$.

Let $\bar{f}: X \longrightarrow Y$ be a shape mapping (we always mean from now on by this a strong shape morphism, i.e. a $\bar{f} \in \overline{\underline{K}}(X,Y)$). Then we have

$$(2) \qquad 1_{E_1} \bar{\wedge} \bar{f}: \ E_1 \bar{\wedge} X \longrightarrow E_1 \bar{\wedge} Y$$

and consequently in the usual way a homomorphism

$$\bar{f}_* = \overline{\underline{E}}_n(\bar{f}): \ \overline{\underline{E}}_n(X) \longrightarrow \overline{\underline{E}}_n(Y).$$

Concerning the details, we refer again to [5] , [6] .

In the category $\overline{\underline{K}} = \overline{\underline{Com}}$ we have:

<u>2.1. Theorem:</u> The functor $\overline{\underline{E}}_*$ has the following properties:

H 1) $\bar{f}_o \simeq \bar{f}_1 \Longrightarrow \overline{\underline{E}}_*(\bar{f}_o) = \overline{\underline{E}}_*(\bar{f}_1)$.

H 2) Let $i: A \subset X$ be any inclusion in <u>Com</u>, $p: X \rightarrow X/A$ the projection, then the sequence

$$\overline{\underline{E}}_*(A) \xrightarrow{i_*} \overline{\underline{E}}_*(X) \xrightarrow{p_*} \overline{\underline{E}}_*(X/A)$$

is exact.[1)]

H 3) There is a natural isomorphism

$$\sigma_n = \sigma : \quad \bar{\underline{E}}_n(X) \approx \bar{\underline{E}}_{n+1}(\Sigma X), \quad n \in Z.$$

We are not going to prove this theorem in this paper but we want to make the following remarks:

1) There exists a functor $h: \underline{K} \longrightarrow \bar{\underline{K}}$, which assigns to each continuous $f: X \to Y$ a $h(f) = \bar{f}: X \longrightarrow Y$ simply by setting

$$h(f)(g) = g\, f\,, \qquad g \in \underline{P}_Y.$$

This allows the following version of H 1):

H 1)' $f_o \simeq f_1 \Longrightarrow \bar{\underline{E}}_*(h(f_o)) = \bar{\underline{E}}_*(h(f_1))$,

$f_i \in \underline{\text{Com}}$.

We denote the functor $\bar{\underline{E}}_*\, h$: $\underline{\text{Com}} \longrightarrow \underline{\text{Ab}}^Z$ by an abuse of notation again by $\bar{\underline{E}}_*$.

2) Instead of H 2) we could have considered the following sequence

$$(3) \quad \bar{\underline{E}}_*(A) \xrightarrow{i_*} \bar{\underline{E}}_*(X) \xrightarrow{j_*} \bar{\underline{E}}_*(X \cup CA),$$

where CA denotes the cone over the space A (with the top vertex of CA as basepoint) and $j: (X,x_o) \subset (X \cup CA, *)$ is the inclusion.

Now we have the following important assertion ([5] theorem A 9): Every inclusion $i: A \subset X$ in $\underline{\text{Com}}$ is a cofibration in $\overline{\underline{\text{Com}}}$ (i.e. h(i) has the homotopy extension property in $\overline{\underline{\text{Com}}}$). - This implies in the usual way that the spaces $(X \cup CA, *)$ and $(X/A, *)$ are homotopy equivalent in $\overline{\underline{\text{Com}}}$. Consequently we have a commutative triangle

1) where we abbreviate $\bar{\underline{E}}_*(h(i))$ resp. $\bar{\underline{E}}_*(h(p))$ by i_* resp. p_*. Concerning the functor $h: \underline{K} \longrightarrow \bar{\underline{K}}$ see the following remark 1).

$$\begin{array}{ccc}
\bar{\underline{E}}_*(X) & \xrightarrow{\quad j_* \quad} & \bar{\underline{E}}_*(X \cup CA) \\
& \searrow^{p_*} & \downarrow \wr\wr\, q_* \\
& & \bar{\underline{E}}_*(X/A) \approx \bar{\underline{E}}_*(X \cup CA\ /\ CA,\ *).
\end{array}$$

This assures us that the sequence in H 2) is exact if and only if the sequence (3) is exact. Hence we can replace H 2) by

H 2)' Let $i\colon A \subset X$ be any inclusion in Com, then (3) is exact.

3) These two preceding remarks relate theorem 2.1. and theorem 3.1. in [5]. In [5] we do not restrict ourselves to CW-spectra, as we do in the present case.

The second kind of homology which we are considering stems from [8] and is defined as follows: Let $\underline{E} \in \underline{\text{Spec}}$ be any spectrum, $X \in \underline{\text{Com}}$ a compactum and $F_n = F(X,S^n) = {S^n}^X$ = the function space. We form the spectrum $\underline{F}(X) = \{F_n\}$ (the bonding maps are obvious) and denote any CW-substitute by the same letter.

Steenrod-homology in the sense of [8] is defined as follows:

(4) $\qquad {}^s\underline{E}_n(X) = {}^s\underline{E}^{-n}(\underline{F}(X)).$

Let $f \in \underline{\text{Com}}(X,Y)$ be continuous, then we have

$$\underline{F}(f)\colon \quad \underline{F}(Y) \longrightarrow \underline{F}(X)$$

and consequently

(5) $\qquad {}^s\underline{E}_n(f) = {}^s\underline{E}^{-n}(\underline{F}(f))\colon\ {}^s\underline{E}_n(X) \longrightarrow {}^s\underline{E}_n(Y).$

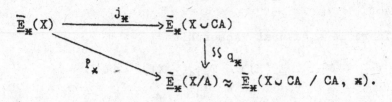

The authors of [8] who introduced this kind of Steenrod homology with coefficents in $\underline{E}$ gave a definition for an arbitrary spectrum (which is not necessarily a CW-spectrum) by applying (4), (5) to any CW-substitute of the given spectrum $\underline{E}$. However we in this paper restrict ourselves to comparing ${}^{s}\underline{E}_{*}(\ )$ with $\bar{\underline{E}}_{*}h = \bar{\underline{E}}_{*}$ only for CW-spectra $\underline{E}$.

Analogously to theorem 2.1. we have:

<u>2.2. Theorem</u>: ${}^{s}\underline{E}_{*}$: $\underline{\text{Com}} \longrightarrow \underline{\text{Ab}}^{Z}$ is a functor, fulfilling H 1)', H 2), H 3) and in addition the following <u>clusteraxiom</u>

H 4) Let $(X_i, x_{io}) \in \underline{\text{Com}}$, $i = 1,2,\ldots$ be a family of spaces and

$$X = \mathop{\text{Cl}}_{i=1}^{\infty} X_i = \varprojlim_{k} (X_1 \vee \ldots \vee X_k)$$

be the cluster of the X_i, then the projections p_i: $X \longrightarrow X_i$ induce a natural isomorphism

$${}^{s}\underline{E}_n(X) \approx \prod {}^{s}\underline{E}_n(X_i).$$

The proof of this theorem is contained in [8](theorem A).

Both homology functors $\bar{\underline{E}}_{*}$ and ${}^{s}\underline{E}_{*}$ have the following property:

<u>2.3. Theorem</u>: On the subcategory $\underline{P}_o \subset \underline{\text{Com}}$ of compact CW-spaces one has natural isomorphisms

$$\varepsilon : \underline{E}_{*} \longrightarrow {}^{s}\underline{E}_{*}$$

resp.

$$\delta : \underline{E}_{*} \longrightarrow \bar{\underline{E}}_{*}$$

where we put as usual

$$\underline{E}_n(X) = \varinjlim \pi_{n+k}(E_1 \wedge X), \quad X \in \underline{P}_o$$

resp. for induced morphisms.

As far as Steenrod homology is concerned, this is again a consequence of theorem A in [8]. The existence of the isomorphism δ follows from [5] theorem 3.4..-

Let $\varphi \in \underline{\text{Spec}}(\underline{E}, \underline{E}')$ be a morphism, then it is immediate that φ induces in a functorial way a natural transformation

$$\varphi_*: \underline{E}_* \longrightarrow \underline{E}'_*$$

as well as a natural transformation:

$${}^s\varphi_*: \quad {}^s\underline{E}_* \longrightarrow {}^s\underline{E}'_* .$$

We notice the following fact:

<u>2.4. Proposition</u>: Let $i: \underline{E} \subset \underline{E}'$ be a dense embedding (cf. §1) then ${}^s\varphi_*$ becomes an isomorphism. So, two spectra $\underline{E}$, $\underline{E}'$ being equivalent in the Boardman category $\underline{B}_h$ give rise to isomorphic homology functors ${}^s\underline{E}_* \approx {}^s\underline{E}'_*$.

3. Relations between $\bar{E}_*$ and ${}^sE_*$:

The following assertions are crucial for our purposes:

3.1. Theorem: Let $\underline{Com}^f \subset \underline{Com}$ be the subcategory of finite dimensional compacta and $\underline{E} \in \underline{Spec}$ a Ω-spectrum, then for $\bar{E}_*$ the clusteraxiom holds on $\underline{Com}^f$. More precisely: Let $X = \overset{\infty}{\underset{i=1}{Cl}} X_i$ be contained in $\underline{Com}^f$, then the projections $p_i : X \longrightarrow X_i$ induce an isomorphism

$$\bar{E}_*(X) \approx \prod_{i=1}^{\infty} \bar{E}_*(X_i) .$$

3.2. Theorem: Let $H_*, H'_* : \underline{Com} \longrightarrow \underline{Ab}^Z$ be two homology functors such that for both of them H 1)', H 2) , H 3), and such that in addition for H'_* the clusteraxiom holds. Assume that a natural transformation $\alpha : H_* \longrightarrow H'_*$ on the subcategory $\underline{P}_o$ of finite polyhedra is defined. Then there exists a unique extension $\bar{\alpha} : H_* \longrightarrow H'_*$ of α which is defined on the whole category $\underline{Com}$. The same conclusion holds for $\underline{Com}$ replaced by $\underline{Com}^f$.

In §5 we will deal with the proofs of these two theorems.- The main theorem of this paper is

3.3. Theorem: a) Let $\underline{E} \in \underline{Spec}$ be any spectrum, then there exists a natural transformation on $\underline{Com}$

$$\beta_{\underline{E}} = \beta : \bar{E}_* \longrightarrow {}^sE_*$$

which is uniquely determined by the property (cf. theorem 2.3.):

$$\beta \,|\, P_o = \varepsilon\, \delta^{-1} .$$

b) Let $\underline{E}$ be an Ω-spectrum in $\underline{Spec}$, then β becomes an isomorphism on $\underline{Com}^f$.

c) Let $\gamma : \underline{E} \to {}_{\Omega}\underline{E}$ be the inclusion of a given $\underline{E} \in$ Spec into the canonical Ω-spectrum ${}_{\Omega}\underline{E}$ (cf. §1), then we have on $\underline{\mathrm{Com}}^f$ a commutative diagram of natural transformations where $\beta_{{}_{\Omega}\underline{E}}$ and ${}^s\gamma_*$ are isomorphisms

$$\begin{array}{ccc}
\bar{\underline{E}}_* & \xrightarrow{\quad \gamma_* \quad} & {}_{\Omega}\bar{\underline{E}}_* \\
\Big\downarrow \beta_{\underline{E}} & & \Big\downarrow \wr\wr \;\; \beta_{{}_{\Omega}\underline{E}} \\
{}^s\underline{E}_* & \underset{\approx}{\xrightarrow{\quad {}^s\gamma_* \quad}} & {}^s_{\Omega}\underline{E}_*
\end{array}$$

Proof: Ad a): Theorem 2.1. assures us that $\bar{\underline{E}}_*$ fulfils H 1)', H 2), H 3) while theorem 2.2. guarantees that ${}^s\underline{E}_*$ fulfils also H 4). Moreover $\alpha = \varepsilon\,\delta^{-1} : \bar{\underline{E}}_* \longrightarrow {}^s\underline{E}_*$ is a natural transformation on $\underline{P}_o$. Hence the conclusion follows from theorem 3.2..-

Ad b): Because $\underline{E}$ is now supposed to be an Ω-spectrum, $\bar{\underline{E}}_*$ is fulfilling a clusteraxiom and we have a unique $\bar{\eta} : {}^s\underline{E}_* \longrightarrow \bar{\underline{E}}_*$ extending $\eta = \delta\,\varepsilon^{-1}$ on $\underline{P}_o$. The uniqueness part of 3.2. guarantees that $\bar{\eta}$ must be an inverse to β.

Ad c): This follows again by applying the uniqueness of $\bar{\alpha}$ stated in theorem 3.2., because we have:

$$\beta_{\underline{E}} | \underline{P}_o = ({}^s\gamma_*)^{-1}\, \beta_{{}_{\Omega}\underline{E}}\; \gamma_* | \underline{P}_o \; .-$$

The fact that ${}^s\gamma_*$ is an isomorphism follows of course by 2.4..-

This completes the proof of theorem 3.3..-

Remarks: 1) theorem 3.3. allows us to describe ${}^s\underline{E}_*$ on $\underline{\mathrm{Com}}^f$ up to an isomorphism as ${}_{\Omega}\bar{\underline{E}}_*$. This works for any $\underline{E} \in$ Spec.

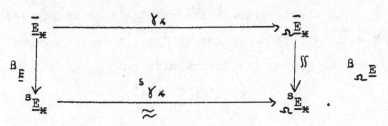

2) It is not known to the author whether all this remains true for <u>Com</u> (instead of $\underline{\text{Com}}^f$). As will be seen in §5, the finite dimensionality enters because we need at one place Alexander duality.

3) For $\underline{E}$ being an arbitrary spectrum we have a definition of $\overline{\underline{E}}_*$ as well as one of ${}^s\underline{E}_*$. In the first case we only have to repeat §2 (1). As for ${}^s\underline{E}_*$, we again replace $\underline{E}$ by any CW-substitute $\underline{E}'$ (cf. [8] §2) and define

$$ {}^s\underline{E}_* = {}^s\underline{E}'_*. $$

Now the relation between $\underline{E}_*$ and ${}^s\underline{E}_*$ becomes a little more delicate. We are not going to pursue this further in this paper.

<u>3.4. Corollary:</u> For any $\underline{E} \in \underline{\text{Spec}}$, there exists an extension of ${}^s\underline{E}_* | \underline{\text{Com}}^f$ over $\overline{\underline{\text{Com}}}^f$, the corresponding shape category. Let in particular X, Y be two spaces, having the same Borsuk-Mardesic shape, then there exists an isomorphism

(1) $$ {}^s\underline{E}_*(X) \approx {}^s\underline{E}_*(Y). $$

Proof: We merely have to extend ${}^s_\Omega\underline{E}_*$ over the shape category, but this is furnished by $\overline{{}_\Omega\underline{E}_*} | \overline{\underline{\text{Com}}}^f$. Moreover let $X, Y \in \underline{\text{Com}}$ have the same Borsuk shape, then [4] theorem 6.1. implies that they are homotopy equivalent in $\overline{\underline{\text{Com}}}$.

Hence we have isomorphisms

$$ \begin{array}{ccc} \overline{{}_\Omega\underline{E}_*}(X) & \approx & \overline{{}_\Omega\underline{E}_*}(Y) \\ \wr\wr & & \wr\wr \\ {}^s\underline{E}_*(X) & & {}^s\underline{E}_*(Y) \end{array} $$

thereby completing the proof of 3.4..-

<u>Remark</u>: A result related to (1) is already contained in [9](5.10)

4. Alexander duality:

Both homology functors appear in an Alexander duality theorem:

Let $X \subset S^{n+1}$ be compact and $\underline{E} \in \underline{\text{Spec}}$ be any spectrum, then there exists a, with respect to inclusions natural, isomorphism

$$(1) \qquad {}^{s}\underline{E}_p(X) \approx {}^{s}\underline{E}^{n-p}(S^{n+1} \backslash X).$$

This is identical with theorem B in [8].

On the other hand we have a, with respect to inclusions natural, isomorphism

$$(2) \qquad \overline{\underline{E}}_p(X) \approx \underline{E}^{n-p}(S^{n+1} \backslash X).$$

This is a special case of theorem 7.1. in [5].

In both cases we are dealing with reduced homology and cohomology for non-based spaces, i.e. we have for example

$$\overline{\underline{E}}_p(X) = \lim_{\overrightarrow{l}} \bar{\pi}_{n+l}(E_l \bar{\wedge} X^+).$$

By an abuse of notation we used in both cases the same symbol for this kind of (co-)homology as before.

According to our results in §3 we can recognize (1) as a special case of (2) in view of the following chain of isomorphisms:

$$\begin{aligned} {}^{s}\underline{E}_p(X) \approx \overline{\underline{E}}_p(X) &\approx {}_{\sim}\underline{E}^{n-p}(S^{n+1} \backslash X) \\ &\approx {}_{\sim}^{s}\underline{E}^{n-p}(S^{n+1} \backslash X) \\ &\approx {}^{s}\underline{E}^{n-p}(S^{n+1} \backslash X). \end{aligned}$$

The first isomorphism appearing in this sequence is (2) while the remaining isomorphisms stem from §1 resp. §3.

On the other hand it can be realized that in general (2) is not a consequence of (1) because there are more homology (and cohomology) functors appearing in (2) than in (1). Taking (1) and (2) to be granted, we have in [5] §8 derived the fact that

(3) $$\bar{\underline{E}}_p(X) \approx {}^s\underline{E}_p(X)$$

for any finite dimensional X and given Ω-spectrum $\underline{E} \in \underline{\text{Spec}}$. This can be deduced from the chain of isomorphisms

$$\begin{aligned}\bar{\underline{E}}_p(X) &\approx \underline{E}^{n-p}(S^{n+1}\setminus X)\\ &\approx {}^s\underline{E}^{n-p}(S^{n+1}\setminus X).\\ &\approx {}^s\underline{E}_p(X).\end{aligned}$$

5. Proofs of theorems 3.1., 3.2.:

a) Ad theorem 3.1.: The proof is modelled closely along the pattern of the proof of proposition 4.4. in [6] (where our assertion is proved for periodic spectra).

We introduce the following property of a spectrum $\underline{E} \in \underline{\text{Spec}}$:

S) There exists a number $p \in \mathbb{Z}$ such that for all i, the mapping $\sigma : \Sigma E_i \longrightarrow E_{i+1}$ induces an isomorphism of the (2i+p)-skeleton.

The following assertion is proved in [6] (proposition 4.1.):

5.1. Proposition: Assume that for $\underline{E} \in \underline{\text{Spec}}$ condition S) holds, then the clusteraxiom holds for $\overline{\underline{E}}_*$.

We do not repeat the proof but refer to [6]. The condition S) is needed in order to bring a sequence

$$\bar{f}_i : S^{n+l(i)} \longrightarrow E_{l(i)} \bar{\wedge} X_i$$

$X = \overset{\infty}{\underset{i=1}{\text{Cl}}} X_i$, of shape mappings $\bar{f}_i \in \overline{\underline{E}}_n(X_i)$ onto the same level

$$\bar{f}_i' : S^{n+l} \longrightarrow E_l \bar{\wedge} X_i$$

with a $l \in \mathbb{Z}$, not depending upon i.-

5.2. Proposition: Let $q \in \mathbb{Z}$ be a fixed number and assume that the Ω-spectrum $\underline{E} = \{E_n\} \in \underline{\text{Spec}}$ has the property that all E_i are (i-q)-connected. Then there exists a spectrum $\underline{E}' = \{E_n'\} \in \underline{\text{Spec}}$, as well as a morphism $\underline{f} \in \underline{\text{Spec}}(\underline{E}, \underline{E}')$ such that $\underline{f}_*: \overline{\underline{E}}_* \longrightarrow \overline{\underline{E}}_*'$ becomes an isomorphism of the homology functors.

Proof: The fact that $\underline{E}$ is a Ω-spectrum yield, together with the stability theorem ([2] Satz 8.7), the following sequence of isomorphisms

$$\bar{\pi}_{i+1}(E_{n+1}) \approx \bar{\pi}_i(\Omega E_{n+1}) \approx \bar{\pi}_i(E_n)$$

$$\approx \bar{\pi}_{i+1}(\Sigma E_n)$$

for $i \leq 2(n-q)$. Now we are able to construct $\underline{E}'$ inductively: We set $E'_n = E_n$ for $n \leq q$ and assume that E'_m has already been constructed for $m \leq n$ together with homotopy equivalences $f_m\colon E_m \longrightarrow E_m'$ (compatible with the related bonding maps σ, σ'). There is an isomorphism

$$(1) \qquad \pi_{i+1}(E_{n+1}) \approx \pi_{i+1}(\Sigma E'_n), \qquad i \leq 2(n-q),$$

which allows us to replace

$$(E_{n+1})^{2(n-q)-1} \quad \text{by} \quad (\Sigma E'_n)^{2(n-q)-1}$$

without altering the homotopy type. This is done in a standard way: We attach the $2(n-q)$-cells of E_{n+1} to $(\Sigma E'_n)^{2(n-q)-1}$ by means of maps

$$f\colon \text{bd}\, D^{2(n-q)} = S^{2(n-q)-1} \longrightarrow (\Sigma E'_n)^{2(n-q)-1}$$

which can be chosen because of (1). Then we continue and construct the k-skeletons of E'_{n+1}, $k > 2(n-q)$ inductively.
So, we finally obtain $E'_{n+1} \simeq E_{n+1}$ with $\sigma'\colon \Sigma E'_n \longrightarrow E'_{n+1}$ which induces an isomorphism on the $2n-(2q+1)$ skeleton.
Consequently $\underline{E}'$ fulfils condition S) with $p = 1-2q$.
Our construction yields a $\underline{f}\colon \underline{E} \longrightarrow \underline{E}'$ such that $f_n\colon E_n \longrightarrow E'_n$ becomes a homotopy equivalence for all n.
We have homotopy inverses $g_n\colon E'_n \longrightarrow E_n$ of f_n which commute according to our construction with the bonding maps σ, σ' up to homotopy. A simple cofiber argument provides us with a family $\langle g'_n \rangle$, such that $g'_n \simeq g_n$ and such that the g'_n strictly commute with the bonding maps σ, σ'. Moreover for each $\bar{a} \in \mathfrak{Z} \in \underline{E}_*(X)$

(resp. $\bar{a}' \in \zeta' \in \tilde{\underline{E}}'_*(X)$) we have a homotopy between $\bar{a}$ and $(g'_1 \wedge 1)(f_1 \wedge 1)\bar{a}$ (resp. $\bar{a}' \simeq (f_1 \wedge 1)(g'_1 \wedge 1)\bar{a}'$). This makes sure that $\underline{g}'_* : \tilde{\underline{E}}'_*(\ ) \longrightarrow \tilde{\underline{E}}_*(\ )$ becomes an inverse to $\underline{f}_*: \tilde{\underline{E}}_* \longrightarrow \tilde{\underline{E}}'_*$.

So we have an isomorphism

$$\tilde{\underline{E}}_* \approx \tilde{\underline{E}}'_* .-$$

Proposition 5.2. enables us to complete the

<u>proof of theorem 3.1.</u>: Let $n \in Z$ be a fixed number and $X = \overset{\infty}{\underset{i=1}{Cl}} X_i \in \underline{Com}^f$ be a given space. There exists a $S^N \supset X \supset X_i$ for all i. Let $\underline{E}$ be any Ω-spectrum. By a well-known procedure we can kill the first (m-q) homotopy groups of E_m (for a $q \in Z$ to be fixed later) without affecting the higher homotopy groups. This is accomplished by taking the fibers of suitable Postnikov decompositions of E_m. The spectrum constructed in this way is called $\underline{E}'$ and it is still an Ω-spectrum. We have for any compact $Y \subset S^N$:

$$\underline{E}'^p(S^N \setminus Y) \approx \tilde{\underline{E}}'_{N-p-1}(Y)$$

$$\underline{E}^p(S^N \setminus Y) \approx \tilde{\underline{E}}_{N-p-1}(Y),$$

according to Alexander duality (cf. §4 (2)).

However since

$$E_p = E'_p \qquad \text{for } p > q$$

we conclude

(2) $$\tilde{\underline{E}}'_{N-p-1}(Y) \approx \tilde{\underline{E}}_{N-p-1}(Y)$$

as long as we have $p > q$, for example with $p = q+1$. Thus by setting $n = N-(q+1)-1 = N-q$ we finally obtain

$$q = N - n.$$

Now we observe that the isomorphism (2) is natural with respect to inclusions (because it stems from Alexander duality). Since the inverse to the cluster isomorphism in 3.1. is induced by the inclusions $X_i \subset X$, the diagram

$$\begin{array}{ccc} \bar{E}_n(X) & \approx & \prod_{i=1}^{\infty} \bar{E}_n(X_i) \\ \wr\wr & & \wr\wr \\ \bar{E}'_n(X) & \approx & \prod_{i=1}^{\infty} \bar{E}'_n(X_i) \end{array}$$

with vertical isomorphisms coming from (2) is commutative. According to proposition 5.2. $\bar{E}'_n$ fulfils the cluster-axiom. All this implies that for $\bar{E}_n$ and our fixed $X = \mathrm{Cl}_{i=1}^{\infty} X_i$ a clusteraxiom holds.

This completes the proof of theorem 3.1..-

<u>b) Ad theorem 3.2.:</u> Theorem 3.2. is <u>almost</u> identical with corollary 5.6. resp. theorem 5.8. in [6]. The difference lies merely in the fact that in [6] <u>both</u> functors H_*, H'_* are required to fulfil the clusteraxiom, while in our present case we merely claim that it will be sufficient to do so for H'_*. However the proof of theorem 5.3. [6] is easily seen to go through with this less restrictive assumption: An element $\bar{\zeta} \in H_n(X)$ is mapped onto $\tilde{\alpha}(\bar{\zeta})$ in the following way: Embed X into a Hilbert cube and replace $\bar{\zeta}$ by $\zeta \in H_{n+1}(Q,X)$ with $\partial\zeta = \bar{\zeta}$. Now ζ is "cut into infinitely many pieces ζ_i" by a procedure explained in [6] lemma 5.5.. For this we do <u>not</u> need the clusteraxiom. Due to the special character of the ζ_i, α can be applied, yielding a family $\{\alpha(\zeta_i)\}$. Finally we put together these $\alpha(\zeta_i)$ obtaining a $\tilde{\alpha}(\zeta)$. Only for this procedure we need lemma 5.4. which of course heavily depends on the validity of a clusteraxiom.

This indicates that we are allowed to remove the assumption of the clusteraxiom for H_*.-

6. Examples:

In this section we are going to realize explicitely that:

1) There are spectra $\underline{E}$ such that for $\bar{\underline{E}}_*$ the clusteraxiom does not hold.

2) there are spectra $\underline{E}$, $\underline{E}'$ being equivalent in $\underline{B}_h$ but having the property that $\bar{\underline{E}}_*$ is not isomorphic to $\bar{\underline{E}}'_*$.

The first example is already included in [6] §4; we have to report it again because we need it in order to establish the second example. It should be observed that the mere existence of spectra $\underline{E}$, $\underline{E}'$ having the property 2) can be deduced from the first example: Take for $\underline{E}$ the spectrum in 1) and for $\underline{E}'$ the spectrum $\underline{\Omega}\underline{E}$. Then the existence of a natural isomorphism between $\bar{\underline{E}}_*$ and $\overline{\Omega\underline{E}}_*$ would imply the existence of a natural isomorphism between $\bar{\underline{E}}_*$ and ${}^s\underline{E}_*$. Since the latter fulfils a clusteraxiom while the first one does not, this leads immediately to a contradiction. - Our aim is to make this more explicit, without involving any properties of ${}^s\underline{E}_*$.

We come to the construction of $\underline{E}$:

We put 1) $E_n = *$ for $n < 0$, 2) $E_o = S^o$ and inductively 3) $E_{n+1} = S^{n+1} \vee \Sigma E_n$. The mapping $\sigma_n: \Sigma E_n \longrightarrow E_{n+1}$ is in this case a cellular inclusion. We set $X_i = S^o = \{x_i, *\}$ and $X = \overset{\infty}{\underset{i=1}{\mathrm{Cl}}}\ X_i$.

As a result of all this we have

$$E_k \bar{\wedge} X_i = E_k \wedge X_i = E_k .$$

Let $(f_i: S^{o+i} \subset S^i \vee \Sigma E_{i-1}) \in \zeta_i \in \underline{E}_o(X_i)$ be the element defined by the inclusion into the first summand. Assume the existence of a $\zeta \in \underline{E}_o(X)$ such that $p_{i*}\zeta = \zeta_i$, with projection $p_i: X \longrightarrow X_i$. Then let $\bar{f}: S^{O+l} \longrightarrow E_l \bar{\wedge} X$ be any representative of ζ. We have $p_i \bar{f} =$

$\bar{f}_i': S^{o+l} \longrightarrow E_l \bar{\wedge} X_i = E_l \wedge X_i = E_l$. This $\bar{f}_i' = f_i'$ can without loss of generality be assumed to be a continuous mapping (since <u>every</u> shape morphism into a CW-space is homotopic to a continuous mapping, cf. [5] 1.3.). On the other hand f_i' must be lying in the same stable class with f_i. However the construction of the f_i guarantees that there is no universal level l independent of i, such that all f_i can be lifted to a $f_i': S^l \longrightarrow E_l$.
This establishes example 1).-

The second example is constructed in the following way:

Let $\underline{E}$ be as before and consider ${}_{\Omega}\underline{E}$, the Ω-spectrum whose construction was indicated in §1, together with the morphism $\gamma : \underline{E} \longrightarrow {}_{\Omega}\underline{E}$. To each $\zeta_i \in \underline{E}_o(X_i)$ which we considered already above, we obtain $\gamma_*(\zeta_i) = \zeta_i' \in$ $\in \underline{E}^o(X_i)$. This ζ_i' admits a representative $h_i: S^1 \longrightarrow {}_{\Omega}E_1 = {}_{\Omega}E_1 \wedge X_i$ which is constructed as follows: Let

$$h_i': S^1 \longrightarrow \Omega^{i-1} E_i$$

be the adjoint of $f_i: S^i \longrightarrow E_i$, $f_i \in \zeta_i$. Due to the explicit construction of ${}_{\Omega}E_1$ in §1 we have an inclusion

$$\Omega^{i-1} E_i \subset {}_{\Omega}E_1.$$

The required h_i is the composite of h_i' with this inclusion. By construction we have $h_i \in \zeta_i'$, where now, differently from the case of the ζ_i, all representatives h_i have a common level, namely $l = 1$.

It can be easily verified that there is only one counterimage to each ζ_i' under γ_* (namely ζ_i): Each h_i has up to homotopy, f_i as representative going into a E_l with lowest possible index l (which is in this case equal to i).

Now these $\zeta_i' \in {}_{\Omega}\underline{E}_o(X_i)$ can be composed to a $\zeta' \in {}_{\Omega}\overline{\underline{E}}_o(X)$ such that $p_{i*}\zeta' = \zeta_i'$ (p_i: $X = \overset{\infty}{\underset{i=1}{\mathrm{Cl}}}\; X_i \longrightarrow X_i$ denoting the projection). This procedure is sufficiently well described in the proof of [6] proposition 4.1.. For this purpose one needs only the fact that all ζ_i' have representatives h_i at the same level l(in our case l= 1).

In order to get a $\bar{h}$: $S^1 \longrightarrow E_1 \bar{\wedge} X$, we approximate X by taking a CW-space $P_m = \bigvee_{i=1}^{m} X_i$ and the natural projections g_m: $X \longrightarrow P_m$, mapping each X_j, $j > m$ into the common basepoint $*$. (Equally well one could replace g_m by all inclusions into a suitable CW-space $P_m' \supset X$) .

We evaluate $\bar{h}$ at g_m by defining

$$\bar{h}(g_m) = (h_1 \vee \ldots \vee h_m)\kappa \quad : S^1 \longrightarrow E_1 \wedge P_m$$

where κ: $S^1 \longrightarrow S^1 \vee \ldots \vee S^1$ (m summands) denotes the comultiplication. By general shape theory (cf. [5] appendix) this turns out to be sufficient for determining a shape mapping $\bar{h}$.

This ζ' cannot have a counter image $\zeta \in \overline{\underline{E}}_o(X)$ because otherwise we would have

$$p_{i*}\zeta = \zeta_i \text{ , } i=1,2,\ldots$$

which contradicts the non-existence of such a ζ which we deduced in the treatment of the first example.-

References:

[1] F. Adams: Stable homotopy and generalized homology
Chicago Lecture Notes in Mathem.
The Univ. of Chicago Press (1974)

[2] F.W.Bauer: Homotopietheorie
B.I. Hochschultaschenbücher,
Bd.475 a/b, Bibliographisches
Institut, Mannheim (1971)

[3] " A shape theory with singular homology
Pacific Journal of Mathem. ,Vol.62
No. 1 p. 25-65 (1976)

[4] " Some relations between shape constructions
Cahiers de Topologie et geom. diff.
Vol. XIX-4 p. 337-367 (1978)

[5] " Duality in manifolds
(mimeographed notes)

[6] " Extensions of generalized homology functors
(mimeographed notes)

[7] J. Boardman: Stable homotopy theory (ch. II)
(mimeographed notes)
The Johns Hopkins Univ. (1970)

[8] D.S.Kahn, J.Kaminker, C.Schochet Generalized homology theories on compact metric spaces
Mich. Math. Journal, Vol. 24
p. 203-224 (1977)

[9] J.Kaminker, C.Schochet — K-Theory and Steenrod homology; Applications to the Brown-Douglas-Fillmore theory of operator algebras Trans. Am. Math. Soc. 227 p. 63-1o7 (1977)

[10] J. Milnor: — On the Steenrod homology theory (mimeographed notes) Berkeley (1960)

[11] K. Morita — Cech cohomology and covering dimension for topological spaces Fund. Math. LXXXVII p. 31-52 (1975)

STRONG SHAPE THEORY

by

Fritz Cathey

The purpose of this paper is to present a brief yet somewhat complete account of the strong shape theory of metric compacta. Various approaches to the subject are surveyed here ([6], [14], [24] and [7]) and some of the central results relating strong shape theory to ordinary shape theory are proved.

The origins of strong shape theory can be traced back to papers by D. Christie [10] and Quigley [32]. Current interest evidently stems from fairly recent developments in "coherent" pro-homotopy theory, sometimes called Steenrod homotopy theory. The relation which strong shape theory bears to Steenrod homotopy theory is entirely analogous to that which shape theory bears to Cech homotopy theory.

One of the principal distinctions between the shape and strong shape theories is the fact that the latter is a generalized homotopy theory. Indeed, the category CM of metric compacta is a model for "left homotopy theory" in the sense of D. W. Anderson [1], where the ensuing homotopy category, ho-CM , is the strong shape category. (Here, embeddings play the role of cofibrations and strong shape equivalences that of weak equivalences.) The shape category is a quotient of the strong shape category, but the classification of spaces by strong shape is the same as by shape. A central problem is describing those shape equivalences which are in fact strong shape equivalences.

The paper is organized as follows. §0 is devoted to preliminaries. In §1 we begin to study strong shape from a moderately homotopy theoretic point of view. The notion of SSDR-map is introduced, a shape version of "cofibration + homotopy equivalence", and some of the important properties of this class of maps are obtained. The strong shape category, sSh(CM) , is then defined as a certain category of "left fractions" which arises by formally inverting the collection of homotopy classes of SSDR-maps of compacta. Even at this level of abstraction one obtains some interesting results of geometric content. (see, e.g., (1.15)-(1.18)).

In §2 we consider several representations of sSh(CM) in other more concrete categories. We look at the approaches to strong shape theory of Calder and Hastings [6], Dydak and Segal [14], and the author [7]. We also look at the approaching homotopy theory of Quigley [32], recently generalized by Kodama and Ono [24], which gives a very useful geometric representation of strong shape theory. (There are several other approaches to strong shape, due to Lisica [27] and Bauer [3], which we do not consider for lack of space).

Finally, in §3 we consider the functor $\theta : sSh(CM) \to Sh(CM)$ that relates strong to ordinary shape theory. The central results here are actually consequences of (coherent) pro-homotopy theory, but are demonstrated here in a more geometrical fashion suitable to the compact setting.

§0 Preliminaries

We make use of a small bit of category theory which is reviewed here for the reader's convenience.

Let $R : \mathcal{C} \to \mathcal{B}$ be a functor. The full image of R is a category fimR and a factorization of R :

$$\mathcal{C} \xrightarrow{E} \text{fimR} \xrightarrow{J} \mathcal{B}$$

where E is the identity on objects and J is fully faithful. Every functor has a full image and it is essentially unique.

Now, suppose $\mathcal{B} \subseteq \mathcal{A}$ is a full subcategory of $\mathcal{A}$. Then a morphism $\tau : X \to Y$ of $\mathcal{A}$ with Y in $\mathcal{B}$ is called a $\mathcal{B}$-reflection of X if the function

$$\tau^{\#} : \mathcal{B}(Y,B) \to \mathcal{A}(X,B)$$

is bijective for each B in $\mathcal{B}$. Note that, if $\tau' : X \to Y'$ is another $\mathcal{B}$-reflection, then there is a unique isomorphism $\eta : Y \to Y'$ such that $\tau' = \eta\tau$. Let $\mathcal{C} \subseteq \mathcal{A}$ be a subcategory of $\mathcal{A}$ and let $\{\tau_X : X \to RX\}_{X\in\mathcal{C}}$ be a collection of $\mathcal{B}$-reflections, where R is some function mapping objects of $\mathcal{C}$ to objects of $\mathcal{B}$. Then R extends to a functor, $R : \mathcal{C} \to \mathcal{B}$, if we define Ru for $u \in \mathcal{C}(A,X)$ by the equation $(Ru)\tau_A = \tau_X u$. Such a functor we call a reflection of $\mathcal{C}$ in $\mathcal{B}$ or, simply, a reflector. It is easily checked that any two such reflectors are naturally equivalent and we draw on this fact repeatedly in the sequel.

We will denote by M (resp. CM) the category of pointed metrizable spaces (resp. compact metrizable pointed spaces) and pointed continuous maps. Henceforth, unless otherwise indicated, spaces will be assumed to be pointed metrizable maps and homotopies, pointed continuous. Note that the assumption of pointedness is not unduly restrictive since the unpointed category can be regarded as a full subcategory of M in the usual way by introducing disjoint basepoints. If $\mathcal{B}$ is a category which admits a homotopy relation, then we will write $h\mathcal{B}$ to denote the quotient by this relation.

§1. SSDR-maps and the Strong Shape Category

Recall that a map $i:A \to X$ is called an SDR-map if i embeds A as a strong deformation retract of X. It is well-known that this is equivalent to saying that i is a cofibration and homotopy equivalence. The following provides a shape version of this notion.

(1.1) Definition: A closed subspace A of a space X is called a shape strong deformation retract of X provided that there exists a space M and a closed embedding $\alpha : X \to M$ such that the following condition holds: for any pair of neighborhoods (U,V) of $(\alpha X, \alpha A)$ in M there is a homotopy $H : X \times I \to M$ rel A such that $H_0 = \alpha$, $\operatorname{im} H \subseteq U$ and $\operatorname{im} H_1 \subseteq V$. A closed embedding $i : A \to X$ will be called an SSDR-map if i embeds A as a shape strong deformation retract of X.

Example: The Warsaw circle $A \subseteq \mathbb{R}^2$ is a shape strong deformation retract of $\mathbb{R}^2 \setminus \{pt.\}$, where pt. is any point in the bounded component of $\mathbb{R}^2 \setminus A$.

Remarks: (1) It is a routine exercise to show A is a shape strong deformation retract of X if and only if the condition stated above holds for arbitrary closed embeddings $\alpha : X \to M$ where M is an ANR. (2) If, in the definition, we drop the requirement that H be rel A and add $H_1|A = \alpha|A$, then we obtain the notion of shape deformation retract. It is known that this amounts to saying that the inclusion map $A \subseteq X$ is a shape equivalence. It follows that every SSDR-map is a shape equivalence. Whether or not the converse holds (for closed embeddings) is an open problem, but for partial results see (1.15), (3.3) and (3.6) of this paper.

(1.2) Theorem: Let $i : A \to X$ be a map. Then the following are equivalent:
(1) i is an SSDR-map;
(2) for any map $f : A \to P$, P an ANR, there is a map $f^* : X \to P$ such that $f^* i = f$ and any two such extensions are homotopic rel iA;
(3) for any commutative square of maps

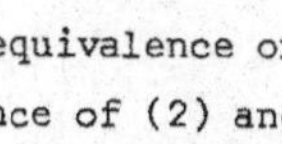

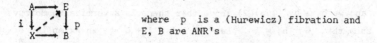

where p is a (Hurewicz) fibration and E, B are ANR's

there is a map represented by the dotted arrow making the resultant diagram commute.

The equivalence of (1) and (2) is a straight forward exercise in ANR theory. The equivalence of (2) and (3) is also routine and is based on the proof of an analogous result for SDR-maps. (See [39] Theorem 8). We remark that here it is necessary to use the fact that fibrations of metrizable spaces are regular (See, e.g., [13] p.397).

(1.3) Remarks: (1) Assuming that $i : A \subseteq X$ is a closed inclusion, we have by the HEP that (1.2) (2) is equivalent to saying that the inclusion-induced functions of homotopy classes

$$[X,P] \to [A,P]$$

$$[X \times I,P] \to [X \times \dot{I} \cup A \times I,P]$$

are surjective for any ANR, P.

(2) Assuming (1.2)(1), (2) and (3), the map represented by the dotted arrow in (3) is in fact unique up to a vertical homotopy. (Again, see e.g. [39]).

Several important consequences of Theorem (1.2) (1) ⇔ (3) are the following:

(1.4) <u>Proposition</u>: Given a pushout (in M)

$$\begin{array}{ccc} A & \longrightarrow & Y \\ {\scriptstyle i}\downarrow & & \downarrow{\scriptstyle i'} \\ X & \longrightarrow & Z \end{array}$$

if i is an SSDR-map, then so is i'. ∎

(1.5) <u>Proposition</u>: Let $i : A \to X$ and $j : X \to Y$ be closed embeddings. If any two of i,j,ji are SSDR-maps, then so is the third.

<u>Proof</u>: We consider only the case were i, ji are SSDR-maps. The other two cases may be treated similarly and are easier in comparison. So, consider a commutative diagram:

$$\begin{array}{ccc} A & & \\ {\scriptstyle i}\downarrow & & \\ X & \xrightarrow{f} & E \\ {\scriptstyle j}\downarrow & & \downarrow{\scriptstyle p} \\ Y & \xrightarrow[g]{} & B \end{array}$$

where p is a fibration and E, B are ANR's .

By (1.2) (1) ⇒ (3) , there is a map $h : Y \to E$ such that $ph = g$ and $hji = fi$. Then $phj = pf$ and we have by (1.3)(2) above that there is a vertical homotopy $G : X \times I \to E$ such that $G : hj \underset{p}{\simeq} f$. Let $G^* : Y \times I \to E$ be an extension of G , i.e., $G^*(j \times id) = G$, such that $G^*_o = h$. Then $G^*_1 j = f$ and $pG^* : g \simeq pG^*_1$ rel jX. Since p is regular, pG^* lifts to a homotopy $H : H_o \simeq G^*_1$ rel jX and we see that H_o satisfies $pH_o = g$, $H_o j = f$. ∎

(1.6) <u>Corollary</u>: Let $i : A \subseteq X$ be closed. Then the inclusion $X \times 0 \cup A \times I \subseteq X \times I$ is an SSDR-map. If in addition, i is an SSDR-map, then so is the inclusion $X \times \dot{I} \cup A \times I \subseteq X \times I$. ∎

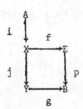

Theorem (1.2) shows how to extend the notion of SSDR-map over the class of all topological spaces. In this extended setting (1.4) and (1.5) remain valid provided that in (1.5) we only consider embeddings $j : X \to Y$ such that (Y, jX) has the HEP with respect to ANR's. We make use of this to prove the following result.

(1.7) <u>Proposition</u>: If $i : A \to X$ is an SSDR-map and $e : A \to X$ is a closed embedding homotopic to i, then e is an SSDR-map.

<u>Proof</u>: Let $H : A \times I \to X$ be a homotopy from i to e and let $Z(G)$ be the mapping cylinder of $G = (H,p) : A \times I \to X \times I$, where p is projection on I. (This space need not lie in M). The inclusions $i_o, i_1 : X \to X \times I$ determine SDR-maps $j_o, j_1 : X \to Z(G)$. Now, regarding I as $0 \times I \cup I \times 0 \cup 1 \times I$, there is an evident closed embedding $k : A \times I \to Z(G)$ such that $k_o = j_o i$ and $k_1 = j_1 e$. Moreover, $(Z(G), k(A \times I))$ has the HEP with respect to ANR's, (consider the inclusions $k(A \times I) \subseteq Z(i) \cup A \times I \cup Z(e) \subseteq Z(G)$). By (1.5), k_o is an SSDR-map, and since $A \times 0 \subseteq A \times I$ is an SDR-map, so is k. Arguing in reverse we conclude that $k_1 = j_1 e$, and hence e, is an SSDR-map. ∎

(1.8) <u>Corollary</u>: If a closed embedding $i : A \to X$ is a homotopy equivalence, then it is an SSDR-map.

<u>Proof</u>: Just consider the mapping cylinder $Z(i)$ and use (1.7). ∎

In practice, SSDR-maps often arise in the following manner.

(1.9) <u>Lemma</u>: Let (X_n, p_n^m) and (Y_n, q_n^m) be towers (i.e., inverse sequences) of spaces and let $(i_n : X_n \to Y_n)$ be a level system of maps. (By "system" we mean that $i_n p_n^m = q_n^m i_m$ for $n \leq m$). Assume that each X_n is compact and each i_n is an SDR-map. Then the induced map of inverse limits, $i : X \to Y$, is an SSDR-map provided either (a) each bond q_n^{n+1} is a fibration, or (b) each Y_n is compact.

<u>Proof</u>: (a) Observe that in this case each projection $q_n : Y \to Y_n$ is a fibration. It follows, therefore, that a deformation of Y_n onto $i_n X_n$ rel $i_n X_n$ lifts to a deformation of Y onto $q_n^{-1} i_n X_n$ rel $q_n^{-1} i_n X_n$. Since the X_n's are compact, the net $(q_n^{-1} i_n X_n)$ refines the basis of neighborhoods of iX in Y, and the result follows. (b) In this case we begin by inductively "turning" each bond q_n^{n+1} into a fibration (see proof of Theorem (2.5)) thus obtaining a tower $(\tilde{Y}_n, \tilde{q}_n^m)$ bonded by fibrations, and a system of SDR-maps $(j_n = Y_n \to \tilde{Y}_n)$. By (a), the induced map of inverse limits $j : Y \to \tilde{Y}$ is an SSDR-map. But each $j_n i_n$ is also an SDR-map, so ji is an SSDR-map as well. The result now follows from (1.5). ∎

<u>Remark</u>: H. Hastings and J. Hollingsworth [21] have shown, using the homotopy theory of pro-spaces, that every SSDR-map of compacta is in fact an inverse limit of PL expansions of compact polyhedra.

We now proceed towards our definition of the strong shape category of compacta $sSh(CM)$. The following theorem is of fundamental importance. Let $\Sigma \subseteq mor(hCM)$ denote the collection of homotopy classes of SSDR-maps in CM.

(1.10) <u>Theorem</u>: Σ admits a calculus of left fractions. That is, the following conditions hold in hCM: (1) Σ is closed under composition and contains all identity morphisms; (2) given $u : A \to Y$ and $s : A \to X$ with $s \in \Sigma$, there exist u' and s' with $s' \in \Sigma$ such that $s'u = u's$:

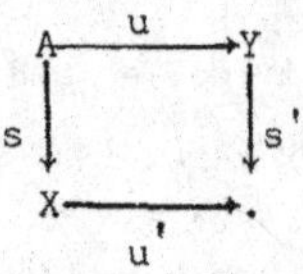

(3) given $u, v: X \rightrightarrows Y$ and $s \in \Sigma$ such that $us = vs$, there exists $t \in \Sigma$ such that $tu = tv$;

$$\bullet \xrightarrow{\ s\ } X \underset{v}{\overset{u}{\rightrightarrows}} Y \dashrightarrow^{t} \dashrightarrow \bullet \ .$$

<u>Proof</u>: (1) follows from (1.5), and (2) follows from (1.4) and the fact that pushouts exist in CM. To see that (3) holds, let $f,g : X \rightrightarrows Y$ be maps representing u,v respectively and let $i : A \to X$ be an SSDR-map representing s. Without loss of generality, assume that i is an inclusion map. Now, using f,g and a homotopy from $f|A$ to $g|A$, define a map $h : X \times \dot{I} \cup A \times I \to Y$ and form the pushout:

$$
\begin{array}{ccc}
X \times \dot{I} \cup A \times I & \xrightarrow{\ h\ } & Y \\
\uparrow & & \downarrow j \\
X \times I & \longrightarrow & X \times I \cup_h Y
\end{array}
$$

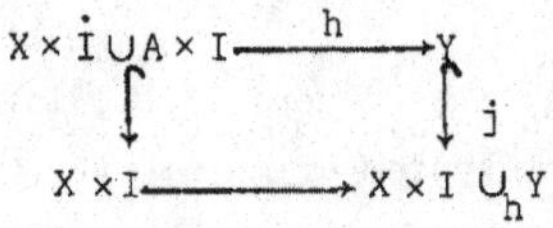

Then by (1.6) and (1.4), j is an SSDR-map. Clearly $jf \simeq jg$ and hence, letting $t = [j]$, $tu = tv$. ∎

Theorem (1.10) permits us to define a "category of left fractions" Σ^{-1}hCM as follows. (See [19] and [34] for details). Objects of Σ^{-1}hCM are those of CM , while a morphism from X to Y in Σ^{-1}hCM is an equivalence class of pairs (u,s) with $s \in \Sigma$ for which there is a diagram $X \xrightarrow{\ u\ } \bullet \xleftarrow{\ s\ } Y$. Two pairs (u,s) and (u',s') are equivalent if there exist morphisms v,v' in hCM and $t \in \Sigma$ so as to yield a commutative diagram:

(1.11)

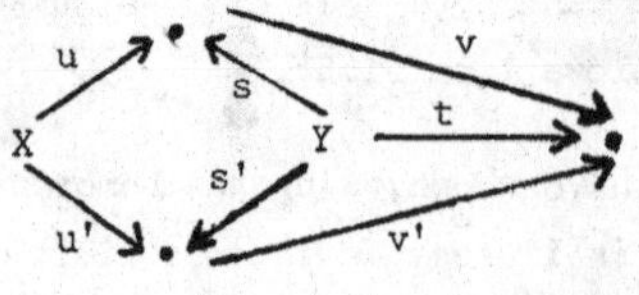

The equivalence class of (u,s) is written $s\backslash u$ and called a (left) fraction. Fractions are composed using (1.10) (2):

(1.12) $(t\backslash v)(s\backslash u) = s't\backslash v'u$

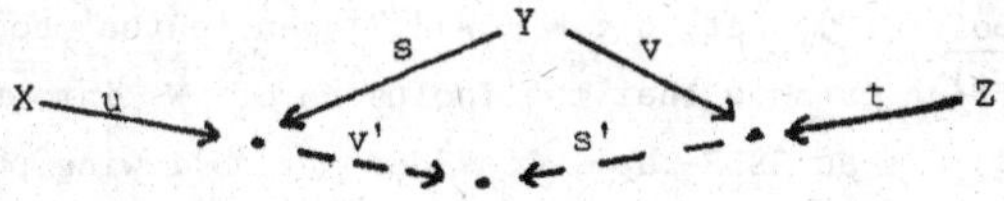

Clearly the identity morphism on X is $1_X\backslash 1_X = s\backslash s$ where s is any morphism of Σ such that domain $s = X$.

There is a functor $P_\Sigma : hCM \to \Sigma^{-1}hCM$ defined by $P_\Sigma(u) = 1_Y\backslash u$ for $u : X \to Y$ in hCM. This functor localizes hCM at Σ. That is, P_Σ inverts Σ (i.e., P_Σ carries Σ into a class of isomorphisms) and for any functor $F : hCM \to \mathcal{B}$ which inverts Σ, there is a unique functor $\overline{F} : \Sigma^{-1}hCM \to \mathcal{B}$ such that $\overline{F}P_\Sigma = F$. (In fact, $\overline{F}(s\backslash u) = F(s)^{-1}F(u)$).

Henceforth we refer to $\Sigma^{-1}hCM$ as the <u>strong shape category</u> and write sSh(CM). We call P_Σ the <u>strong shape functor</u> and write $S^* : hCM \to sSh(CM)$.

Several important consequences of the preceding development are the following facts. Call a map f a strong shape equivalence if $S^*[f]$ is invertible.

(1.13) The composite functor $CM \to hCM \to sSh(CM)$ localizes CM at the class of strong shape equivalences. $\blacksquare$

(1.14) Every morphism of sSh(CM) is of the form $S^*[i]^{-1}S^*[j]$, where j is an embedding and i is an SSDR-map. $\blacksquare$

(1.15) A closed embedding $i : A \to X$ is a strong shape equivalence if and only if it is an SSDR-map.

<u>Proof</u>: Using (1.11), (1.12) and (1.7) one can show that if $S^*[i]$ is invertible, then there is a commutative diagram of embeddings

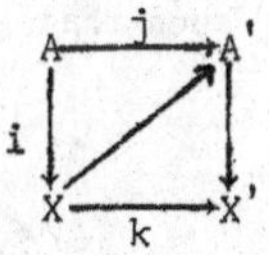

where j,k are SSDR-maps.

Now use (1.2) (1) $\Leftrightarrow$ (2) and (1.6). $\blacksquare$

(1.16) A map $f : X \to Y$ is a strong shape equivalence if and only if the inclusion of X into the base of the mapping cylinder $Z(f)$ is an SSDR-map. An unpointed strong shape equivalence $f : X \to Y$ induces a strong shape equivalence $f : (X,x) \to (Y,f(x))$ for any choice of basepoint $x \in X$. $\blacksquare$

(1.17) (Dydak and Segal) A map $f : X \to Y$ is a strong shape evidence if and only if the projection map $W \to W \cup_f Y$ is a shape equivalence for every compactum W with $X \subseteq W$.

<u>Proof</u>: ($\Rightarrow$) Let $q : W \to W \cup_f Y$ denote the projection map. Then it will more than suffice to show that the inclusion of W into the base of the mapping cylinder $Z(q)$ is an SSDR-map. Consider the following pushout diagrams:

$$\begin{array}{ccc} X & \hookrightarrow & W \\ \downarrow & & \downarrow i \\ Z(f) & \longrightarrow & W \cup_X Z(f) \end{array} \qquad\qquad \begin{array}{ccc} W \times 0 \cup X \times I & \longrightarrow & W \cup_X Z(f) \\ \downarrow & & \downarrow j \\ W \times I & \longrightarrow & Z(q) \end{array}$$

By (1.16), (1.6) and (1.4) both i and j are SSDR-maps. It follows that the composite $ji : W \hookrightarrow Z(q)$ is one as well.

($\Leftarrow$) By (1.16) and Remark (1.3) (1) it will suffice to show that the inclusions $X \hookrightarrow Z(f)$ and $Z(f) \times \dot{I} \cup X \times I \hookrightarrow Z(f) \times I$ are shape equivalences. By assumption f is a shape equivalence and therefore so is $X \hookrightarrow Z(f)$. Also, the projection $q : Z(f) \to Z(f) \cup_f Y$ is assumed to be a shape equivalence. Since $Z(f) \cup_f Y = Z(f) \times I \cup X \times I$, the inclusion $Y \hookrightarrow Z(f) \overset{q}{\to} Z(f) \times \dot{I} \cup X \times I$ is a shape equivalence, and therefore so is $Z(f) \times \dot{I} \cup X \times I \longrightarrow Z(f) \times I$ by (1.5). $\blacksquare$

Kozlowski [26] has shown that the criterion of (1.17) is satisfied by any hereditary shape equivalence. Thus we have the following result: (1.18) Hereditary shape equivalences are strong shape equivalences.

In the next section we shall need a criterion for deciding when a functor on sSh(CM) is fully faithful. This is provided by the following lemma of a purely categorical nature.

(1.19) <u>Lemma:</u> Let $R : hCM \to \mathcal{B}$ be a functor which inverts Σ and let $\overline{R} : sSh(CM) \to \mathcal{B}$ satisfy $\overline{R}S^* = R$. Then:

(1) $\overline{R}$ is full provided that for any two compacta X, Y and morphism $\phi : RX \to RY$, there are homotopy classes u, s with $s \in \Sigma$ such that $\phi = (Rs)^{-1}Ru$.

(2) $\overline{R}$ is faithful provided that for any two compacta X, Y and homotopy classes $u, v : X \to Y$ such that $Ru = Rv$, there exists a $t \in \Sigma$ such that $tu = tv$. $\blacksquare$

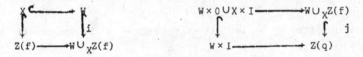

§2. Representations of sSh(CM)

The first representation we consider, due to the author [7], can be regarded as a natural homotopy theoretic outgrowth of the theory of SSDR-maps presented in the previous section. Recall that in axiomatic homotopy theory, c.f. [33], an object Z of a model category is called fibrant if the terminal morphism $Z \to *$ is a fibration, i.e., has the right lifting property with respect to all morphisms which are both cofibrations and weak equivalences.

(2.1) <u>Definition:</u> A space Z will be called <u>fibrant</u> provided that for every SSDR-map $i : A \to X$ and map $f : A \to Z$, there is a map $f^* : X \to Z$ such that $f^*i = f$. The full subcategory of M whose objects are fibrant spaces will be denoted by F.

(2.2) <u>Examples:</u> By Theorem (1.2) every ANR and every inverse limit of a tower of ANR's bonded by fibrations is a fibrant space. In particular, every compact metric topological group is fibrant. (See, e.g., Theorem 54 of [31]). The Warsaw circle is not fibrant.

As consequences of (1.6) and the above definition we have

[2.3) Every closed pair (X,A) has the HEP with respect to the class of fibrant spaces.

(2.4) If $i : A \to X$ is an SSDR-map, then $[i]^\# : [X,Z] \to [A,Z]$ is bijective for every fibrant space Z.

Note that, if in (2.4) the space X happens to be fibrant, then $[i]$ is an hF-reflection.

(2.5) <u>Theorem:</u> Every compactum X has an hF-reflection. In fact X embeds in a fibrant space $|X|$ which admits a "deformation" $D : |X| \times [0,\infty) \to |X|$ satisfying:

(1) $D_0 = \mathrm{id}$ and $D_t(x) = x$ for $x \in X$;

(2) for each neighborhood U of X in $|X|$ there is a $T \geq 0$ such that $D_t(|X|) \subseteq U$ for $t \geq T$;

(3) $D_t D_{t'} = D_{\max\{t,t'\}}$ for $t,t' \in [0,\infty)$.

<u>Proof:</u> Suppose first that X is given as the inverse limit of a tower of compact ANR's $(X_n;P_n^m)$. Then by inductively turning each bond into a fibration (c.f. [36] p. 99) :

$$\begin{array}{ccccccc}
X_o & \longleftarrow \cdots \longleftarrow & X_n & \xleftarrow{P_n^{n+1}} & X_{n+1} & \longleftarrow & \cdots \\
\| & & \downarrow i_n & & \downarrow i_{n+1} & & \\
\tilde{X}_o & \longleftarrow \cdots \longleftarrow & \tilde{X}_n & \xleftarrow{\tilde{P}_n^{n+1}} & \tilde{X}_{n+1} & &
\end{array}$$

one obtains a tower $(\tilde{X}_n;\tilde{P}_n^m)$ bonded by fibrations and a system of SDR-maps $(i_n : X_n \to \tilde{X}_n)$. Passing to the limit we obtain a fibrant space $\tilde{X}$ by (2.2) and an SSDR-map $i : X \to \tilde{X}$ by (1.9). Then $[i]$ is an hF-reflection of X .

Now let $X_0 \supseteq X_1 \supseteq X_2 \supseteq \ldots$ be a net of compact ANR's shrinking to X (so $X = \cap X_n$). Define a space $|X|$ and maps $i : X \to |X|$ and $D : |X| \times [0,\infty) \to |X|$ as follows:

$$|X| = \{\omega \in X_0^{[0,\infty)} | \omega[n,\infty) \subseteq X_n \text{ for } n \geq 0\} \quad ;$$

$$i(x)(\lambda) = x \text{ for } x \in X \text{ and } \lambda \in [0,\infty) \quad ;$$

$$D_t(\omega)(\lambda) = \omega(\max\{t,\lambda\}) \text{ for } \omega \in |X| \text{ and } t,\lambda \in [0,\infty) .$$

Give $|X|$ the (metrizable) topology inherited from the compact open topology on $X_0^{[0,\infty)}$ (which is the same as the topology of uniform convergence on compacta). Use the embedding i to regard $|X|$ as a space containing X. Then it is easy to verify that D has the required properties. It is also easy to verify that $i : X \to |X|$ is an SSDR-map and that $|X|$ is fibrant. However, we will check this last fact after the manner of the first half of this proof. Let

$$|X|_n = \{\omega \in X_0^{[0,n]} | \omega[k,n] \subseteq X_k \text{ for } 0 \leq k \leq n\}$$

and let $\rho_n^m : |X|_m \to |X|_n$, $n \leq m$,

be the map induced by restriction. Then $(|X|_n ; \rho_n^m)$ is a tower of ANR's bonded by fibrations whose inverse limit is $|X|$. Thus $|X|$ is fibrant. Moreover, for each n, the natural embedding $i_n : X_n \to |X|_n$ is an SDR-map (with left inverse $e_n : |X|_n \to X_n$, the "evaluation at n" map). Thus $i = \lim i_n$ is an SSDR-map. ∎

(2.7) <u>Corollary</u>: Let $|X|$ be as in Theorem (2.5). Then for every compactum A and map $f : A \to |X|$ there is a commutative diagram:

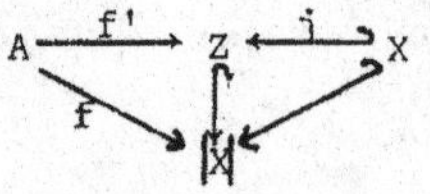

where Z is compact and the inclusion j is an SSDR-map. The deformation D of (2.5) induces a deformation of Z on itself.

<u>Proof</u>: Just let $Z = X \cup D(f(A) \times [0,\infty)) \subseteq |X|$. Condition (3) of (2.5) insures that D deforms Z on itself. ∎

By Theorem (2.5) we have (as remarked in §0) that there exist reflections of hCM in hF.

(2.8) <u>Theorem</u>: Any reflector $R : hCM \to hF$ inverts Σ and induces a fully faithful functor $\bar{R}: sSh(CM) \to hF$.

<u>Proof</u>: Let $\{\tau_X : X \to RX\}$ be the collection of reflections defining R. Then given $s \in \Sigma$, $s : A \to X$, we have by (2.1) and (2.4) a unique $S^* : X \to RA$ such

that the following diagram commutes:

$$\begin{array}{ccc} A & \xrightarrow{s} & X \\ {\scriptstyle \tau_A}\downarrow & \swarrow{\scriptstyle s^*} & \downarrow{\scriptstyle \tau_X} \\ RA & \xrightarrow[Rs]{} & RX \end{array}$$

It follows that Rs is invertible. Now let $\overline{R} : sSh(CM) \to hF$ satisfy $\overline{R}S^* = R$. To show that $\overline{R}$ is fully fiathful we shall verify that R satisfies the conditions of Lemma (1.19). Without loss of generality, we assume that $RX = |X|$ and that τ_X is represented by the inclusion map $X \subseteq |X|$ for each compactum X.

<u>$\overline{R}$ is full</u>: Given $w^* : RX \to RY$ let $w = w^*\tau_X : X \to RY$. Then by (2.7) there is a commutative diagram:

(2.9)
$$\begin{array}{ccccc} X & \xrightarrow{w'} & Z & \xleftarrow{s} & Y \\ & {\scriptstyle w}\searrow & \downarrow & \swarrow{\scriptstyle \tau_Y} & \\ & & RY & & \end{array} \qquad \text{with } s \in \Sigma .$$

A simple argument using the reflective property of R shows that $w^* = (Rs)^{-1}Rw'$.

<u>$\overline{R}$ is faithful</u>: Let $u,v : X \rightrightarrows Y$ be such that $Ru = Rv$. Then $\tau_y u = \tau_y v$. Let f, g be maps representing u,s respectively and, regarding f, g as maps into $|Y| = RY$, let $H : X \times I \to |Y|$ be a homotopy from f to g. Again, by (2.7) there is a commutative diagram of maps:

(2.10)
$$\begin{array}{ccccc} X \times I & \longrightarrow & Z & \xleftarrow{j} & Y \\ & {\scriptstyle H}\searrow & \downarrow & \swarrow & \\ & & |Y| & & \end{array}$$

with Z compact and j an SSDR-map. Letting $S = [j]$ we infer that $su = sv$. ∎

<u>Remark</u>: It would be nice if $sSh(CM)$ could in fact be represented in hCM. In this direction, S. Ferry [18] has proved that if a compactum X is approximatively 1-connected (UV^1), then X has the shape of some compactum X' with $sSh(A,X) \approx [A,X']$ for each finite dimensional compactum A.

Let X be a compactum and let $|X|$ be as in (2.6). Then each map $f:A \to |X|$ determines an "adjoint" map $\hat{f} : A \times [0,\infty) \to X_0$ such that $\hat{f}_t(A) \subseteq X_n$ for $t \geq n$. This observation allows us to make the connection between strong shape theory and the approaching homotopy theory introduced by Quigley [32] (later generalized to fine-shape theory by Kodama and Ono [24]).

(2.11) <u>Definition</u>: Let (X,A) and (Y,B) be pointed pairs of compacta and let $\Phi : X \times [0,\infty) \to Y$ be a map such that $\Phi_t : X \to Y$ is pointed for each $t \geq 0$. Then Φ is called an <u>approaching map</u> (write $\Phi : (X,A) \to (Y,B)$) if the following condition holds: for every neighborhood U of B in Y there is a neighborhood V of

A in X and a $T \geq 0$ such that $\Phi_t(V) \subseteq U$ for $t \geq T$.

There is a category $\mathcal{A}$ whose objects are pairs of compacta and whose morphisms are approaching maps. Composition of morphisms is defined "levelly": $(\Phi\Psi)_t = \Phi_t\Psi_t$, $t \geq 0$. Observe that any map of pairs determines a "stationary" approaching map and so we may (and do) regard the category of pairs (of compacta) and maps of pairs as a subcategory of $\mathcal{A}$. There is an embedding, $CM \to \mathcal{A}$, which sends X to (X,X).

Two approaching maps $\Phi,\Psi : (X,A) \rightrightarrows (Y,B)$ are called approaching homotopic if there is an approaching map $H : (X \times I, A \times I) \to (Y,B)$ which is a homotopy from Φ to Ψ. Denote by $h\mathcal{A}$ the quotient of $\mathcal{A}$ by the relation of approaching homotopy, and write $[\Phi]$ for the approaching homotopy class of Φ. It is not hard to see that the embedding, $CM \to \mathcal{A}$, induces a full embedding, $I : hCM \to h\mathcal{A}$.

Now, let $\mathcal{B}$ denote the full subcategory of $\mathcal{A}$ whose objects are pairs (M,X) with M an AR. We will see that there are fully faithful representations of $sSh(CM)$ in $h\mathcal{B}$. (Note that $h\mathcal{B}$ is equivalent to C_f, the fine shape category defined by Kodama and Ono). We will need the following lemma.

(2.12) <u>Lemma:</u> Let (N,Y) be an object of $\mathcal{B}$ and let $i : (X',X) \subseteq (M,X)$ be an inclusion map regarded as a morphism of $\mathcal{A}$. Then the induced function

$$[i]^{\#} : h\mathcal{A}((M,X),(N,Y)) \to h\mathcal{A}((X',X),(N,Y))$$

is bijective. In particular, if M is an AR, then $[i]$ is an $h\mathcal{B}$-reflection of (X',X).

<u>Proof:</u> Following roughly the manner of proof of [4], IX, (4.9), but with extensive use of the HEP, one can show the following: if $A \subseteq A' \subseteq A''$ are compacta, then each approaching map, $(A',A) \to (N,Y)$, extends to an approaching map $(A'',A) \to (N,Y)$. The result follows from this upon considering the sequences $X \subseteq X' \subseteq M$ and $X \times I \subseteq M \times \dot{I} \cup X' \times I \subseteq M \times I$. ∎

Now each compactum lies in a compact AR and so it follows from (2.12) that there exists a reflection of $h\mathcal{A}$ in $h\mathcal{B}$.

(2.13) <u>Theorem:</u> Let $R : h\mathcal{A} \to h\mathcal{B}$ be a reflector. Then the composite functor RI inverts Σ and induces a fully faithful representation of $sSh(CM)$ in $h\mathcal{B}$.

<u>Proof:</u> Let $s : A \to X, s \in \Sigma$, and assume that s is represented by an inclusion map. Let $(M,M) \supseteq (X_1,A_1) \supseteq (X_2,A_2) \supseteq \ldots$ be a net of pairs of compact ANR's shrinking to (X,A) such that M is an AR. Then we may also assume without loss of generality that $R(A,A) = (M,A)$ and $R(X,X) = (M,X)$. Now, using the nets above we obtain a pair of fibrant spaces $(|X|,|A|)$ with $|A|$ closed in $|X|$ and, since $s \in \Sigma$, the inclusion $|A| \subseteq |X|$ is a homotopy equivalence. By (2.3) and ([36] Cor. 10 and Thm. 11, p.31), we infer that there is a deformation $D : |X| \times I \to |X| \text{ rel} |A|$ such that D_1 defines a retraction $r : |X| \to |A|$. Consideration of the adjoint maps $\hat{D}$ and $\hat{r}$ yields an approaching homotopy commutative diagram:

$$\begin{array}{ccc} (A,A) & \longrightarrow & (X,X) \\ \downarrow & \overset{\hat{r}}{\swarrow} & \downarrow \\ (M,A) & \longrightarrow & (M,X) \end{array}$$

where all unlabeled arrows are inclusions. Thus the following diagram commutes:

$$\begin{array}{ccc} IA & \xrightarrow{Is} & IX \\ \downarrow & \overset{[\hat{r}]}{\swarrow} & \downarrow \\ RIA & \xrightarrow[RIs]{} & RIX \end{array}$$

and we infer, exactly as in the proof of (2.8), that RIs is invertible.

It remains to show that RI satisfies the conditions of Lemma (1.19). So, let us choose for each compactum X a fibrant space $|X|$ as in (2.6) such that X_0 is an AR. Then it is not hard to see that for compacta X,Y there is a bijection, $[X,|Y|] \approx hA(IX,RIY)$, induced by the correspondence $f \to \hat{f}$. We now argue as in the proof of (2.8), using the diagrams (2.9) and (2.10). ∎

Next we consider the approach to strong shape theory of Dydak and Segal [14] which gives rise to representations of sSh(CM) in the proper homotopy category of locally compact AR's, or, of locally finite contractible polyhedra (or CW complexes). In light of the work of Chapman and Siebenmann [9], this approach is seen to establish an equivalence between sSh(CM) and the proper homotopy category of contractible Q-manifolds which admit boundary. In particular, sSh(CM) is equivalent to the proper homotopy category of complements of Z-sets in Q.

Dydak and Segal develop strong shape theory for pairs of compacta. For simplicity, we only consider the theory for pointed compacta. Let $\mathcal{P}$ denote the category of locally compact separable spaces and proper maps. We will make the restrictive assumptions that each object of $\mathcal{P}$ contains a proper ray (a properly embedded copy of $[0,\infty)$) emanating from the base point, and that maps in $\mathcal{P}$ carry rays onto rays. Let $h\mathcal{P}$ denote the associated proper homotopy category. (Homotopies are required to map rays on rays). If f is a proper map, let $[f]_p$ denote the proper homotopy class of f.

Recall that for any tower of compacta $\underline{X} = (X_1 \xleftarrow{P_1^2} X_2 \xleftarrow{P_2^3} X_3 \leftarrow \dots)$ there is an object CTel $\underline{X}$ in $\mathcal{P}$ called the contractible telescope (or contractible infinite mapping cylinder) of $\underline{X}$ obtained by glueing together copies of the mapping cylinders $Z(P_n^{n+1})$ in an obvious way and then attaching a cone to the top of $Z(P_1^2)$:

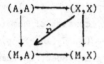

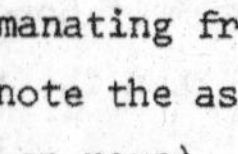

... CTel $\underline{X}$

(The basepoints of the tower $\underline{X}$ determine the ray in CTel $\underline{X}$). If $\underline{Y}$ is another tower and $\underline{f} = (f_n : X_n \to Y_n)$ is a level system of maps (commuting with bonds), then there is an obvious map CTel $\underline{f}$: CTel $\underline{X} \to$ CTel $\underline{Y}$ and one quickly sees that CTel provides a functor from the category of towers of compacta and level systems of maps to the category $\mathcal{P}$. Moreover, if $\underline{f}$ happens to consist of homotopy equivalences, then CTel $\underline{f}$ is a proper homotopy equivalence (c.f. [16] Prop. (3.7.4)). Indeed, by factoring each map f_n through its mapping cylinder, one can produce a factorization of $\underline{f}$:

$$\underline{X} \xrightarrow{\underline{i}} Z(\underline{f}) \xrightarrow{\underline{r}} \underline{Y}$$

such that $\underline{i}$ is a level system of SDR-maps and such that $\underline{r}$ has a right inverse $\underline{j}$ which is also a system of SDR-maps. It is easy to see that CTel $\underline{i}$ and CTel $\underline{j}$ are proper SDR-maps and hence that CTel $\underline{f}$ is a proper homotopy equivalence.

Now, identify each compactum X with the tower whose bonds are all the identity map 1_X. Observe that CTel yields a full embedding, $I : hCM \to h\mathcal{P}$, with $IX \approx X \times [0,\infty)/X \times 0$ (see [14] Lemma 3.1).

Let $\underline{X}$ be a tower of compact ANR's whose limit is X and let $\underline{P} = (P_n : X \to X_n)$ be the cone of projection maps. Then there is a proper map CTel $\underline{P}$: $IX \to$ CTel $\underline{X}$ with CTel $\underline{X}$ an AR. Let $hANR_p$ denote the full subcategory of $h\mathcal{P}$ whose objects are ANR's.

(2.14) Theorem: $[\text{CTel } \underline{P}]_p$ is an $hANR_p$-reflection of IX. If $R : imI \to hANR_p$ is the induced reflector, then the composite functor RI inverts Σ and induces a fully faithful functor from sSh(CM) to $hANR_p$.

Proof: In case $\underline{X}$ is bonded by inclusion maps, then by [14], Lemma 3.3, $\text{CTel}_{\underline{P}}$ embeds IX as a proper, shape strong deformation retract of CTel $\underline{X}$. That $[\text{CTel } \underline{P}]_p$ is an $hANR_p$-reflection now follows from the proper homotopy theory of ANR's. (c.f. [2]). In the general case, let $\text{CTel}^+\underline{X}$ denote the "completed" contractible telescope of $\underline{X}$ obtained by adjoining X as a "boundary" to CTel $\underline{X}$ (c.f. [9] and [18]). Then $\text{CTel}^+\underline{X}$ is a compact AR which supports a net (nested sequence) of compact ANR's, $\underline{Z}$, shrinking to X, and a level system of homotopy equivalences $\underline{r} = (r_n : Z_n \to X_n)$ such that $r_n | X = P_n$, $n \geq 1$. Then the following diagram commutes:

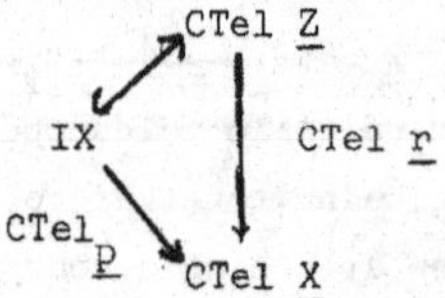

and, as noted earlier, CTel $\underline{r}$ is a proper homotopy equivalence. It follows that $[\text{CTel } \underline{P}]_p$ is an $hANR_p$-reflection.

Let $R : imI \to hANR_p$ be the induced reflector. It remains to show that RI inverts Σ and satisfies the conditions of Lemma (1.19). But this can be argued exactly as in the proofs of (2.8) and (2.13) using the following fact: Let

$X_0 \supseteq X_1 \supseteq \dots$ be a net of compact ANR's shrinking to X, and define $\underline{X}$ as in (2.6). Then $\mathrm{CTel}\ \underline{X} = \bigcup_{n\geq 0} X_n \times [0,n+1]/X_0 \times 0$ is proper homotopy equivalent to RIX, and for each compactum A, there is a bijection $[A,|X|] \approx hP(IA,\mathrm{CTel}\ \underline{X})$ induced by the correspondence $f \to \overline{f}$ where $\overline{f}(a,t) = (f(a)(t),t)$ for $a \in A$, $t \geq 0$. $\blacksquare$

(2.15) <u>Remarks:</u> (1) That RI induces a full functor from $sSh(CM)$ to $hANR_p$ also follows from ([14] Theorems 5.2 and 5.3). (2) Dydak and Segal define the strong shape category as the full image of the composite functor $hCM \xrightarrow{I} hP \xrightarrow{S} Sh_J(hP)$ where S is the shape functor of the inclusion $J : hANR_p \subseteq hP$ (in the sense of Deleanu and Hilton [15]). By Theorem (2.14) and the Yoneda Lemma, this category is isomorphic to $sSh(CM)$. (3) It is an interesting fact that if $f : \mathrm{CTel}\ \underline{X} \to \mathrm{CTel}\ \underline{Y}$ is a proper map representing a strong shape isomorphism $\alpha : X \to Y$, then f is an infinite simple homotopy equivalence. This follows from the "Peripheral Homeomorphism Paradox" of Chapman and Siebenmann [9]. (4) Steenrod [37] used the CTel construction in his development of (Steenrod) homology theory for compacta. In a roughly similar fashion one can geometrically define Steenrod bordism theories for compacta. How to do this for finite dimensional compacta is briefly indicated in a paper by Dold [12].

We turn now to a consideration of the strong shape theory of Calder and Hastings [6] which yields a representation of $sSh(CM)$ in ho(pro-Top). First, we degress briefly to discuss this latter category.

Edwards and Hasings [16] have shown that the usual (see [38]) closed model structure on Top (the category of topological spaces and maps) extends over the category pro-Top. The associated homotopy category, ho-pro-Top, is then obtained by localizing pro-Top at the class of morphisms represented by level systems of homotopy equivalences. This localization can be realized as follows. First, note that the ordinary relation of homotopy of maps extends naturally to the pro setting, and yields a quotient category h(pro-Top). Denote by $h(\text{pro-Top})_f$ the full subcategory whose objects are fibrant systems. (A tower is fibrant for instance if its bonds are fibrations). Now every object has an $h(\text{pro-Top})_f$ - reflection represented by an isomorphism followed by a level system of homotopy equivalences. (See [16] §4.3). ho-pro-Top is, up to isomorphism, the full image of any reflector $R : h(\text{pro-Top}) \to h(\text{pro-Top})_f$.

Note that the inverse limit functor, $\lim : \text{pro-Top} \to \text{Top}$, respects the homotopy relation and hence determines a functor $\text{hlim} : h(\text{pro-Top}) \to h\text{Top}$. The composition $\text{hlim} \circ R$ factors across the full image of R and therefore defines a functor $\text{holim} : \text{ho-pro-Top} \to \text{ho-Top} \equiv h\text{Top}$. It is easily seen that holim is right adjoint to the inclusion $\text{ho-Top} \subseteq \text{ho-pro-Top}$, ie, $[X,\text{holim}\ \underline{Y}] \approx \text{ho-pro-Top}(X,\underline{Y})$ for X in ho-Top, $\underline{Y}$ in ho-pro-Top.

Now, let PL denote the piecewise linear category of compact polyhedra and let ho-pro-PL denote the full subcategory of ho-pro-Top whose objects come from pro-PL . In the approach to strong shape theory of Calder and Hastings each topological space X is associated to an object $sSh(X)$ in ho-pro-PL , namely the universal system $(X \downarrow PL) \to PL$. If X is compact, then the canonical morphism $X \to sSh(X)$ in ho-pro-Top is a ho-pro-PL-reflection of X , [6] Proposition (3.6). The full image of the induced reflector $sSh : hCM \to$ ho-pro-PL is defined by Calder and Hastings to be the strong shape category of compact metric spaces. The following shows that this definition is consistent with that given in §1.

(2.16) <u>Theorem:</u> $sSh : hCM \to$ ho-pro-PL induces a full embedding of $sSh(CM)$ in ho-pro-PL .

<u>Proof:</u> For each compactum X , let $\underline{X}$ be a tower (inverse sequence) in pro-PL whose inverse limit is X and let $\underline{P} : X \to \underline{X}$ be the cone of projections, which is a morphism of pro-Top . Then $[\underline{P}]$, the image of $\underline{P}$ in ho-pro-Top , is a ho-pro-PL-reflection of X . (This is Proposition (3.6) of [6]; see also Proposition (2.10) of [7]). Thus we may assume that $sSh(X) = \underline{X}$ for each compactum X .

Now, the functor holim carries $[\underline{P}]$ into an $h\mathcal{F}$-reflection $\text{holim}[\underline{P}] : X \to \text{holim}\,\underline{X}$. This follows from the proof of (2.15) since $\text{holim}\,\underline{X} = \lim \underline{\tilde{X}}$ where $\underline{\tilde{X}}$ is obtained from $\underline{X}$ by turning the bonds into fibrations. Thus for any two compacta X,Y there is a commutative diagram:

$$\begin{array}{ccc} \text{ho-pro-PL}(\underline{X},\underline{Y}) & \xrightarrow[\approx]{[\underline{P}]^{\#}} & \text{ho-pro-Top}(X,\underline{Y}) \\ \Big\downarrow \text{holim} & & \approx \Big\downarrow \text{holim} \\ [\text{holim}\,\underline{X},\text{holim}\,\underline{Y}] & \xrightarrow[(\text{holim}[\underline{P}])^{\#}]{\approx} & [X,\text{holim}\,\underline{Y}] \end{array}$$

It follows that holim induces a fully faithful functor, $\text{fim}\, sSh \to h\mathcal{F}$, such that the composite functor, $hCM \to \text{fim}\, sSh \to h\mathcal{F}$, is an $h\mathcal{F}$-reflector. The result now follows from (2.8). ∎

<u>Remarks:</u> In light of results by Heller [22], and Edwards and Hastings [16] (concerning inj-sapces), there is strong evidence for regarding ho-pro-PL as the dual of the homotopy category of polyhedra (where "polyhedron" is taken to mean "simplicial polytope with the weak topology"). With respect to this duality, $sSh(CM)$ corresponds to the homotopy category of countable polyhedra.

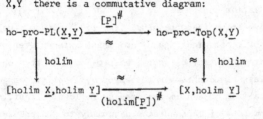

§3. The Functor $\theta : sSh(CM) \to Sh(CM)$

Let $Sh(CM)$ denote the shape category of pointed compacta and let $S : hCM \to Sh(CM)$ be the shape functor. Since S inverts Σ there is a unique functor $\theta : sSh(CM) \to Sh(CM)$ such that $\theta S^* = S$. This functor is full. Indeed, if $F : A \to X$ is a shape morphism, then by a result of Moszyńska [28], $F = S[j]^{-1}S[i]$ where j is the inclusion of X into the base of a "shape mapping cylinder" for F. But j is constructed as a limit of SDR-maps in CM and hence is an SSDR-map by (1.9). Thus $F = \theta([j]\backslash[i])$.

In general, little is known about the function $\theta_{A,X} : sSh(A,X) \to Sh(A,X)$ for arbitrary A,X in CM. However, it is possible to characterize the fiber $\ker \theta_{A,X} = \theta^{-1}_{A,X}(*)$ where $*$ denotes the base point of the pointed set $Sh(A,X)$. Note that the assignment $A,X \to \ker \theta_{A,X}$ determines a bifunctor on $sSh(CM)$, covariant in X and contravariant in A. The characterization of this functor that we are about to describe stems from a fairly widely known $\lim^1$ short exact sequence of Bousfield and Kan [5]. (See, also, the papers by Cohen [11], Grossman [20] and Vogt [40], and Theorem (5.21) of the book by Edwards and Hastings [16]). Our approach, although not the most general, is quite geometric and hence permits a fairly simple proof of certain naturality conditions only alluded to in the literature.

Let tow-hANR denote the full subcategory of pro-hM whose objects are towers of ANR's (bonded by homotopy classes), and for each compactum X let $\underline{P}_X : X \to \underline{X} = (X_n;P^m_n)$ be a tow-hANR-reflection of X. (For example, let $\underline{X}$ be determined by some inverse sequence of compact ANR's in CM whose limit is X). Then it is well known that the associated reflector $hCM \to$ tow-hANR induces a fully faithful functor $Sh(CM) \to$ tow-hANR. This is essentially the inverse systems approach to shape theory of Mardešić and Segal. We make the identification $Sh(A,X) = \text{tow-hANR}(\underline{A},\underline{X})$ and observe that $\underline{P}_A$ induces a bijection $(P_A)^{\#}:Sh(A,X) \approx \lim([A,X_n])$.

Now the assignment $A,X \to ([\Sigma A,X_n]) \approx (\text{tow-hANR}(\Sigma\underline{A},X_n))$, where Σ denotes the reduced suspension functor, determines a bifunctor from $Sh(CM)$ to tow-Gr = (tower category of groups) which is covariant in X and contravariant in A. Composing with θ yields a tow-Gr valued bifunctor on $sSh(CM)$.

Recall that there is a functor

$$\lim{}^1 : \text{tow-Gr} \to S_* = (\text{category of pointed sets})$$

defined as follows. For a tower of groups $\underline{G} = (G_n,p^m_n)$, $\lim^1\underline{G} = \Pi G_n/\sim$ where $(x_n) \sim (y_n)$ if and only if there exists $(g_n) \in \Pi G_n$ such that $x_n = g_n y_n p^{n+1}_n(g^{-1}_{n+1})$ for each n. If $\underline{h} : \underline{G} \to \underline{H}$ is a morphism of tow-Gr, represented by a system of homomorphisms $(h_{\emptyset(n)} : G_{\emptyset(n)} \to H_n)$ with $\emptyset$ monotone and the $h_{\emptyset(n)}$'s commuting with the bonding homomorphisms, then $\lim^1\underline{h}[(x_n)] = [(y_n)]$ where

$$y_n = h_{\emptyset(n)}(p_{\emptyset(n)}^{\emptyset(n)} x_{\emptyset(n)} \cdot \dots p_{\emptyset(n)}^{\emptyset(n+1)-1} x_{\emptyset(n+1)-1})$$

(Here $[(x_n)]$ denotes the equivalence class of (x_n), etc.). Note that if $\emptyset$ is the identity, then $\lim^1 \underline{h}[(x_n)] = [(h_n x_n)]$.

(3.1) <u>Theorem</u>: The S_* valued bifunctors $\ker \theta_{A,X}$ and $\lim^1([\Sigma A, X_n])$ defined on sSh(CM) are naturally equivalent. In particular, there is a bifunctorial short exact sequence of pointed sets:

(3.2) $\lim^1([\Sigma A, X_n]) \rightarrowtail sSh(A,X) \twoheadrightarrow Sh(A,X)$

<u>Proof</u>: Without loss of generality, we may assume that for each compactum X, $\underline{P}_X : X \to \underline{X}$ is represented by a system of inclusion maps $X \subseteq X_n$ where (X_n) is a net of compact ANR's shrinking to X. Then, with $|X|$ defined as in (2.6), one checks that the following diagram commutes:

(3.3)
$$\begin{array}{ccc} sSh(A,X) & \xrightarrow{\theta_{A,X}} & Sh(A,X) \\ \approx \downarrow & & \approx \downarrow (\underline{P}_A)^{\#} \\ [A,|X|] & \xrightarrow[([e_n]_{\#})]{} & \lim_n([A,X_n]) \end{array}$$

where $e_n : |X| \to X_n$ is evaluation at n and where the first isomorphism is the result of (2.8).

The obvious embedding $\Pi\Omega X_n \to |X|$, where Ω is the loop functor, determines a function $\Pi[\Sigma A, X_n] \approx \Pi[A, \Omega X_n] \to [A,|X|]$ whose image is $\ker([e_n]_{\#})$. This function is readily seen to induce an injection, $\lim^1([\Sigma A, X_n]) \to [A,|X|]$, and using (3.3) we obtain the desired bijection, $\lim^1([\Sigma A, X_n]) \approx \ker\theta_{A,X}$. Alternatively, one can proceed as follows, c.f. Cohen [11]. It is well known that given a fibration sequence $F \to E \xrightarrow{P} B$, then $\pi_1 B$ acts on $\pi_o F$ so as to yield an exact sequence of pointed sets

$$\pi_o F/\text{action} \rightarrowtail \pi_o E \to \text{im}\, \pi_o p .$$

Now, $\Pi\Omega X_n \to |X| \xrightarrow{(e_n)} \Pi X_n$ is a fibration sequence and "raising to the exponent A" gives a fibration sequence which yields:

(3.4) $\qquad \pi_o(\Pi\Omega X_n)^A/\text{action} \rightarrowtail \pi_o|X|^A \twoheadrightarrow \text{im}\, \pi_o(P_n)^A .$

Making the adjoint identifications

$\pi_o(\Pi\Omega X_n)^A = \Pi[\Sigma A, X_n] = \pi_1(\Pi X_n)^A$

$\pi_o|X|^A = [A,|X|]$ and $\text{im}\, \pi_o(e_n)^A = \lim([A,X_n])$,

we see that (3.4) yields the exact sequence

$$\lim^1([\Sigma A, X_n]) \rightarrowtail [A,|X|] \twoheadrightarrow \lim([A,X_n]) .$$

We shall now verify that the bijection $\lim^1([\Sigma A, X_n] \approx \ker\theta_{A,X}$ is natural in X. The naturality in A is easier to see and is left to the reader. In each case we use the functorial property of $\lim^1$ and the characterization of strong shape morphisms as left fractions. Given $\alpha \in sSh(X,Y)$ we must verify the commutivity of the square below:

$$\begin{array}{ccc} \lim^1([\Sigma A, X_n]) & \rightarrow & sSh(A,X) \\ {\scriptstyle \lim^1((\theta\alpha)_\#)} \downarrow & & \downarrow {\scriptstyle \alpha_\#} \\ \lim^1([\Sigma A, Y_n]) & \rightarrow & sSh(A,Y) \end{array}$$

By (1.14) we may assume that $\alpha = s\backslash u$ where s,u are represented by inclusions $Y \subseteq Z$, $X \subseteq Z$ respectively. Then one can construct a net of compact ANR's (Z_n) shrinking to Z such that $X_n \subseteq Z_n$ and $Y_n \subseteq Z_n$ for each n. These inclusions determine morphisms $\underline{u} : \underline{X} \to \underline{Z}$ and $\underline{s} : \underline{Y} \to \underline{Z}$ such that $\theta\alpha = \underline{s}^{-1}\underline{u}$. They also determine inclusions $|X| \subseteq |Z|$ and $|Y| \subseteq |Z|$ the homotopy classes of which we denote by $\underline{\underline{u}}$ and $\underline{\underline{s}}$ respectively. Then it is easy to see that the following diagram commutes:

$$\begin{array}{ccccc} \lim^1([\Sigma A, X_n]) & \rightarrowtail & [A,|X|] & \approx & sSh(A,X) \\ {\scriptstyle \lim^1(\underline{u}_\#)} \downarrow & & \downarrow {\scriptstyle \underline{\underline{u}}_\#} & & \downarrow {\scriptstyle S^*(u)_\#} \\ \lim^1([\Sigma A, Z_n]) & \rightarrowtail & [A,|Z|] & \approx & sSh(A,Z) \\ {\scriptstyle \lim^1(\underline{s}_\#)} \uparrow & & \uparrow {\scriptstyle \underline{\underline{s}}_\#} & & \uparrow {\scriptstyle S^*(s)_\#} \\ \lim^1([\Sigma A, Y_n]) & \rightarrowtail & [A,|Y|] & \approx & sSh(A,Y) . \end{array}$$

The result follows since $\lim^1(\underline{s}_\#)^{-1}\lim^1(\underline{u}_\#) = \lim^1((\theta\alpha)_\#)$ and $S^*(s)^{-1}S^*(u) = s\backslash u = \alpha$. ∎

<u>Remark</u>: In case $A = S^k$, (3.2) reduces to the exact sequence

$$\lim^1 \pi_{k+1} X_n \rightarrowtail \underline{\underline{\pi}}_k X \twoheadrightarrow \check{\pi}_k X$$

where $\check{\pi}_k X$ is the kth shape group of X and $\underline{\underline{\pi}}_k X$ is the kth approaching homotopy group of X in the sense of Quigley [32]. Kodama and Koyama [25] use this sequence to prove an Hurewicz isomorphism theorem for Steenrod homology.

Theorem (3.1) is particularly interesting when $X(A)$ is a group (cogroup) object in $sSh(CM)$. For then the same is true in $Sh(CM)$ and $\theta_{A,X}$ is a homomorphism of groups. Note that this will be the case if $X(A)$ is an H-group (H-cogroup). We thus have the following easy consequence of (3.1).

(3.3) <u>Corollary</u>: Let $\alpha : X \to Y$ be a strong shape morphism and suppose that $\theta(\alpha)$ is invertible. Then

$$\alpha_\# : sSh(A,X) \to sSh(A,Y)$$

is bijective for all cogroup objects A, and

$$\alpha^{\#} : sSh(Y,B) \to sSh(X,B)$$

is bijective for all group objects B. If in addition X and Y are both group objects, cogroup objects, or X is a group object and Y is a cogroup object, then α is invertible. ■

Example: There is a 2-sphere-like space X such that $\ker\theta_{X,X}$ is uncountable:

$$\ker\theta_{X,X} \to sSh(X,X) \twoheadrightarrow Sh(X,X) \approx \mathbb{Z} .$$

In particular, the family of strong shape automorphisms of X, $\theta^{-1}_{X,X}\{-1,1\}$, is uncountable.

To define X, we choose an increasing sequence of prime numbers (p_n) and set $X = \lim(X_n, f_n^{n+1})$ where $X_n = S^2$ and f_n^{n+1} is a map of degree p_n. Then

$$[X,X_n] \approx \operatorname{colim}(\mathbb{Z} \xrightarrow{p_1} \mathbb{Z} \xrightarrow{p_2} \cdots) \approx G ,$$

where G is the additive subgroup of the rationals generated by $\left\langle \frac{1}{p_n} \right\rangle$, and $([X,X_n])$ is isomorphic in tow-Gr to the net $(p_1p_2\cdots p_nG)$. Thus $Sh(X,X) \approx \lim([X,X_n]) \approx \bigcap p_1p_2\cdots p_nG = \mathbb{Z}$. Similarly, $[\Sigma X,X_n] \approx G$ and $([X,X_n])$ is isomorphic to $(p_1^2p_2^2\cdots p_n^2G)$. (Here we make use of Theorem 3.6 p.86 of [23], from which we infer that $[f_n^{n+1}]_{\#} : [\Sigma X,X_{n+1}] \to [\Sigma X,X_n]$ corresponds to multiplication by p_n^2).

Now, there is an exact sequence of towers

$$0 \to (p_1^2p_2^2\cdots p_n^2G) \to (G) \to (\mathbb{Z}/p_1^2p_2^2\cdots p_n^2\mathbb{Z}) \to 0$$

and passing to the limit we have by [5] proposition (2.3), an exact sequence

$$0 \to G \to \lim(\mathbb{Z}/p_1^2p_2^2\cdots p_n^2\mathbb{Z}) \to \lim{}^1(p_1^2p_2^2\cdots p_n^2G) \to 0 .$$

It follows that $\lim^1(p_1^2p_2^2\cdots p_n^2G) \approx \ker\theta_{X,X}$ is uncountable.

We remark that the space X above can be used to show that $Sh(CM)$ is not the category CM localized at the class of shape equivalences. Indeed, denoting the latter category by $CM\{s.e.\}^{-1}$ we have by (3.3) that the functor $sSh(X,S^*\_)$ factors across $CM\{s.e.\}^{-1}$. It clearly does not factor across $Sh(CM)$.

The next result is of considerable importance since it implies that θ induces a bijection between the collections of shape types and strong shape types.

(3.4) Theorem: Let $F : A \to X$ be a shape isomorphism. Then there is a strong shape isomorphism $\alpha : A \to X$ such that $\theta(\alpha) = F$.

This result is proved by Dydak and Segal in [14]. Moreover, in light of Chapman's representation of $Sh(CM)$ in the weak proper homotopy category (see [8]), it also follows from a result of Edwards and Hastings [17] which states that every weak proper homotopy equivalence is weakly proper homotopic to a proper homotopy equivalence. The method of proof used by Edwards and Hastings (see [16], Theorem 5.2.9) allows us to describe α in case F is represented by an inclusion map. Suppose then that $X \subseteq Q$, the Hilbert cube, and that $i : A \subseteq X$ is a shape equivalence representing F. Then we can find a net (X_n, A_n), $n \geq 0$, of compact ANR pairs shrinking to (X,A) and a sequence of deformations $D_t^n : X_{n+1} \to X_n$ such that $\mathrm{im}\, D_1^n \subseteq A_n$ and $D_1^n(a) = a$ for $a \in A_{n+1}$.

(3.5) The approaching map $\Phi : (A,A) \to (Q,X)$ defined by $\Phi_t(a) = D^n_{n+1-t}(a)$ for $n \leq t \leq n+1$, $n \geq 0$, represents a strong shape isomorphism $\alpha : A \to X$ such that $\theta(\alpha) = F$. In fact, the associated map $\hat{\Phi} : A \to |X|$ is an SSDR-map such that $e_n \hat{\Phi} = \text{inclusion}$, $n \geq 0$.

Proof: Note that $A = \lim(X_n; r_n^{n+1})$ where $r_n^{n+1} = D_1^n : X_{n+1} \to X_n$. Now, with $(|X|_n; \rho_n^{n+1})$ defined as in the proof of(2.5)we inductively define a system of SDR-maps $(j_n : X_n \to |X|_n)$ as follows: Let $j_o = id : X_o \to |X|_o$ and for $n \geq 0$ let

$$j_{n+1}(x)(t) = \begin{cases} j_n r_n^{n+1}(x)(t) & , 0 \leq t \leq n \\ D^n_{n+1-t}(x) & , n \leq t \leq n+1 \end{cases}$$

for $x \in X_{n+1}$. Observe that for each $n \geq 0$, $e_n j_n = id_{X_n}$ so that j_n is a homotopy equivalence and hence an SDR-map. By (1.9) $\hat{\Phi} = {}^n\lim j_n : A \to |X|$ is an SSDR-map and the result follows.

An important unsolved problem is whether or not $\theta(\alpha)$ an isomorphism implies α an isomorphism. Partial results are (3.3) and the following important theorem of Dydak and Segal [14].

(3.6) Theorem: If A,X are (pointed) continua of finite shape dimension and if $\alpha \in sSh(A,X)$ is such that $\theta(\alpha)$ is an isomorphism, then α is an isomorphism.

Calder and Hastings [6] remark that for finite dimensional A,X this follows from a result of L. C. Siebenmann [35]. The general case follows, since, by [30] Theorem (1.6), each continuum of finite shape dimension has the shape and hence (by (3.4)) strong shape of a finite dimensional continuum.

Theorem (3.6) also follows from the next somewhat more general result.

(3.7) Theorem: Let A,X be continua and $\alpha : A \to X$ a strong shape morphism such that $\theta(\alpha)$ is an n-equivalence, i.e., such that the morphism of pro groups $\text{pro} - \pi_k A \to \text{pro} - \pi_k X$ induced by $\theta(\alpha)$ is an isomorphism for $k \leq n-1$ and is an epimorphism for $k = n$. Then for any compactum Z,

$$\alpha_{\#} : sSh(Z,A) \to sSh(Z,X)$$

is surjective if $\text{shdim}\, Z \leq n-1$ and is injective if $\text{shdim}\, Z < n-1$. (Here shdim denotes shape dimension).

<u>Proof</u>: By (1.14) we may assume without loss of generality that α is represented by an inclusion $A \subseteq X$. Then the pair (X,A) is shape n-connected, i.e., $\text{pro} - \pi_k(X,A) = 0$ for $1 \leq k \leq n$. (see, e.g., [29]). Assume that $X \subseteq Q$, the Hilbert cube. Then it follows that, given any ANR neighborhood pair (U,V) of (X,A) in Q there exists an ANR neighborhood pair (U',V') of (X,A) in (U,V) with the following property:

(*) For any CW pair (K,L) with $\dim K \leq n$ and map $f : (K,L) \to (U',V')$ there is a map $f' : K \to V$ such that the following diagram commutes up to a homotopy rel L

$$\begin{array}{ccc} (K,L) & \xrightarrow{f} & (U',V') \\ {\scriptstyle f'}\downarrow & & \downarrow \\ (V,V) & \longrightarrow & (U,V) \end{array}$$

Now, given any compact pair (Z,B) with $\dim Z \leq n$ and map $f : (Z,B) \to (U',V')$, one can factor f, up to homotopy, across a CW pair (K,L) with $\dim K \leq n$. We infer that (*) holds with "compact pair (Z,B)" substituted for "CW pair (K,L)". (By exercise D4 p. 57 of [36], the homotopy from f to f' can be assumed stationary on B).

Let (X_n,A_n) be a net of compact ANR neighborhoods of (X,A) in Q shrinking to (X,A) and let $\phi : \mathbb{N} \to \mathbb{N}$ be a monotone function such that $(X_{\phi(n)},A_{\phi(n)}) \subseteq (X_n,A_n)$ has property (*).

Now, as we noted earlier each compactum Z with $\text{shdim}\, Z \leq k$ has the strong shape of a compactum Z' with $\dim Z' \leq k$. Thus it suffices to prove the theorem with "dim" in place of "sh dim." Assume that $\dim Z \leq n-1$ and that $\Phi : (Z,Z) \to (Q,X)$ is an approaching map representing some strong shape morphism from Z to X. By reparameterizing if necessary we may assume that $\Phi(Z \times [n,\infty)) \subseteq X_{\phi^2(n)}$ for $n \geq 0$. Then by (*), Φ extends to a map $\Phi' : Z \times ([0,\infty) \times 0 \cup \mathbb{N} \times I) \to Q$ such that $\Phi'(Z \times n \times I) \subseteq X_{\phi(n)}$ and $\Phi'(Z \times n \times 1) \subseteq A_{\phi(n)}$ for $n \in \mathbb{N}$. Again by (*), Φ' extends to a map $\Phi'' : Z \times [0,\infty) \times I \to Q$ such that $\Phi''(Z \times [n,n+1] \times I) \subseteq X_n$ and $\Phi''(Z \times [n,n+1] \times 1) \subseteq A_n$ for $n \geq 0$. Thus Φ'' determines an approaching map, $\Phi''_1 : (Z,Z) \to (Q,A)$ such that $[i\, \Phi''_1] = [\Phi] : (Z,Z) \to (Q,X)$, where $i : (Q,A) \to (Q,X)$ is the approaching map determined by the inclusion $(Q,A) \subseteq (Q,X)$. Since i represents α we infer that $\alpha_\#$ is surjective.

One proves in similar fashion that if $\dim Z < n-1$, then $\alpha_\#$ is injective. ∎

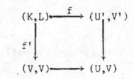

REFERENCES

[1] D. W. Anderson, Fibrations and geometric realizations, Bull. Amer. Math. Soc. 84(5) (1978), 765-788.

[2] B. J. Ball and R. Sher, A theory of proper shape for locally compact metric spaces, Fund. Math. 86(1974), 163-192.

[3] F. W. Bauer, A characterization of movable compacta, J. Reine Angew Math. 293/294 (1977), 394-417.

[4] K. Borsuk, Theory of Shape, Monografie Matematycne 59 Warsawa, 1975.

[5] A. K. Bousfield and D.M. Kan, Homotopy limits, completions and localizations, Lecture Notes in Math. 304, Springer, Berlin-Heidelberg-New York, 1973.

[6] A. Calder and H.M. Hastings, Realizing strong shape equivalences, J. of Pure and Applied Algebra (to appear).

[7] F. Cathey, Strong shape theory, Ph.D. thesis, University of Washington, 1979.

[8] T. A. Chapman, On some applications of infinite-dimensional manifolds to the theory of shape, Fund. Math. 76(1972), 181-193.

[9] T. A. Chapman and L. Siebenmann, Finding a boundary for a Hilbert cube manifold, Acta. Mathematica 137(1976), 171-208.

[10] D. E. Christie, Net homotopy for compacta, Trans. Amer. Math. Soc. 56(1944), 275-308.

[11] J. Cohen, Inverse limits of principal fibrations, Proc. London Math. Soc. (3) 27(1973), 178-192.

[12] A. Dold, Geometric cobordism and the fixed point transfer, in Alg. Top. Proc. Vancouver, Lecture Notes in Math. 673, Springer, Berlin-Heidelberg-New York (1977), 32-87.

[13] J. Dugundji, Topology, Allyn and Bacon, Inc., Boston, 1966.

[14] J. Dydak and J. Segal, Strong shape theory, Dissertationes Mathematica, (to appear).

[15] A. Deleanu and P. Hilton, Generalized shape theory, in General topology and its relations to modern analysis and algebra IV, ed., - J. Novak, Lecture Notes in Math. 609, Spring, Berlin-Heidelberg-New York (1977), 59-65.

[16] D. A. Edwards and H.M. Hastings, Cech and Steenrod homotopy theory, with applications to geometric topology, Lecture Notes in Math. 542, Springer, Berlin-Heidelberg-New York, 1976.

[17] ____________ and ____________, Every weak proper homotopy equivalence is weakly properly homotopic to a proper homotopy equivalence, Trans. Amer. Math. Soc. (to appear).

[18] S. Ferry, A stable converse to the Victoris-Smale theorem with applications to shape theory, Trans. Amer. Math. Soc. 261(2)(1980, 369-386.

[19] P. Gabriel and M. Zisman, Calculus of fractions and homotopy theory, Ergebnisse der Mathematik 35, Springer, Berlin-Heidelberg-New York, 1967.

REFERENCES (continued)

[20] J. Grossman, Homotopy classes of maps between pro-spaces, Mich. Math. J. 21 (1974), 355-362.

[21] H. M. Hastings and J. Hollingsworth, Shape equivalences, CE-equivalences, and the Siebenmann obstruction, Topology and its Applications (to appear).

[22] A. Heller, Completions in abstract homotopy theory, Trans. Amer. Math. Soc. 147(1970), 573-602.

[23] P. J. Hilton, An introduction to homotopy theory, Cambridge Univ. Press (43), 1953.

[24] Y. Kodama and J. Ono, On fine shape theory, Fund. Math. 105(1979), 29-39.

[25] Y. Kodama and A. Koyama, Hurewicz isomorphism theorem for Steenrod homology, Proc. Amer. Math. Soc. 74(2) (1979), 363-367.

[26] G. Kozlowski, Images of ANR's, Trans. Amer. Math. Soc. (to appear).

[27] J. T. Lisica, Exactness of sequences of spectral homotopy grsups for shape theory, Soviet Math. Dok. 18 (5) (1977), 1186-1190.

[28] M. Moszynska, On shape and fundamental deformation retracts II, Fund. Math. 77 (1973), 235-240.

[29] ___________, The Whitehead Theorem in the theory of shapes, Fund. Math. 80 (1973), 221-263.

[30] S. Nowak, Some properties of the fundamental dimension, Fund. Math. 85(1974), 211-227.

[31] L. S. Pontrjagin, Topological groups, Princeton Univ. Press, 1939.

[32] J. Quigley, An exact sequence from the nth to the (n-1)st fundamental group, Fund. Math. 77(1973), 195-210.

[33] D. G. Quillen, Homotopical algebra, Lecture Notes in Math. 43, Springer, Berlin-Heidelberg-New York, 1967.

[34] H. Schubert, Categories, Springer, Berlin-Heidelberg-New York, 1972.

[35] L. C. Siebenmann, Infinite simple homotopy types, Indag. Math. 32,(1970), 479-495.

[36] E. H. Spanier, Algebraic topology, McGraw-Hill, New York, 1966.

[37] N. E. Steenrod, Regular cycles on compact metric spaces, Ann. of Math. (2) 41(1940), 833-851.

[38] A. Strøm, The homotopy category is a homotopy category, Arch. Math. (Basel) 23 (1973), 435-441.

[39] ________, Note on cofibrations II, Math. Scand. 22(1968), 130-142.

[40] R. M. Vogt, On the dual of a lemma of Milnor, Proc. Adv. Study Inst. Alg. Top., Aarhus Universitet, Denmark, 1970.

University of Kansas
Lawrence, Kansas 66045

INVERSE LIMITS AND RESOLUTIONS

Sibe Mardešić, Zagreb

1. Introduction

In 1971 S.Mardešić and J.Segal [9] defined the shape category $\mathrm{Sh}|\mathrm{Cpt}$ of compact Hausdorff spaces by means of inverse systems of compact polyhedra (compact ANR's) as follows (see also [3]). Every compact Hausdorff space X admits an inverse system $\underline{X} = (X_\lambda, p_{\lambda\lambda'}, \Lambda)$ of compact polyhedra and (continuous) maps, indexed by a directed set Λ and a collection of maps $p_\lambda : X \to X_\lambda$, $\lambda \in \Lambda$, such that $\underline{p} = (p_\lambda) : X \to \underline{X}$ is an inverse limit of $\underline{X}$ in the category Top. If $\underline{Y} = (Y_\mu, q_{\mu\mu'}, M)$ is another inverse system of compact polyhedra with an inverse limit $\underline{q} = (q_\mu) : Y \to \underline{Y}$, then every map $f : X \to Y$ induces a unique morphism $\underline{f} : \underline{X} \to \underline{Y}$ of the pro-category pro-HTop such that the following diagram commutes

$$
(1) \qquad \begin{array}{ccc}
\underline{X} & \xleftarrow{\ \underline{p}\ } & X \\
{\scriptstyle \underline{f}}\downarrow & & \downarrow{\scriptstyle f} \\
\underline{Y} & \xleftarrow[\ \underline{q}\ ]{} & Y
\end{array} .
$$

Here HTop denotes the homotopy category of spaces, and f, $\underline{p}$ and $\underline{q}$ are also interpreted as morphisms of pro-HTop. In particular, if $\underline{X}$ and $\underline{X}'$ are two inverse systems of compact polyhedra and $\underline{p} : X \to \underline{X}$, $\underline{p}' : X \to \underline{X}'$ are inverse limits, then the identity map $1_X : X \to X$ determines a unique morphism $\underline{i} : \underline{X} \to \underline{X}'$ of pro-HTop. Similarly, different inverse limit expansions $\underline{q} : Y \to \underline{Y}$, $\underline{q}' : Y \to \underline{Y}'$ of a space Y determine a unique morphism $\underline{j} : \underline{Y} \to \underline{Y}'$. Two morphisms $\underline{f} : \underline{X} \to \underline{Y}$, $\underline{f}' : \underline{X}' \to \underline{Y}'$ of pro-HTop are considered equivalent provided the following diagram commutes in pro-HTop

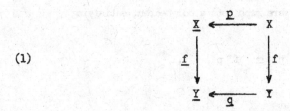

$$\begin{array}{ccc} \underline{X}' & \xleftarrow{\underline{i}} & \underline{X} \\ {\scriptstyle \underline{f}'}\downarrow & & \downarrow{\scriptstyle \underline{f}} \\ \underline{Y}' & \xleftarrow[\underline{j}]{} & \underline{Y} \end{array} \tag{2}$$

We now define the category $Sh|Cpt$ by taking as its objects all compact Hausdorff spaces and as its morphisms $F:X \to Y$ the equivalence classes of the morphisms $\underline{f}:\underline{X} \to \underline{Y}$ of pro-HTop.

There are two crucial properties of inverse limits $\underline{p}:X \to \underline{X}$ of inverse systems of compacta, which make the above contructions possible ([9], Lemmas 3 and 4).

(E1) For every polyhedron P and every map $h:X \to P$, there exists a $\lambda \in \Lambda$ and a map $f:X_\lambda \to P$ such that

$$h \simeq f\, p_\lambda \;. \tag{3}$$

(E2) If $f', f'' : X_\lambda \to P$ are maps into a polyhedron, satisfying

$$f'\, p_\lambda \simeq f''\, p_\lambda \;, \tag{4}$$

then there is a $\lambda' \geq \lambda$ such that

$$f'\, p_{\lambda\lambda'} \simeq f''\, p_{\lambda\lambda'} \tag{5}$$

Since every polyhedron has the homotopy type of an ANR and conversely, every ANR has the homotopy type of a polyhedron, one is allowed to replace in (E1), (E2) the requirement that P be a polyhedron by the requirement that P be an ANR.

In 1975 K.Morita showed [11] that one can define in an analogous way the shape category Sh of arbitrary topological spaces (originally introduced in a different manner in [4]). More precisely, given a space X, Morita considers inverse systems $\underline{X} = (X_\lambda, [p_{\lambda\lambda'}], \Lambda)$ in pro-HTop, where all X_λ are polyhedra (not necessarily compact), and morphisms $([p_\lambda]):X \to \underline{X}$ in pro-HTop satisfying conditions (E1) and

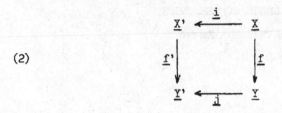

(E2) (with P ranging over arbitrary polyhedra). We will refer to such morphisms as to HPol-expansions of X . The only assertion, which requires a new proof concerns the existence of HPol-expansions for arbitrary spaces X . Morita has shown that the usual Čech system, which consists of the nerves N_λ of all the locally finite normal coverings λ of X , and the canonical maps $p_\lambda : X \to N_\lambda$, yield an HPol-expansion as required.

However, the analogy of Morita's construction with the construction in the compact case is not complete, because in the latter case the system $\underline{X}$ associated with X is an inverse systems in Top , while the Čech system is only a system in HTop. An alternate approach associates with X its Vietoris system, which is an inverse system in Top, but in this case too the projections $X \to X_\lambda$ are defined only up to their homotopy class so that $\underline{p}: X \to \underline{X}$ is again a morphism of pro-HTop.

In this paper we will describe the notion of a resolution of a space, which gives a genuine generalization of the above described construction of Sh | Cpt . Essentially, resolutions are special inverse limits $\underline{p}: X \to \underline{X}$ in Top. Application of the homotopy functor H to a polyhedral resolution always yields an HPol-expansion (in the sense of Morita). Furthermore, all spaces X admit polyhedral resolutions. Thus polyhedral resolutions can be used to define the shape category Sh . We will also define resolutions of maps and we will use polyhedral resolutions of maps to define shape fibrations for maps between arbitrary topological spaces. Detailed proofs of our assertions will appear in [6] (and some also in [10]).

2. Resolutions of spaces

Let $\underline{X} = (X_\lambda, p_{\lambda\lambda'}, \Lambda)$ be an inverse system in Top. A morphism $\underline{p} = (p_\lambda): X \to \underline{X}$ of pro-Top is called a resolution of the space X provided the following conditions are satisfied :

(R1) For every polyhedron P , open covering $\underline{V}$ of P and map $h: X \to P$ there is a $\lambda \in \Lambda$ and a map $f: X_\lambda \to P$ such that the maps h and $f p_\lambda$ are $\underline{V}$-near.

(R2) Every polyhedron P and open covering $\underline{V}$ of P admit an open covering

$\underline{V}$ of P admit an open covering $\underline{V}'$ of P with the following property. Whenever $\lambda \in \Lambda$ and $f, f': X_\lambda \to P$ are $\underline{V}'$-near maps, then there exists a $\lambda' \geq \lambda$ such that the maps $f p_{\lambda\lambda'}$ and $f' p_{\lambda\lambda'}$ are $\underline{V}$-near.

An equivalent formulation is obtained if one replaces the requirement that P be a polyhedron by the requirement that P be an ANR (for metric spaces). This is an immediate consequence of the following fact.

Proposition 1. Let P be a polyhedron and $\underline{U}$ an open covering. Then there exists an ANR Q and maps $f: P \to Q$, $g: Q \to P$ such that $g f$ and 1_P are $\underline{U}$-near maps. Similarly, let Q be an ANR and $\underline{V}$ an open covering of Q. Then there exist a polyhedron P and maps $g: Q \to P$, $f: P \to Q$ such that $f g$ and 1_Q are $\underline{V}$-near.

Proof. Let K be a triangulation of P so fine that the closed simplexes of K refine $\underline{U}$. Let $Q = K$ be endowed with the strong (metric) topology. Then Q is an ANR and the identity map $i: P \to Q$ is a homotopy equivalence with a homotopy inverse $j: Q \to P$, which preserves closed simplexes. Consequently, $j \cdot i$ and 1_P are $\underline{U}$-near. The usual proof that every ANR is dominated by a polyhedron actually establishes the second assertion of Proposition 1.

The usefulness of resolutions in shape theory is based on the following theorem.

Theorem 1. Every resolution $\underline{p}: X \to \underline{X}$ satisfies conditions (E1) and (E2). In particular, if $\underline{p}$ is a polyhedral resolution (i.e. each X_λ is a polyhedron), then $\underline{p}$ is an HPol-expansion of X.

We will establish properties (E1) and (E2) for P an ANR, making use of § 2, Lemma 2 from [5], which reads as follows.

Lemma 1. Let P and P' be ANR's (for metric spaces), let X be a topological space and let $g', g'': P' \to P$, $h': X \to P'$ be maps satisfying

$$g' h' \simeq g'' h' \quad . \tag{6}$$

Then there is an ANR P'' and there are maps $g: P'' \to P'$, $h'': X \to P''$ such that

$$g h'' = h' \quad , \tag{7}$$

$$g' g \simeq g'' g \quad . \tag{8}$$

Proof of Theorem 1. (E1) Let $\underline{V}$ be an open covering of $P \in$ ANR such that any two $\underline{V}$-near maps into P are homotopic. By (R1) there is a $\lambda \in \Lambda$ and a map $f:X_\lambda \to P$ such that the maps h and $f p_\lambda$ are $\underline{V}$-near and therefore homotopic.

(E2) Let P be an ANR and let $f', f'':X_\lambda \to P$ be maps satisfying (4). We choose $\underline{V}$ as in the proof of (E1) and then we choose $\underline{V}'$ according to (R2). The maps $f'p_\lambda$, $f''p_\lambda:X \to P$ determine a map $h':X \to P' = P \times P$ such that

$$g'h' = f'p_\lambda \tag{9}$$

$$g''h' = f''p_\lambda \quad , \tag{10}$$

where $g', g'': P \times P \to P$ denote the two projections. Since g' and g'' satisfy (6), Lemma 1 yields an ANR P'' and maps $g:P'' \to P'$, $h'':X \to P''$ satisfying (7) and (8). Let $\underline{V}''$ be an open covering of P'', which refines $(g'g)^{-1}(\underline{V}')$ and $(g''g)^{-1}(\underline{V}')$. Property (R1) yields a $\lambda'' \geq \lambda$ and a map $f:X_{\lambda''} \to P''$ such that fp_λ and h'' are $\underline{V}''$-near maps. Consequently, the maps $g'g f p_{\lambda''}$ and $g'g h'' = g'h' = f'p_\lambda$ are $\underline{V}'$-near. Hence, by the choice of $\underline{V}'$, for sufficiently large $\lambda' \geq \lambda''$ the maps $g'g f p_{\lambda''\lambda'}$ and $f'p_{\lambda\lambda'}$ are $\underline{V}$-near and therefore

$$g'g f p_{\lambda''\lambda'} \simeq f' p_{\lambda\lambda'} . \tag{11}$$

Similarly,

$$g''g f p_{\lambda''\lambda'} \simeq f'' p_{\lambda\lambda'} . \tag{12}$$

Finally, (8), (11) and (12) yield the desired homotopy (5).

3. Characterization of resolutions

Resolutions can be conveniently characterized by means of the following two conditi-ons, which are applicable to an arbitrary morphism $\underline{p} = (p_\lambda):X \to \underline{X} = (X_\lambda, p_{\lambda\lambda'}, \Lambda)$ of pro-Top.

(B1) Let $\lambda \in \Lambda$ and let U be an open neighborhood of $\overline{p_\lambda(X)}$ in X_λ. Then there exist a $\lambda' \geq \lambda$ such that

(13) $$p_{\lambda\lambda'}(X_{\lambda'}) \subseteq U \ .$$

(B2) For every normal (i.e. numerable) covering $\underline{U}$ of X there is a $\lambda \in \Lambda$ and a normal covering $\underline{V}$ of X such that $(p_\lambda)^{-1}(\underline{V})$ refines $\underline{U}$.

The characterization is given by the next two theorems.

Theorem 2. If a morphism $\underline{p}:X \to \underline{X}$ has properties (B1) and (B2), then $\underline{p}$ is a resolution.

Theorem 3. Every resolution $\underline{p}:X \to \underline{X}$ with all X_λ normal spaces has properties (B1) and (B2).

Corollary 1. Let $\underline{X} = (X_\lambda, p_{\lambda\lambda'}, \Lambda)$ be an inverse system of polyhedra (ANR's) and let X be a topological space. A morphism $\underline{p} = X \to \underline{X}$ of pro-Top is a resolution if and only if $\underline{p}$ satisfies conditions (B1) and (B2).

Remark 1. P.Bacon [1] has considered morphisms $\underline{p}:X \to \underline{X}$ of pro-Top, which satisfy (B2) and a slightly stronger form of (B1) under the name complement of $\underline{X}$. He has proved that for paracompact X complements satisfy (E1) and (E2), which is a weaker form of our Theorem 1. Closely related is the work of K.Morita [12] , who also proved a weaker version of Theorem 1. Note that the papers of Bacon and Morita do not contain any condition similar to our condition (R2).

Proof of Theorem 2. We will verify conditions (R1) and (R2) with P an ANR.

(R1) Let $\underline{V}$ be an open covering of P and $h:X \to P$ a map. One can assume that P is a closed subset of a convex set K of a normed vector space. Let G be an open neigborhood of P in K , which admits a retraction $r:G \to P$. Let $\underline{V}' = r^{-1}(\underline{V})$ and let $\underline{V}''$ be an open covering of G , which refines $\underline{V}'$ and consists of convex sets. Then $\underline{U} = h^{-1}(\underline{V}'')$ is a normal covering of X . Hence, by (B2), there is a $\mu \in \Lambda$ and a normal covering $\underline{W}$ of X such that $(p_\mu)^{-1}(\underline{W})$ refines $\underline{U}$. Let $\underline{W}'$ be a normal covering of X_μ , which is a star-refirement of $\underline{W}$. We next choose an open neighborhood N of $\overline{p_\mu(X)}$ in X_μ and a normal locally finite covering $\underline{H}$ of N such that $\underline{H}$ refines $\underline{W}'|N$ and each $H \in \underline{H}$ meets $\overline{p_\mu(X)}$. We then assign to each $H \in \underline{H}$ a point $y_H \in h((p_\mu)^{-1}(H))$. Let $(\phi_H, H \in \underline{H})$ be a partition of unity on N subordinated to the cover $\underline{H}$. We define a map $g:N \to K$ by the formula

(14) $$g(z) = \sum_{H \in \underline{H}} \phi_H(z)\, y_H \ , \qquad z \in N .$$

g is actually a map into G and the maps $g\,p_\mu$ and h are $\underline{V}'$-near. Consequently, the maps $r\,g\,p_\mu$ and h are $\underline{V}$-near.

We now apply property (B1) and obtain a $\lambda \geq \mu$ such that

(15) $$p_{\mu\lambda}\ (X_\lambda) \subseteq N .$$

Finally, the map $f = r\,g\,p_{\mu\lambda} : X_\lambda \to P$ has the desired property that $f\,p_\lambda$ and h are $\underline{V}$-near.

(R2). Let P be an ANR and $\underline{V}$ an open covering of P. We take for $\underline{V}'$ any star-refinement of $\underline{V}$. Let $\lambda \in \Lambda$ and let $f', f'' : X_\lambda \to P$ be maps such that the maps $f'\,p_\lambda, f''\,p_\lambda$ are $\underline{V}'$-near. Then it is readily seen that any point $y \in \overline{p_\lambda(X)}$ admits a $V \in \underline{V}$ such that $f'(y), f''(y) \in V$.

This enables us to choose for each $y \in \overline{p_\lambda(X)}$ an open neighborhood U_y in X such that the maps $f' \mid U_y$, $f'' \mid U_y$ are $\underline{V}$-near. Consequently, the union U of all U_y, $y \in \overline{p_\lambda(X)}$, is a an open neighborhood of $\overline{p_\lambda(X)}$ such that $f' \mid U$ and $f'' \mid U$ are $\underline{V}$-near maps. By (B1), there is a $\lambda' \geq \lambda$ such that $p_{\lambda\lambda'}(X_{\lambda'}) \subseteq U$. Hence, the maps $f'\,p_{\lambda\lambda'}$ and $f''\,p_{\lambda\lambda'}$ are also $\underline{V}$-near.

<u>Proof of Theorem 3</u>. Let $\underline{p}: X \to \underline{X}$ be a resolution and let $\underline{U}$ be a normal covering of X. A locally finite partition of unity subordinated to $\underline{U}$ determines a canonical map $h: X \to P$ into the nerve P of the covering $\underline{U}$. The open stars of the vertices form an open covering $\underline{W}$ of P such that $h^{-1}(\underline{W})$ refines $\underline{U}$. Let $\underline{W}'$ be a star-refinement of $\underline{W}$. By (R1) there is a $\lambda \in \Lambda$ and a map $f: X_\lambda \to P$ such that the maps $f\,p_\lambda$ and h are $\underline{W}'$-near. Clearly, $\underline{V} = f^{-1}(\underline{W}')$ is a normal covering of X_λ. It is now readily seen that $(p_\lambda)^{-1}(\underline{V})$ refines $\underline{U}$. Indeed, if $V = f^{-1}(W'), W' \in \underline{W}'$, and if $St(W', \underline{W}') \subseteq W$, $W \in \underline{W}$, then

(16) $$(p_\lambda)^{-1}(V) \subseteq h^{-1}(W) .$$

This establishes property (B2) (without any assumptions on the spaces X_λ).

Proof of (B1). Let $\lambda \in \Lambda$ and let U be an open set in X_λ containing $\overline{p_\lambda(X)}$. Choose a map $f': X_\lambda \to I = [0,1]$ such that

(17) $$f'|\overline{p_\lambda(X)} = 0 \quad ,$$

(18) $$f'|X_\lambda \setminus U = 1 ,$$

Also consider the constant map $f'' = 0 : X_\lambda \to I$. Furthermore, let $\underline{V}$ be the open covering $\underline{V} = \{[0,1),(0,1]\}$ of I. Clearly, $f'p_\lambda = f''p_\lambda$. Hence, by (R2), there is a $\lambda' \geq \lambda$ such that the maps $f'p_{\lambda\lambda'}$ and $f''p_{\lambda\lambda'}$ are $\underline{V}$-near and therefore,

(19) $$f'p_{\lambda\lambda'}(X_{\lambda'}) \subseteq [0,1) .$$

However, this implies

(20) $$p_{\lambda\lambda'}(X_{\lambda'}) \subseteq U .$$

Example 1. Let M be a metric space and $Y \subseteq M$ an arbitrary subset. Let $\underline{X} = (X_\lambda, p_{\lambda\lambda'}, \Lambda)$ be the inverse system, which consists of all open neighborhoods of X in M, and let all $p_{\lambda\lambda'}$ and all $p_\lambda : X \to X_\lambda$ be inclusions. Then $\underline{p}: X \to \underline{X}$ is a resolution. Moreover, if M is an ANR, then $\underline{p}$ is an ANR-resolution of X. This is easily proved by verifying conditions (B1) and (B2).

4. Resolutions and inverse limits

The next theorem shows that in the most important cases resolutions are inverse limits.

Theorem 4. Let $\underline{p}: X \to \underline{X} = (X_\lambda, p_{\lambda\lambda'}, \Lambda)$ be a resolution. If all X_λ are normal spaces and X is a topologically complete space, then $\underline{p}$ is an inverse limit of $\underline{X}$.

Corollary 2. Let $\underline{p}: X \to \underline{X}$ be a polyhedral resolution (ANR-resolution). If X is paracompact, then $\underline{p}$ is an inverse limit of $\underline{X}$.

Proof of Theorem 4. Let X be topologically complete, i.e. it admits a complete uniformity, or equivalently, X is complete with respect to its maximal uniformity. Let Y be a space and $\underline{g} = (g_\lambda): Y \to \underline{X}$ a morphism of pro-Top. We must show that there is a unique map $g:Y \to X$ such that

(21) $$g_\lambda = p_\lambda g \quad , \lambda \in \Lambda \quad .$$

Let $y \in Y$. An easy argument based on property (B1) shows that

(22) $$(p_\lambda)^{-1}(W) \neq \emptyset$$

for any $\lambda \in \Lambda$ and any neighborhood W of $g_\lambda(y)$ in X_λ. As a consequence of this fact one proves that the set $\underline{C} = \{(p_\lambda)^{-1}(W) : \lambda \in \Lambda$, W a closed neighborhood of $g_\lambda(y)\}$ is a centered collection of closed subsets of X. Furthermore, if $\underline{U}$ is a normal covering of X, then property (B2) yields a $\lambda \in \Lambda$ and a normal covering $\underline{V}$ of X_λ such that $(p_\lambda)^{-1}(\underline{V})$ refines $\underline{U}$. Hence, if W is a closed neighborhood of $g_\lambda(y)$, contained in a member of $\underline{V}$, then $(p_\lambda)^{-1}(W)$ is contained in an element of $\underline{U}$. Since the normal coverings $\underline{U}$ of X generate the maximal uniformity of X, it follows that $\underline{C}$ is a Cauchy collection, with respect to this uniformity. By the topological completeness of X, one concludes that the intersection of the elements of $\underline{C}$ is a unique point $x \in X$. It readily follows that

(23) $$p_\lambda(x) = g_\lambda(y)$$

for all $\lambda \in \Lambda$. Moreover, the point x is characterized by this property. This proves that there is a unique function $g:Y \to X$ satisfying (21). Finally, one proves that g is continuous.

Remarks 2. A slightly weaker form of our Theorem 4 was proved by K.Morita in [12].

We will now show by an example that inverse limits need not be resolutions so that the converse of Theorem 4 generally does not hold.

Example 2. Let $I_n \subseteq R^2$ denote the 2-cell $[n-1,n]\times[0,1]$, $n \in N$. Let

(24) $$X_n = (\bigcup_{k=1}^{n} \partial I_k) \cup [n,\infty)\times[0,1] \quad ,$$

(25) $$X = \bigcup_{k=1}^{\infty} \partial I_k \quad ,$$

and let $p_{nn+1}:X_{n+1} \to X_n$, $p_n:X \to X_n$ be inclusion maps. Then $\underline{X} = (X_n, p_{nn+1})$ is an inverse sequence of polyhedra and $\underline{p} = (p_n):X \to \underline{X}$ is an inverse limit, because

$\cap X_n = X$. Nevertheless, $\underline{p}$ is not a resolution, because property (B1) is not fulfilled. Moreover, property (E1) is not fulfilled, because the identity map $h=1_X$ does not factor up to homotopy through any of the terms X_n ($H_1(X_n)$ is finitely generated and $H_1(X)$ is not). Such examples show that inverse limits cannot be used for developing shape theory in the non-compact case. However, they are suitable in the compact case because of the next theorem.

Theorem 5. Let $\underline{p}:X \to \underline{X} = (X_\lambda, p_{\lambda\lambda'}, \Lambda)$ be a morphism in the pro-category of compact Hausdorff spaces. Then $\underline{p}$ is a resolution if and only if $\underline{p}$ is an inverse limit.

Proof. The necessity of the condition follows from Theorem 4. To obtain the sufficiency, it suffices to recall the well-known fact that conditions (B1) and (B2) hold in the compact case.

5. Existence of polyhedral resolutions

P.Bacon [1] has proved the following theorem.

Theorem 6. Every topological space X admits a polyhedral resolution $\underline{p}:X \to \underline{X}$.

For completeness we sketch a proof. Let A be the set of all normal coverings α of X . For each $\alpha \in A$ one chooses a locally finite partition of unity $\phi_\alpha = (\phi_V, V \in \alpha)$ subordinated to α. We denote by X_α the nerve of α. The partition ϕ_α determines a canonical map $p_\alpha : X \to X_\alpha$. Next one considers the set Λ of all finite subsets $\lambda = \{\alpha_1, \dots, \alpha_n\}$ of A ordered by inclusion. By X_λ one denotes the nerve of the covering $\alpha_1 \wedge \dots \wedge \alpha_n = \{V_1 \cap \dots \cap V_n : (V_1, \dots, V_n) \in \alpha_1 \times \dots \times \alpha_n\}$. If $\lambda \le \lambda' = \{\alpha_1, \dots, \alpha_n, \dots, \alpha_m\}$, one obtains a natural simplicial map $p_{\lambda\lambda'}: X_{\lambda'} \to X_\lambda$ which takes the vertex $(V_1, \dots, V_n, \dots, V_m)$ of $X_{\lambda'}$ into the vertex $(V_1, \dots, V_n)$ of X_λ. One obtains thus a polyhedral inverse system $\underline{X} = (X_\lambda, p_{\lambda\lambda'}, \Lambda)$. One now defines maps $p_\lambda : X \to X_\lambda$ as canonical maps determined by the partition of unity $(\phi_{(V_1, \dots, V_n)}, (V_1, \dots, V_n) \in \alpha_1 \wedge \dots \wedge \alpha_n)$, given by

$$\phi_{(V_1, \dots, V_n)} = \phi_{V_1} \cdot \dots \cdot \phi_{V_n} . \tag{26}$$

It is readily seen that $\underline{p} = (p\ )$ is a morphism $X \to \underline{X}$ of pro-Top with property (B2).

Unfortunately,(B1) need not be satisfied and one must extend the system $\underline{X}$ to a new polyhedral system $\underline{Y} = (Y_\mu, q_{\mu\mu'}, M)$ and $\underline{p}$ to a morphism $\underline{q}:X \to \underline{Y}$ as follows. For each $\lambda \in \Lambda$ one chooses a basis of open neighborhoods $\underline{V}_\lambda$ of the set $\overline{p_\lambda(X)}$. Then M consists of pairs $\mu=(\lambda,V)$, where $\lambda \in \Lambda$ and $V \in \underline{V}_\lambda$. One puts $Y_\mu = V$ and $q_\mu = p_\lambda : X \to Y_\mu$. Furthermore, one puts $\mu \leq \mu' = (\lambda', V')$ provided $\lambda \leq \lambda'$ and $p_{\lambda\lambda'}(V') \subseteq V$ and one puts $q_{\mu\mu'} = p_{\lambda\lambda'} : V' \to V$. It is readily seen that $\underline{q}$ has properties (B1) and (B2) and is thus a polyhedral resolution of X.

The next theorem is proved in [6].

Theorem 7. Every topological space X admits an ANR-resolution $p:X \to \underline{X}$.

6. Resolutions of maps and shape fibrations

Let $f:X \to Y$ be a map of topological spaces. A resolution of the map f is a triple $(\underline{p}, \underline{q}, \underline{f})$, where $\underline{p}:X \to \underline{X} = (X_\lambda, p_{\lambda\lambda'}, \Lambda)$ and $\underline{q}:Y \to \underline{Y} = (Y_\mu, q_{\mu\mu'}, M)$ are resolutions of the spaces X and Y respectively, and $\underline{f} = (f_\mu, \phi)$ is a map of systems satisfying

(27) $$\underline{f}\,\underline{p} = \underline{q}\,f\ ,$$

i.e. for every $\mu \in M$ one has

(28) $$f_\mu p_\mu = q_\mu f\ .$$

If $\underline{p}$ and $\underline{q}$ are polyhedral (ANR) resolutions, then we call $(\underline{p}, \underline{q}, \underline{f})$ a polyhedral (ANR) resolution of f.

Generalizing theorems 6 and 7 one has this result [6].

Theorem 8. Every map $f:X \to Y$ between topological spaces admits a polyhedral (ANR) resolution.

For maps of systems $(f_\mu, \phi):\underline{X} \to \underline{Y}$ one defines the approximate homotopy lifting property (AHLP) with respect to a class of spaces $\underline{S}$ as follows. A pair

$(\lambda, \mu) \in \Lambda \times M$ is called admissible provided $\pi(\mu) \leq \lambda$. If (λ', μ') is another admissible pair, we put $(\lambda, \mu) \leq (\lambda', \mu')$ provided $\lambda \leq \lambda'$, $\mu \leq \mu'$ and the following diagram commutes

$$
(29) \qquad \begin{array}{ccc}
X_\lambda & \xleftarrow{p_{\lambda\lambda'}} & X_{\lambda'} \\
\downarrow f_{\mu\lambda} & & \downarrow f_{\mu'\lambda'} \\
Y_\mu & \xleftarrow[q_{\mu\mu'}]{} & Y_{\mu'}
\end{array} ,
$$

where $f_{\mu\lambda} = f_\mu p_{\phi(\mu)\lambda}$, $f_{\mu'\lambda'} = f_{\mu'} p_{\phi(\mu')\lambda'}$.

We say that $\underline{f}:\underline{X} \to \underline{Y}$ has the AHLP with respect to $\underline{S}$ provided for every admissible pair (λ, μ) and for normal coverings $\underline{U}_\lambda$ and $\underline{V}_\mu$ of X_λ and Y_μ respectively, there exist an admissible pair $(\lambda', \mu') \geq (\lambda, \mu)$ and a normal covering $\underline{V}_{\mu'}$ of $Y_{\mu'}$ such that for an arbitrary $S \in \underline{S}$ and for maps $h:S \to X_\lambda$, $H:S \times I \to Y_{\mu'}$ such that $f_{\mu'\lambda'} h$ and H_0 are $\underline{V}_{\mu'}$-near maps, there is an $\tilde{H}:S \times I \to X_\lambda$ such that the maps $p_{\lambda\lambda'}h$ and $\tilde{H}_0$ are $\underline{U}_\lambda$-near and the maps $f_{\mu\lambda}\tilde{H}$ and $q_{\mu\mu'}H$ are $\underline{V}_\mu$-near.

A special case of maps of systems are level-maps $\underline{f} = (f_\mu, \phi):\underline{X} \to \underline{Y}$, where $\Lambda = M$, $\phi = 1_\Lambda$ and $f_\lambda p_{\lambda\lambda'} = q_{\lambda\lambda'} f_{\lambda'}$, whenever $\lambda \leq \lambda'$. In this case the AHLP assumes this simpler form. For every $\lambda \in \Lambda$ and for normal coverings $\underline{U}_\lambda$ and $\underline{V}_\lambda$ of X_λ and Y_λ respectively, there exists a $\lambda' \geq \lambda$ and a normal covering $\underline{V}_{\lambda'}$ of $Y_{\lambda'}$ such that whenever $S \in \underline{S}$ and maps $h:S \to X_{\lambda'}$,$H:S \times I \to Y_{\lambda'}$ are such that $f_{\lambda'} h$ and H_0 are $\underline{V}_{\lambda'}$-near, then there is an $\tilde{H}:S \times I \to X_\lambda$ such that the maps $p_{\lambda\lambda'} h$ and $\tilde{H}_0$ are $\underline{U}_\lambda$-near and the maps $f_\lambda \tilde{H}$ and $q_{\lambda\lambda'} H$ are $\underline{V}_\lambda$-near .

Maps of systems can always be reduced to level-maps by a well-known procedure (see [5] and also [10]).

One can now define shape fibrations as maps $f:X \to Y$ between topological spaces, which admit a polyhedral resolution $(\underline{p}, \underline{q}, \underline{f})$ with the property that the map of systems $\underline{f}:\underline{X} \to \underline{Y}$ has the AHLP with respect to all spaces. We quote without proof a theorem from [6] (see also [2]).

Theorem 9. If for a polyhedral (ANR) resolution $(\underline{p}, \underline{q}, \underline{f})$ of f the map of systems $\underline{f}$ has the AHLP, then for any other polyhedral resolution $(\underline{p}', \underline{q}', \underline{f}')$ of f the map of systems $\underline{f}'$ also has the AHLP.

In particular, one can always assume that $\underline{f}$ is a level map of systems, which enables one to use the simpler form of the AHLP. Theorem 9 also shows that for compact metric spaces X,Y the above defined notion of shape fibration coincides with the one introduced in [7].

Several results known for shape fibrations between compact metric spaces hold also for closed shape fibrations between topological spaces. To this effect we quote here a theorem of Q.Haxhibeqiri [2].

Theorem lo. Let $f:X \to Y$ be a closed mapping, which is a shape fibration and let Y be a normal space. Furthermore, let $*$ be a base-point of Y and let the fiber $F = f^{-1}(*)$ be P-embedded in X. Then for any base point $*$ of F the map f induces an isomorphism of the homotopy pro-groups

$$\text{pro-}\ \pi_n(X,F,*) \to \text{pro-}\ \pi_n(Y,*) .$$

As a consequence, one has an exact sequence of pro-groups

$$\ldots \to \text{pro-}\pi_n(F,*) \to \text{pro-}\pi_n(X,*) \to \text{pro-}\pi_n(X,F,*) \to \text{pro-}\pi_{n-1}(Y,*) \to \ldots ,$$

which generalizes the corresponding results from [8].

REFERENCES:

[1] P.Bacon : Continuous functors, General Topology and its Appl. 5(1975), 321-331.

[2] Q.Haxhibeqiri: Shape fibrations for topological spaces, Ph.D.Thesis, Univ.of Zagreb, 198o.

[3] S.Mardešić: Retracts in shape theory, Glasnik Mat. Ser.III 6(26)(1971),153-163.

[4] S.Mardešić: Shapes for topological spaces, General Topology and its Appl. 3(1973), 265-282.

[5] S.Mardešić: The foundations of shape theory, Lecture Notes, Univ. of Kentucky,1978.

[6] S.Mardešić: Approximate polyhedra, resolutions of maps and shape fibrations, Fund. Math. (to appear).

[7] S.Mardešić and T.B.Rushing: Shape fibrations, I, General Topology and its Appl. 9 (1978), 193-216.

[8] S.Mardešić and T.B.Rushing:Shape fibrations,II, Rocky Mountain J.Math.9(1979), 283-298.

[9] S.Mardešić and J.Segal: Shapes of compacta and ANR-systems, Fund. Math. 72(1971), 41-59.

[10] S.Mardešić and J.Segal: Shape Theory, submitted.

[11] K.Morita : On shapes of topological spaces, Fund. Math. 86(1975), 251-259.

[12] K.Morita : On expansions of Tychonoff spaces into inverse systems of polyhedra, Sci.Rep. Tokyo Kyoiku Daigaku 13(1975), 66-74.

APPLICATION OF THE SHAPE THEORY IN THE CHARACTERIZATION OF EXACT HOMOLOGY THEORIES AND THE STRONG SHAPE HOMOTOPIC THEORY

L. D. Mdzinarishvili
Tbilisi, USSR

The paper deals with two topics: 1) the characterization of exact homology theories of compact spaces and 2) the strong shape homotopic theory and its relation to homology.

The first topic is connected with the results of the research carried out jointly by H. N. Inasaridze and the author [1], while the second topic with the results of Z. R. Miminoshvili [2]. Both topics are closely interconnected.

The characterization of exact homologies of compact spaces proposed by H. N. Inasaridze and the author is interesting for the reasons as follows: in the first place, it provides a positive answer to J. Milnor and W. Massey's question whether it is possible to characterize exact homologies of compact spaces; in the second place, it is the natural continuation of the research started by S. Eilenberg and N. Steenrod in the axiomatization of homologic theories; finally, the shape theory is used in proving the uniqueness theorem.

In the category A_c of compact spaces S. Eilenberg and N. Steenrod [3] obtained the uniqueness theorem for continuous theories of partially exact homology, which characterizes the Čech theory. The Čech theory, not being exact, satisfies the continuity axiom, i.e. the Čech homology groups are commutative with the inverse limit of compact spaces.

In the category A_{cm} of compact metric spaces J. Milnor [4] obtained the uniqueness theorem, which characterizes the Steenrod theory. The Steenrod theory, being exact, satisfies, in addition to seven Eilenberg-Steenrod axioms, two Milnor axioms: 8) invariance under relative homeomorphism; 9) (cluster axiom) If X is the union of countably many compact subsets $X_1, X_2, \ldots$ which intersect pairwise at a single point b and which have diameters tending to zero, then $H_n(X,b)$ is naturally isomorphic to the direct product of the groups $H_n(X_i,b)$.

Milnor showed that for homologic theories H, satisfying these nine axioms, the modified continuity axiom holds, i.e. if the sequence

$$X_1 \leftarrow X_2 \leftarrow \dots \quad ,$$

where $X_i \in \mathcal{A}_{cm}$, is given, then we have the exact sequence

$$0 \to \varprojlim{}^{(1)} H_{n+1}(X_i) \to H_n(\varprojlim X_i) \to \varprojlim H_n(X_i) \to 0 \quad .$$

Another characterization of the Steenrod homology theory in the category $\mathcal{A}_{cm}$ was obtained by E. Skljarenko [5], while that of the Čech homology by K. Kaul [6].

Besides, on the category $\mathcal{A}_{cm}$ the characterization of the extraordinary Steenrod homology was given by H. N. Inasaridze [7], J. Kaminker and C. Schochet [8]. This characterization consisted of two Milnor axioms and all Eilenberg-Steenrod axioms except for the dimension axiom.

On the category $\mathcal{A}_c$ of compact spaces (and on more general categories) the homologies of Kolmogorov [9], Borel-Moore [10], Milnor [4], Chogoshvili [11], Lefshetz [12], Skljarenko [5], Inasaridze [7], Kuzminov-Shvedov [13], W. Massey [14] are defined; the property of these homologies is that, on the one hand, they are exact theories, while, on the other hand, on the category $\mathcal{A}_{cm}$ they are isomorphic to Steenrod homologies. It is therefore interesting to characterize on the category $\mathcal{A}_c$ those exact homologies which would coincide on the category $\mathcal{A}_{cm}$ with Steenrod homologies; this was indicated by J. Milnor [4], W. Massey [14] and V. I. Kuzminov [15].

V. I. Kuzminov [15] proposed the characterization of exact homologies defined from the co-chains or, speaking more exactly, the isomorphism of Borel-Moore homologies with Milnor homologies was proved.

Recently, N. A. Berikashvili [16] obtained in the category $\mathcal{A}_c$ the axiomatics of exact homology theories by adding to the Eilenberg axioms three axioms, expressing, respectively, the properties of relative homeomorphism, continuity with respect to the union of spheres and isomorphisms of the homology of a compact space with the direct limit of homologies of the natural filtration of the associated canonical limit space.

As shown by S. Eilenberg and N. Steenrod, in the category A_c exact homologies do not satisfy the continuity axiom. Therefore to prove the uniqueness theorem for exact homologies it is necessary to replace the continuity axiom by such an axiom that would express partial continuity. The relation between the exactness axiom and the continuity axiom would thereby be established. Namely, those homology theories which are continuous are partially exact, while those homology theories which are partially continuous are exact. This approach allowed us to characterize homologies both as exact and partially exact on the category A_c as well as on the category A_{cm}. In this case, using the finite polyhedrons only, in our investigation we use the shape theory. We thereby think that we have completed the investigation started by S. Eilenberg and N. Steenrod to study the relation between continuity and exactness in the theory of compact space homology.

Let $h_* = \{h_n, n \in Z\}$ be a homotopic invariant homologic sequence from the category A_{cm} into the category of Abelian groups.

The property of partial continuity in the category A_{cm} is: if $(X,A) = \lim\limits_{\leftarrow i} (K_i, L_i)$, where each (K_i, L_i), $i = 1,2,\dots,$ is a pair of finite polyhedrons, then for all n we have the natural exact sequence

$$0 \to \lim_{\leftarrow}{}^{(1)} h_{n+1}(K_i, L_i) \to h_n(X,A) \overset{\Phi_n}{\to} \lim_{\leftarrow} h_n(K_i, L_i) \to 0$$

The naturality is understood in the following sense: If $(X',A') = \lim\limits_{\leftarrow} (K_i', L_i')$ and the morphism $f: (X,A) \to (X',A')$ and the morphisms $\varphi_i: (K_i, L_i) \to (K_i', L_i')$, $i = 1,2, \dots,$ compatible with spectrum morphisms and projections with an accuracy of homotopy, are given, then for all n the commutative diagram

$$\begin{array}{ccc}\lim\limits_{\leftarrow}{}^{(1)} h_{n+1}(K_i,L_i) & \longrightarrow & h_n(X,A) \\ \Big\downarrow \lim\limits_{\leftarrow}{}^{(1)} h_{n+1}(\varphi_i) & & \Big\downarrow h_n(f) \\ \lim\limits_{\leftarrow}{}^{(1)} h_{n+1}(K_i',L_i') & \longrightarrow & h_n(X',A')\end{array}$$

holds.

If the canonical mapping Φ_n is isomorphism for all n, then we have the continuity property (in this case naturality always holds) in the category A_{cm}.

In the category A_{cm} the homology theory, satisfying Eilenberg-Steenrod axioms, is said to be the partially continuous theory of exact homologies, if it possesses the property of partial continuity. If in A_{cm} the theory of partially exact homology possesses the continuity property, then we call it the continuous theory of partially exact homologies.

Theorem I. On the category A_{cm} any partially continuous theory of exact homologies is isomorphic to the Steenrod homology theory and any continuous theory of partially exact homologies is isomorphic to the Čech homology theory with the same group of coefficients.

When proving uniqueness theorems in the categories A_{cm} and A_c, for exact homology theories we establish the isomorphism with the exact homology theory $H_*(X,A,G)$ constructed in [7]. A simplified version of its construction will now be given.

Let $Cov(X,A) = \{(U_\alpha, V_\alpha)\}$ be a set of finite open multiplicative coverings of the compact pair $(X,A) \in A_c$. Note that the elements of each covering U_α must not be repeated and if $u_\alpha \cap v_\alpha \neq \emptyset$, then $u_\alpha \cap v_\alpha \in V_\alpha$, $v_\alpha \in V_\alpha$. We introduce in $Cov(X,A)$ the strong refinement $>>: (U_\beta, V_\beta) >> (U_\alpha, V_\alpha)$, if U_β and V_β are refined, respectively, in U_α and V_α; in this case if $u_\alpha \cap u_\beta \neq \emptyset$, then $u_\alpha \cap u_\beta \in U_\beta$, where $u_\alpha \in U_\alpha$, $u_\beta \in U_\beta$. Then $(Cov(X,A), >>)$ is a partially ordered directed set. If (K_α, L_α) and (K_β, L_β) are the nerves of the coverings (U_α, V_α) and (U_β, V_β), respectively, then we have the simplicial mapping

$p^\alpha_\beta:(K_\beta,L_\beta)\to(K_\alpha,L_\alpha)$, where $p^\alpha_\beta(u_\beta)=\bigcap\limits_{u_\alpha\supset u_\beta} u_\alpha$.

Let $C_*(X,A,G)=\varprojlim\,[C_*(K_\alpha,L_\alpha,G),q^\alpha_\beta]$, where $C_n(K_\alpha,L_\alpha,G)$ is the group of n-dimensional chains with coefficients in the Abelian group G and the homomorphism q^α_β is induced by the simplicial mapping p^α_β. Then $H_n(X,A,G)=H_n(C_*(X,A,G))$. The morphism $f:(X,A)\to(Y,B)$ induces the mapping $\bar{f}:\mathrm{Cov}\,(Y,B)\to\mathrm{Cov}\,(X,A)$ of the directed sets and if $\bar{f}(U_\beta,V_\beta)=(U_\alpha,V_\alpha)$, then we have simplicial mappings $f^\beta_\alpha:(K_\alpha,L_\alpha)\to(K_\beta,L_\beta)$, where $f^\beta_\alpha(u_\alpha)=\bigcap\limits_{f^{-1}(u_\beta)=u_\alpha} u_\beta$, which induce the homomorphism $H_n(f):H_n(X,A,G)\to H_n(Y,B,G)$.

<u>Theorem 2</u>. On the category A_{cm} the homology theory $H_*(X,A,G)$ satisfies the Milnor axiomatics of the Steenrod homology theory. On the category A_c it satisfies the Steenrod-Eilenberg axiomatics and for it we have the following exact sequences

$$0\to \mathrm{Ext}(\check{H}^{n+1}(X,A),G)\to H_n(X,A,G)\to \mathrm{Hom}\,(\check{H}^n(X,A),G)\to 0,$$

which is natural with respect to (X,A) and G,

$$0\to \varprojlim{}^{(1)}\, H_{n+1}(K_\alpha,L_\alpha,G)\to H_n(X,A,G)\to \check{H}_n(X,A,G)\to 0,$$

where the limit $\varprojlim$ is taken with respect to $(\mathrm{Cov}\,(X,A),>>)$, $n\in Z$. The homomorphism $\varprojlim{}^{(1)}\, H_{n+1}(K_\alpha,L_\alpha,G)\to H_n(X,A,G)$ is defined as the composition

$$\varprojlim{}^{(1)}\, H_{n+1}(K_\alpha,L_\alpha,G)\tilde{\to}\varprojlim{}^{(1)}\;\mathrm{Hom}(\check{H}^{n+1}(K_\alpha,L_\alpha),G)\to$$
$$\to\mathrm{Ext}(\check{H}^{n+1}(X,A),G)\to H_n(X,A,G).$$

We give the scheme of the proof of Theorem I for the absolute case. Let $X\in A_{cm}$; then by virtue of Lemma 2 [4] we have $X=\varprojlim\{N_i,p_i\}$ where N_i is the realization of the finite open covering, while $p_i:N_{i+1}\to N_i$ is the continuous mapping, not induced by the simplicial mapping, $N_i=|V_i|$.

From the Milnor sequence

$$V_1<V_2<V_3<\ldots$$

we construct with usual refinement a new sequence of multiplicative coverings with strong refinement

$$W_1 << W_2 << W_3 << \dots \ .$$

Each W_i is constructed from V_i by adding $V_i \cap (V_{i_1} \cap \dots \cap V_{i_s})$, where $v_i \in V_i$, $V_{i_k} \in V_j$, $1 \le j \le i$ (without repetition).

Taking the realizations of the respective nerves, we obtain the sequence

$$M_1 \xleftarrow{q_1} M_2 \xleftarrow{q_2} M_3 \longleftarrow \dots)$$

where q_i are continuous mappings induced by the simplicial mappings obtained from refinement.

Let $K(X) = \lim\{M_i, q_i\}$. Then there exists a continuous mapping "on"

$$\begin{array}{c} \varprojlim\{M_i, q_i\} = K(X) \\ \downarrow \theta \\ \varprojlim\{N_i, p_i\} = X \end{array}$$

for each i there exist simplicial mappings

$$M_i \xrightarrow{\varphi_i} N_i \xrightarrow{\psi_i} M_i \ ,$$

such that $\varphi_i \psi_i = 1$, $\psi_i \varphi_i \sim 1$ (adjacent) and the diagram

$$\begin{array}{ccc} X & \xrightarrow{s_i} & N_i \\ \theta \uparrow & & \uparrow \varphi_i \\ K(X) & \xrightarrow{t_i} & M_i \end{array}$$

is commutative with an accuracy of homotopy and $s_i \theta$ and $\varphi_i t_i$ are canonical mappings, since $|\theta^{-1}(V_i)| = N_i$

In the space $K(X)$ the filtration

$$K^0(X) \subset K^1(X) \subset K^2(X) \subset \dots$$

is defined in the natural manner; here $K^p(X) = \varprojlim M_i^p$ and $\cup K^p(X) \neq K(X)$.

This approach is used in the works of D. Edwards and H. Hastings [17], N. A. Berikashvili [16].

Using the partial continuity property, we obtain a chain of isomorphisms

$$h_n(X) \stackrel{\approx}{\leftarrow} h_n(K(X)) \stackrel{\approx}{\leftarrow} h_n(K^{n+2}(X)) \stackrel{\approx}{\leftarrow} h_n(K^{n+1}(X)) \approx$$
$$\approx H_n(\lim_{\leftarrow} C_*(M_i)) \approx H_n(X).$$

A sequence of the functors $h_* = \{h_n, n\in Z\}$ on $\mathcal{A}_{cm}$, satisfying the first six Steenrod-Eilenberg axioms and the partial continuity axiom is called the extraordinary theory of Steenrod homology.

Example. 1) The Steenrod homology theory. 2) K-homologies of Ext (Brown-Douglas-Fillmore [18]).

Let h_* and k_* be extraordinary theories of Steenrod homology and $\varphi_*: h_* \to k_*$ be a natural transformation of these theories which is commutative with the $\lim_{\leftarrow}$-mapping, i.e. if $X = \lim_{\leftarrow} K_i$, then for all $n\in Z$ we have the commutative diagram

$$\begin{array}{ccc} \lim\limits_{\leftarrow}{}^{(1)} h_{n+1}(K_i) & \longrightarrow & h_n(X) \\ \Big\downarrow \lim\limits_{\leftarrow}{}^{(1)}\{\varphi^i_{n+1}\} & & \Big\downarrow \varphi_n \\ \lim\limits_{\leftarrow}{}^{(1)} k_{n+1}(K_i) & \longrightarrow & k_n(X) \end{array}$$

Theorem 3. If $\varphi_*(p): h_*(p) \to k_*(p)$ is an isomorphism, then φ_* is an equivalence (p-point).

The property of partial continuity in the category A_c is: if $\{(K_\alpha, L_\alpha), p^\alpha_\beta\}, \pi_\alpha, (X,A))$ is the inverse system of finite polyhedron pairs (K_α, L_α) associated with the compact pair (X,A) [19], then for all $n\in Z$ the exact sequence

$$0 \to \lim_{\leftarrow}{}^{(1)} h_{n+1}(K_\alpha, L_\alpha) \to h_n(X,A) \xrightarrow{\Phi_n} \lim_{\leftarrow} h_n(K_\alpha, L_\alpha) \to 0$$

holds, which is natural with respect to the morphisms of the associated pairs with the same set of indexes.

If the cannonical mapping Φ_n is always an isomorphism for all n, then we have the continuity property in the category (in this case naturality always holds).

In the category A_c the homology theory, satisfying the Steenrod-Eilenberg axioms, is termed as the partially continuous theory of exact homologies, if it possesses the partial continuity property. If in A_c the theory of partially exact homology possesses the continuity property, then it is called the continuous theory of partially exact homologies.

Example. 3) The homology theories [9], [20], [11], [4], [13], [7] are partially continuous theories of exact homologies. 4) The Čech theory is a continuous theory of partially exact homologies.

Theorem 4. a) On the category A_c there exists and is unique with an accuracy of an isomorphism a partially continuous theory of exact homologies with the given group of coefficients. b) On A_c there exists and is unique with an accuracy of an isomorphism a continuous theory of partially exact homologies with the given group of coefficients.

In this case the proof of the uniqueness theorems is similar to the metric case with the only difference being that instead of the Milnor polyhedron sequence we consider the Morita associated system of finite polyhedron pairs.

From the obtained results and G. S. Chogoshvili's work [21] it follows that the homologies, participating in the duality laws, are either partially continuous and exact or continuous and partially exact. Any partially continuous theory of exact homologies is isomorphic to the Steenrod homology theory on the subcategory A_{cm}.

The second topic, as said at the beginning, is connected with the strong shape theory. Z. R. Miminoshvili proved

Theorem 5. If X is a compact space, then the exact sequence

$$0 \longrightarrow \lim_{\leftarrow}{}^{(1)} \pi_{n+1}(K_\alpha) \longrightarrow \overset{B}{\pi}_n(X) \longrightarrow \check{\pi}_n(X) \longrightarrow 0$$

holds, where $\overset{B}{\pi}_n(X)$ are Bauer homotopies, $\check{\pi}_n(X)$ are Borsuk-Mardešić homotopies and the first derivative is defined by means of strong refinement.

Besides, Z. R. Miminoshvili succeeded in defining for an arbitrary space the strong homologic group $\tilde{H}_n(X)$, which is invariant with respect to strong shapes, and in proving for the case of a compact space the commutativity of the diagram

$$\begin{array}{ccccccccc} 0 \to & \lim\limits_{\leftarrow}{}^{(1)} \pi_{n+1}(K_\alpha) & \to & {}^{B}\pi_n(X) & \to & \pi_n(X) & \to 0 \\ & \downarrow & & \downarrow & & \downarrow \approx & \\ 0 \to & \lim\limits_{\leftarrow}{}^{(1)} H_{n+1}(K_\alpha) & \to & \tilde{H}_n(X) & \to & \check{H}_n(X) & \to 0 \end{array}$$

with exact rows. This enabled him to obtain: 1) the isomorphism $\tilde{H}_n(X) \approx \overset{sk}{H}_n(X) \approx \overset{k}{H}_n(X)$; 2) the Gurevich theorem for strong homotopic and homologic groups in terms of the Gurevich theorem for homotopic and homologic pro-groups.

The strong homology group $\tilde{H}_n(X)$ is defined in the following manner: $\tilde{H}_n(X) = \tilde{Z}_n(X)/\tilde{B}_n(X)$, where $\tilde{Z}_n(X)$ is a subgroup of the group $\prod\limits_{\alpha \in M} \prod\limits_{\alpha<\beta} (Z_n^\alpha \times C_{n+1}^{\alpha\beta})$, consisting of $\{(z_n^\alpha, c_{n+1}^{\alpha\beta})\}$ such that the following conditions are fulfilled: a) if $\alpha<\beta$, in M we have $z_n^\alpha - p_n^{\alpha\beta} z_n^\beta = \partial c_{n+1}^{\alpha\beta}$; b) if $\alpha<\beta<\gamma$, in M we have $c_{n+1}^{\alpha\beta} + p_{n+1}^{\alpha\beta} c_{n+1}^{\beta\gamma} - c_{n+1}^{\alpha\gamma} \in B_{n+1}^\alpha$ while $\tilde{B}_n(X)$ is the subgroup $\tilde{Z}_n(X)$ consisting of $\{(b_n, c_{n+1})\} \in \tilde{Z}_n(X)$ such that the following conditions are fulfilled:

a) there exists an element $\{c_{n+1}^\alpha\} \in \prod\limits_{\alpha \in M} C_{n+1}^\alpha$ such that $b_n^\alpha = \partial c_{n+1}^\alpha$;

b) if $\alpha<\beta$, in M we have $c_{n+1}^\alpha - c_{n+1}^{\alpha\beta} - p_{n+1}^{\alpha\beta} c_{n+1}^\beta \in B_{n+1}^\alpha$ ($c_n^{\alpha\beta} \in C_n^{\alpha\beta} \equiv C_n^\alpha$ if $\alpha<\beta$ in M, where M is a partially ordered directed set of finite multiplicative open coverings with strong refinement).

Литература

1. Х. Н. Инасарилзе, Л. Д. Мдзинаришвили, Сообщения АН ГССР, 99, № 2, 1980.

2. З. Р. Миминошвили, Сообщения АН ГССР, 98, №2, 1980.

3. Н. Стинрод, С. Эйленберг, Основания алгебраической топологии, Москва, 1958.

4. J. Milnor, On the Steenrod Homology Theory (mimeographed), Berkeley, 1960.

5. Е. Г. Склярнко, Успехи матем. наук, 24, №5, 1969.

6. K. Kaul, Colloquium Math., 21, No. 2, 1970

7. Х. Н. Инасаридзе, Труды Тбил. матес. ин-та, X I, 1972.

8. J. Kaminker, C. Schochet. Bull. Amer. Math., Soc., 81, No. 2, 1975.

9. A. N. Kolmogoroff, C. R. de Paris, 202, 1936.

10. A. Borel, J. C. Moore, Mich. Math. J. 7, 1960.

11. Г. С. Чогошвили, Сообщения АН ГССР, I, 1940.

12. С. Лефшец, Алгебраическая топология, М., 1949.

13. В. И. Кузьминов, И. А. Шведов, Сиб. мат. журн., 15, №5, 1974.

14. W. S. Massey, Homology and Cohomology Theory, New York, Dekker 1978.

15. В. И. Кузьминов, Международная топ. конф. тезисы докладов, М., 1979.

16. Н. А. Берикашвили, Довлады АН СССР, 254, № 6, 1980.

17. D. A. Edwards, H. M. Hastings, Čech and Steenrod Homology Theories with Applications to Geometric Topology, Lecture Notes in Mathematics, 542.

18. L. G. Brown, R. G. Douglas, P. A .Fillmore, Bull. Amer. Math. Soc., 79, No .5, 1973.

19. K. Morita, Fund. Math „ 86, No. 3, 1975.

20. Л. Д. Мдзинаришвили, Труды Тбил. мат. ин-та, IX, 1978.

21 . G. Chogoshvili, On homology theory of non-closed sets, Proceedings of the Symposium held in Prague, 1961.

LIST OF PARTICIPANTS

Vlada Baković
VI Crnogorske T-5/I
81400 NIKŠIĆ
Yugoslavia

Friedrich-Wilhelm Bauer
6 FRANKFURT a.M.
Kurhessenstrasse 65
West Germany

Frederick W. Cathey
213 Hanover Place 4
LAWRENCE, KS 66044
USA

Donald S. Coram
Department of Mathematics
Oklahoma State University
STILLWATER, OK 74074
USA

Zvonko Čerin
Department of Mathematics
University of Zagreb
P.P. 187
41001 ZAGREB
Yugoslavia

Jerzy Dydak
University of Warsaw
Department of Mathematics
00-901 WARSAW PKiN
Poland

Minir Efendis
Ulj. "Borisa Kidriča" 304
38320 DJAKOVICA
Yugoslavia

Steven C. Ferry
Department of Mathematics
University of Kentucky
LEXINGTON, KY. 40506
USA

Hiroo Fukaishi
Faculty of Education
Kagawa University
TAKAMATSU, KAGAWA 760
Japan

Ross Geoghegan
Department of Mathematics
SUNY Binghamton
BINGHAMTON, NY 13901
USA

Eraldo Giuli
Via Antica Arischia 21
67110 L'AQUILA
Italy

Harold M. Hastings
Department of Mathematics
Hofstra University
HEMPSTEAD, NY 11550

Qamil Haxhibeqiri
Rr. "Asim Vokshi" no. 86
38320 DJAKOVICA
Yugoslavia

Damir Henč
Vinkovićeva 24
41000 ZAGREB
Yugoslavia

Kreso Horvatić
Department of Mathematics
p.p. 187
Marulićev trg 19/I
41001 ZAGREB
Yugoslavia

Lawrence S. Husch
Department of Mathematics
University of Tennessee
KNOXVILLE, TN 37916
USA

Ivan Ivanšić
Hanamanova 1
41000 ZAGREB
Yugoslavia

Mahendra Jani
William Paterson College
of New Jersey
WAYNE, NJ 07470
USA

Akira Koyama
Osaka Kyoiku University
Department of Mathematics
Tennoji
OSAKA
Japan

Jozef Krasinkiewicz
ul. Krepowieckiego 11/100
01-456 WARSAW
Poland

Miljenko Lapaine
4100 ZAGREB
Kačićeva 26
Yugoslavia

Vera Mardešić
Faculty of Mechanical
Engineering and Shipbuilding
Djure Salaja 5
41000 ZAGREB
Yugoslavia

Sibe Mardešić
Department of Mathematics
University of Zagreb
pp. 187
41001 ZAGREB
Yugoslavia

Gordana Matić
Končareva 242
41000 ZAGREB
Yugoslavia

Leonard Mdzinarishvili
TBILISI
st. Saburtalinsray kor. 3 45
USSR

Uroš Milutinović
Department of Mathematics
University of Zagreb
pp. 187
41001 ZAGREB
Yugoslavia

Maria Moszyńska
Institute of Mathematics
Warsaw University
PKin IX p
00-901 WARSAW
Poland

Darko Mrakovčić
M. Pijade b
51521 PUNAT
Yugoslavia

Peter Mrozik
Schwälmersträsse 3
D-6000 FRANKFURT 90
West Germany

Rubiano Ortegon
Studentsko naselje,
blok lo (714)
61000 LJUBLJANA
Yugoslavia

Jelena Pešut
Gredička 8
41000 ZAGREB
Yugoslavia

Zlata Petričević
J. Cagarina 52/6
11070 NOVI BEOGRAD
Yugoslavia

Riccardo Piergallini
Department of Mathematics
University of Perugia
via Vanvitelli 1
06100 PERUGIA
Italy

Nebojša Ralević
11000 BEOGRAD
Franca Rozmana 22/x
Yugoslavia

Dušan Repovš
Fakulteta za strojništvo
61001 LJUBLJANA
pp 394
Yugoslavia

Jose Rodriguez Sanjurjo
Departamento de Topologia
y Geometria
Facultat de Matematicas
Universidad Complutense
MADRID (3)
Spain

Jack Segal
Department of Mathematics
University of Washington
SEATTLE, WA 98195
USA

Feliciana Serrano
c/Velayos 20
MADRID 35
Spain

Richard B. Sher
Department of Mathematics
University of North Carolina
at Greensboro
GREENSBORO, NC 37412
USA

Luciano Stramaccia
Department of Mathematics
University of Perugia
via Vanvitelli no. 1
06100 PERUGIA
Italy

Dragutin Svrtan
Department of Mathematics
University of Zagreb
p.p. 187
41001 ZAGREB
Yugoslavia

Jože Šrekl
Fakulteta za strojništvo
Murnikova 2
61000 LJUBLJANA
Yugoslavia

Anna Tozzi
Strada 63, 13/A
67100 L'AQUILA
Italy

Šime Ungar
Department of Mathematics
University of Zagreb
p.p. 187
41001 ZAGREB
Yugoslavia

Darko Veljan
Department of Mathematics
University of Zagreb
p.p. 187
41001 ZAGREB
Yugoslavia

Gerard A. Venema
Department of Mathematics
Calvin College
GRAND RAPIDS, MI 49506
USA

Vlasta Vitaljić
Nikole Tesle 12
58000 SPLIT
Yugoslavia

John J. Walsh
Department of Mathematics
University of Tennessee
KNOXVILLE, TN 37917
USA

Tadashi Watanabe
Department of Mathematics
Faculty of Education
1677-1 Yoshida
YAMAGUCHI CITY
Japan

Acknowledgment. The organizers of the Dubrovnik school gratefully acknowledge the financial help received from the University of Zagreb, from SIZ za znanstveni rad SRH and from Savez republičkih i prokrajinskih SIZ-ova za znanost.